This book gives a detailed account of recent work on relations between commutative algebra and intersection theory, with a particular emphasis on applications of the theory of local Chern characters. This theory is the result of many years of development, having originated in topology and been introduced in algebraic geometry about thirty years ago. Building on the algebraic form described in *Intersection Theory* by W. Fulton, Paul Roberts presents further developments and important algebraic applications that were not known at the time Fulton's book was written. Some of these applications come from the author's own work. He also discusses the background in commutative algebra and related questions in homological algebra and describes the relations between these subjects, including extensive discussions of the homological conjectures and of the use of the Frobenius map.

Students and researchers specializing in commutative algebra will find access to a wide range of new ideas in this book.

T0276016

CAMBRIDGE TRACTS IN
MATHEMATICS

General Editors
B. BOLLOBAS, F. KIRWAN, P. SARNAK, C. T. C. WALL

133 Multiplicities and Chern Classes in Local Algebra

PAUL C. ROBERTS
University of Utah

Multiplicities and Chern Classes in Local Algebra

CAMBRIDGE UNIVERSITY PRESS
Cambridge, New York, Melbourne, Madrid, Cape Town, Singapore, São Paulo

Cambridge University Press
The Edinburgh Building, Cambridge CB2 8RU, UK

Published in the United States of America by Cambridge University Press, New York

www.cambridge.org
Information on this title: www.cambridge.org/9780521473163

First published 1998
This digitally printed version 2008

A catalogue record for this publication is available from the British Library

Library of Congress Cataloguing in Publication data
Roberts, Paul, Ph.D.
Multiplicities and Chern classes in local algebra / Paul C.
Roberts.
p.   cm. – (Cambridge tracts in mathematics; 133)
Includes bibliographical references (p. –) and index.
ISBN 0-521-47316-0 (hb)
1. Multiplicity (Mathematics)   2. Chern classes.   3. Commutative
algebra.   I. Title.   II. Series.
QA251.38.R64   1998.
512′.24 – dc21   97-46135
                                                    CIP

ISBN 978-0-521-47316-3 hardback
ISBN 978-0-521-06583-2 paperback

Contents

Preface

The study of multiplicities has been of major importance in commutative algebra since the beginnings of the subject. It developed from the notion of the multiplicity of a root of a polynomial and from counting multiplicities of intersections in algebraic geometry, and it has influenced the development of the subject in many ways. On the other hand, the notion of Chern classes was originally developed in topology, and it is comparatively recently that it has been used in algebraic geometry and even more recently that it has made its way into commutative algebra. It is the aim of this book to present the theories of multiplicities and of Chern classes in an algebraic setting and to describe their relations with each other and with other topics in the field.

There are two somewhat different notions of multiplicities that will be discussed at length in this book. They both originate from algebraic geometry but from somewhat different sources. The first notion of multiplicity that we consider comes from the multiplicity of a variety at a point. This is best illustrated by an example. Suppose we have a curve in a plane that crosses itself, perhaps more than once, at a point. It then makes sense to take the multiplicity of the curve at the point to be the number of components locally at the point. But this straightforward definition in this case already raises problems: For example, if the curve is tangent to itself, it is not immediately clear how the multiplicity should be counted. In addition, the more complicated possibilities in higher dimension make a general definition less obvious. A good notion of multiplicity in this sense was introduced by Samuel, and we will refer to it as *Samuel multiplicity*. Since we will mostly be concerned with algebraic notions here, we will define multiplicities for local rings; in the geometric situation above, the local ring would be the local ring of functions at the point. We also discuss generalizations of this concept to ideals, modules, and sets of ideals.

The second notion is that of the multiplicity of intersection of two subvarieties of a variety. This can be considered a direct generalization of the multiplicity of roots of a polynomial, which is the case of the intersection of the graph of the polynomial with the x-axis. The question of defining intersection multiplicities algebraically in general has been a central topic in algebraic geometry for many years, and several definitions have been proposed. The one we consider here was introduced by Serre, and it involves extensive use of homological algebra. This homological definition has led to a considerable number of purely algebraic questions and conjectures relating to properties of modules and complexes of free modules, and particularly to properties of modules of finite projective dimension.

The first part of this book deals with the various concepts of multiplicity, as well as related ideas in the field of commutative algebra. We prove some of the basic theorems in this field, including properties of prime ideals and dimension, which can be found in the fundamental books on the subject such as Matsumura's *Commutative Rings* or the *Introduction to Commutative Algebra* of Atiyah and Macdonald. On the other hand, we assume that the reader has a knowledge of the basic results that can be found in these or similar books. The elementary results that we prove in this book are those closely related to the rest of the theory; we give references for others.

The other topics we cover revolve around the so-called homological conjectures, which arose in part from questions on multiplicities. We include sections on homological algebra, Cohen-Macaulay properties, Koszul complexes, and dualizing complexes and discuss their relations to questions on multiplicities and to these conjectures. We also discuss some of the methods that have been used to solve these problems. Some of these involve Cohen-Macaulay modules and duality and are presented in the first five chapters. We also discuss the use of the Frobenius map for rings of positive characteristic, which was introduced into this subject by Peskine and Szpiro to answer some of these questions. The Frobenius map is discussed in Chapter 7 after a discussion of the homological conjectures in Chapter 6.

Another important method was made possible by the development of theory of local Chern characters of Baum, Fulton, and MacPherson [5]. This made it possible to solve some of the problems in mixed characteristic, and it also sheds new light on the theory of intersection multiplicities and of Samuel multiplicities. Most of this theory is worked out in a geometric setting in Fulton [17], which is a source of many of the results in this book.

The main aim of the second half of the book is to present the algebraic theory of local Chern characters in a self-contained manner. Although it is not possible to make it completely self-contained, this presentation comes quite close, and the necessary results for proving the properties needed in the applications are proven in detail. In addition, we have consistently emphasized algebraic methods in the constructions.

The background includes the theory of Chow group, which is introduced in the first chapter together with basic properties of prime ideals, functorial properties of projective maps, the basic theory of Chern classes of locally free sheaves, and some of the fundamental properties of the Grassmannian. In some cases our presentation is not totally standard; in particular, projective schemes are defined for multigraded rings, which arise naturally in this situation and which are related to the theory of Hilbert polynomials and multiplicities presented in Chapter 2. In addition, coherent sheaves on projective schemes are almost exclusively treated in terms of the graded modules that are associated to them. In any case, the proofs of theorems are essentially the same, as they usually come down to the case of ordinary projective schemes defined by graded rings.

I would like to thank the students in a course I gave in these topics at the University of Utah last year, using preliminary versions of part of this material. In particular, I want to thank Ionut Ciocan-Fontanine, Elizabeth Jones, Saule Zhoshina, Chin-yi Chan, and Sean Sather-Wagstaff, who made numerous suggestions and contributions to the readability and correctness of the book. And finally, I want to give special thanks to my wife Anne, whose unending patience was essential in bringing this project to completion.

Part One

Multiplicities

1
Prime Ideals and the Chow Group

The main purpose of this chapter is to define the Chow group of a Noetherian ring and prove several of its basic properties. The Chow group is a quotient of the free group whose generators are the prime ideals of the ring, and we devote the first section to a summary of some of the main properties of prime ideals, and in particular the prime ideals associated to a finitely generated module. Much of the material in the first section can be found in several books on commutative algebra, such as Matsumura [44] or Atiyah and Macdonald [1]. While we prove some of these basic facts, we use others without proof, giving a reference to a place in one of these books where a proof can be found. Most of the results of Section 2, in a more general setting, can be found in Fulton [17].

1.1 Prime Ideals in Noetherian Rings

All rings will be assumed to be commutative and have an identity element. A module M over a commutative ring A is *Noetherian* if it has the ascending chain condition on submodules, or, equivalently, if every submodule is finitely generated. The ring A is Noetherian if it is Noetherian as an A-module, which means that it has the ascending chain condition on ideals or that every ideal is finitely generated. We recall that every finitely generated module over a Noetherian ring is Noetherian.

Unless specifically stated otherwise, all rings in this book will be assumed to be Noetherian. By the Hilbert basis theorem (see [1, Theorem 7.5]), this class of rings includes all rings that are finitely generated as algebras over a field or over the ring of integers, and the class is closed under the operations of localization and completion.

Let A be a ring. An ideal $\mathfrak{p}$ of A is *prime* if $\mathfrak{p} \neq A$ and if whenever $ab \in \mathfrak{p}$, we have either $a \in \mathfrak{p}$ or $b \in \mathfrak{p}$. The zero ideal of A is prime if and only if A

3

is an integral domain. While this definition is valid in any commutative ring, whether Noetherian or not, the set of prime ideals in Noetherian rings has many special properties. We denote the set of prime ideals of A by $\mathrm{Spec}(A)$.

An ideal $\mathfrak{a}$ of A is an *annihilator ideal* for the module M if there exists a nonzero element $m \in M$ such that $\mathfrak{a} = \{x \in A \mid xm = 0\}$. The next proposition shows the existence of prime ideals that are annihilator ideals.

Proposition 1.1.1 *Let M be an A-module. Then every ideal that is maximal in the set of annihilator ideals for M is prime.*

Proof Let $\mathfrak{a}$ be a maximal annihilator ideal for M, and suppose that $\mathfrak{a}$ is the annihilator of the element m in M. Let a and b be elements of A such that neither a nor b is in $\mathfrak{a}$. Since a is not in $\mathfrak{a}$, and $\mathfrak{a}$ is the annihilator of m, we have $am \neq 0$. On the other hand, $\mathfrak{a}$ annihilates am, so, by the maximality of $\mathfrak{a}$, we must have that $\mathfrak{a}$ is equal to the annihilator of am. Since $b \notin \mathfrak{a}$, this means that $abm \neq 0$, so $ab \notin \mathfrak{a}$. Thus, $\mathfrak{a}$ is prime. $\square$

In general there is no reason for maximal annihilator ideals to exist (they do not satisfy the conditions of Zorn's lemma, for example), but if the ring is Noetherian the ascending chain condition implies that every annihilator ideal is contained in a maximal annihilator ideal and hence in a prime annihilator ideal.

Let A be a Noetherian ring, and let M be a finitely generated A-module. We say that a prime ideal $\mathfrak{p}$ is *associated* to M, or is an associated prime of M, if there is an element of M whose annihilator is exactly $\mathfrak{p}$, or, equivalently, if there is an embedding of $A/\mathfrak{p}$ into M. Proposition 1.1.1 implies that every nonzero module has at least one associated prime. In fact, it shows that the union of all associated primes of M is exactly the set of zero-divisors on M, where a zero-divisor is an element $a \in A$ such that there exists a nonzero element $m \in M$ with $am = 0$. In fact, every associated prime ideal consists of zero-divisors, and if a is a zero-divisor, then it is contained in the annihilator of a nonzero element and hence by Proposition 1.1.1 in an associated prime ideal.

The *support* of M is the set of prime ideals $\mathfrak{p}$ for which the localization $M_{\mathfrak{p}}$ is not zero. The only associated prime ideal of $A/\mathfrak{p}$ is $\mathfrak{p}$, and the support of $A/\mathfrak{p}$ consists of all primes containing $\mathfrak{p}$. The localization of $A/\mathfrak{p}$ at $\mathfrak{p}$ is the quotient field of the integral domain $A/\mathfrak{p}$, which we denote $k(\mathfrak{p})$. For any module M, every associated prime ideal of M is in the support of M.

Let S be a multiplicatively closed subset of A, and let A_S and M_S denote the localizations of A and M at S. Every prime ideal $\mathfrak{p}$ generates an ideal $\mathfrak{p}_S$ of A_S, and it is not hard to check that $\mathfrak{p}_S$ is prime if $\mathfrak{p}$ does not meet S.

We summarize the behavior of prime ideals under localization in the following proposition.

Proposition 1.1.2 *Let M be an A-module as above, and let S be a multiplicatively closed subset of A. Then*

(i) *The map that sends $\mathfrak{p}$ to $\mathfrak{p}_S$ defines a one-one correspondence between the set of prime ideals of A that do not meet S and the set of prime ideals of A_S.*

(ii) *The support of M_S is the subset of $\operatorname{Spec}(A_S)$ consisting of those $\mathfrak{p}_S$ for which $\mathfrak{p}$ is in the support of M.*

(iii) *The set of associated primes of M_S is the subset of $\operatorname{Spec}(A_S)$ consisting of those $\mathfrak{p}_S$ for which $\mathfrak{p}$ is associated to M.*

Proof We note first that $\mathfrak{p}_S = A_S$ if and only if $\mathfrak{p} \cap S$ is not empty. Assume that $\mathfrak{p} \cap S = \emptyset$. Then if $a/s \in \mathfrak{p}_S$, we have $ta \in \mathfrak{p}$ for some $t \in S$, and since t cannot be in $\mathfrak{p}$ and $\mathfrak{p}$ is prime, we have $a \in \mathfrak{p}$. Thus, a/s is in $\mathfrak{p}_S$ if and only if $a \in \mathfrak{p}$.

Let f be the map from A to A_S that sends a to $a/1$. If $\mathfrak{q} \in \operatorname{Spec}(A_S)$, then $\mathfrak{q} = (f^{-1}(\mathfrak{q}))_S$. But we just proved that if $\mathfrak{p}$ is a prime ideal in A, then $\mathfrak{p} = f^{-1}(\mathfrak{p}_S)$, so it now follows that the map that sends $\mathfrak{q}$ to $f^{-1}(\mathfrak{q})$ is an inverse to the map that sends $\mathfrak{p}$ to $\mathfrak{p}_S$.

The second statement follows from the fact that if $\mathfrak{p} \cap S = \emptyset$, then $(M_S)_{\mathfrak{p}_S} \cong M_{\mathfrak{p}}$.

If $\mathfrak{p}$ is an associated prime of M, then we have an embedding of $A/\mathfrak{p}$ into M, and localizing gives an embedding of $(A/\mathfrak{p})_S$ into M_S, so $\mathfrak{p}_S$ is an associated prime of M_S. Conversely, suppose that $\mathfrak{p}_S$ is an associated prime of M_S and is the annihilator of m/s. By multiplying by the unit s, we may assume that $s = 1$. The annihilator of m will then be contained in $\mathfrak{p}$, and we claim that some multiple tm will have annihilator exactly $\mathfrak{p}$. In fact, if $x \in \mathfrak{p}$, then $(x/1)(m/1) = xm/1 = 0$ in M_S, so $tmx = 0$ for some $t \in S$, and thus x annihilates tm. Since $\mathfrak{p}$ is finitely generated, we may find a t such that every element of the ideal $\mathfrak{p}$ annihilates tm. Since the annihilator of tm is still contained in $\mathfrak{p}$, we have that $\mathfrak{p}$ is the annihilator of tm and is an associated prime of M. $\square$

An important consequence of Proposition 1.1.2 is that every prime ideal that is minimal in the support of M is associated to M. In fact, if $\mathfrak{p}$ is minimal in the support of M, then $\mathfrak{p}_{\mathfrak{p}}$ is the only prime ideal in the support of $M_{\mathfrak{p}}$ over the localization $A_{\mathfrak{p}}$, so $\mathfrak{p}_{\mathfrak{p}}$ must be associated to $M_{\mathfrak{p}}$, and Proposition 1.1.2 implies that $\mathfrak{p}$ is associated to M.

The basic theorem on prime ideals associated to modules is the following.

Proposition 1.1.3 *If*

$$0 \to M' \to M \to M'' \to 0$$

is a short exact sequence, then

(i) *The support of M is the union of the supports of M' and of M''.*
(ii) *Any associated prime of M is an associated prime of M' or of M''.*

Proof The first statement follows from the exactness of localization.

If $\mathfrak{p}$ is an associated prime of M, let m be an element of M whose annihilator is $\mathfrak{p}$. If the submodule generated by m meets M' only in 0, then $\mathfrak{p}$ is associated to M''. If not, there is a nonzero element in $Am \cap M'$, and its annihilator must be $\mathfrak{p}$, so $\mathfrak{p}$ is associated to M'.   □

The next theorem is one of the main tools for studying modules over Noetherian rings.

Theorem 1.1.4 *Let M be a finitely generated module. There is a finite filtration*

$$0 = M_0 \subseteq M_1 \subseteq \cdots \subseteq M_n = M$$

of M such that

$$M_i/M_{i-1} \cong A/\mathfrak{p}_i$$

for some prime ideal $\mathfrak{p}_i$ for each $i = 1, \ldots n$. The set of prime ideals $\mathfrak{p}_i$ thus obtained contains all the associated primes of M. In particular, the set of associated primes of M is finite.

Proof Proposition 1.1.1 implies that there exists a prime ideal $\mathfrak{p}_1$ and an embedding of $A/\mathfrak{p}_1$ into M. Let M_1 be the image of $A/\mathfrak{p}_1$ in M. This gives the first module in the filtration. We may then apply the same process to M/M_1 to get $M_2, \ldots M_n$ together with prime ideals $\mathfrak{p}_2, \ldots \mathfrak{p}_n$. Since M is a Noetherian module this process must eventually stop, and we have a filtration of the required type.

To show that the set of prime ideals $\mathfrak{p}_1, \ldots, \mathfrak{p}_n$ contains every associated prime of M, we use induction on n. If $n = 1$, $M \cong A/\mathfrak{p}_1$, and the only associated prime ideal of $A/\mathfrak{p}_1$ is $\mathfrak{p}_1$. If $n > 1$, we apply Proposition 1.1.3 to the short exact sequence

$$0 \to M_1 \to M \to M/M_1 \to 0.$$

Every associated prime of M must be associated to M_1 or M/M_1, and hence by induction it must be one of the $\mathfrak{p}_i$. □

One of the uses of this theorem is to show the existence of elements that are not zero-divisors on a module. We define a ring to be *local* if it has a unique maximal ideal. Let A be a local ring, and let M be a finitely generated A-module. If the maximal ideal $\mathfrak{m}$ of A consists of zero-divisors on M, then it must be the union of the finite set of associated primes of M. The next proposition shows that then $\mathfrak{m}$ must be an associated prime ideal of M.

Proposition 1.1.5 *If an ideal $\mathfrak{a}$ is contained in a finite union of prime ideals $\mathfrak{p}_1, \ldots \mathfrak{p}_n$, then we have $\mathfrak{a} \subset \mathfrak{p}_i$ for some i.*

Proof We prove this result by induction on n; if $n = 1$ it is clear. If $n > 1$, we may assume that we have no containment relations $\mathfrak{p}_i \subseteq \mathfrak{p}_j$ for $i \neq j$. Suppose that $\mathfrak{a} \not\subset \mathfrak{p}_i$ for all i. By induction we can find, for each i, an element $x_i \in \mathfrak{a}$ with $x_i \notin \mathfrak{p}_j$ for all $j \neq i$; since we are assuming that $\mathfrak{a}$ is contained in $\mathfrak{p}_1 \cup \cdots \cup \mathfrak{p}_n$, we must have $x_i \in \mathfrak{p}_i$. Let $y_i = \prod_{j \neq i} x_j$ for each i. Then y_i is in all the $\mathfrak{p}_j$ except $\mathfrak{p}_i$, and $\sum y_i$ is in $\mathfrak{a}$ but in none of the $\mathfrak{p}_i$. □

We state explicitly the consequence mentioned above for the maximal ideal of a local ring.

Corollary 1.1.6 *Let A be a local ring with maximal ideal $\mathfrak{m}$, and let M be a finitely generated A-module. Then exactly one of the following holds:*

(i) *The ideal $\mathfrak{m}$ is an associated prime of M.*
(ii) *There exists an element $x \in \mathfrak{m}$ that is not a zero-divisor on M.*

Proof If $\mathfrak{m}$ is associated to M, then clearly every element of $\mathfrak{m}$ is a zero-divisor on M. On the other hand, if $\mathfrak{m}$ consists of zero-divisors on M, it is contained in the union of associated primes of M. By Proposition 1.1.5 it must therefore be contained in one of the associated primes, and since $\mathfrak{m}$ is maximal it must itself be an associated prime of M. □

1.2 Cycles and the Chow Group

Let A be a Noetherian ring. The *dimension* (or Krull dimension) of A, denoted $\dim(A)$, is the supremum of integers i for which there exists a chain of prime ideals $\mathfrak{p}_0 \subset \mathfrak{p}_1 \subset \cdots \subset \mathfrak{p}_i$ such that $\mathfrak{p}_j$ is properly contained in $\mathfrak{p}_{j+1}$ for each

j. For any module M the dimension of M is the supremum of lengths of chains of prime ideals in the support of M. In most cases we will consider, the dimension of a module or ring will be finite, but this is not true in general for Noetherian rings (see Nagata [48, Appendix A]). The dimension of a local ring in particular is finite; we will prove this statement in the next chapter. We note that the dimension of any quotient $A/\mathfrak{a}$ of A is the same whether considered as a ring or as an A-module, since the support of $A/\mathfrak{a}$ is equal to $\mathrm{Spec}(A/\mathfrak{a}) = \{\mathfrak{p} \in \mathrm{Spec}(A) \mid \mathfrak{p} \supseteq \mathfrak{a}\}$.

If $\mathfrak{p}$ is a prime ideal of A, the *height* of $\mathfrak{p}$ is the dimension of the localization $A_\mathfrak{p}$, or, equivalently, the supremum of lengths of chains of prime ideals contained in $\mathfrak{p}$.

Some of the basic definitions in this chapter make sense for all Noetherian rings, but for much of the theory to work properly it will be necessary to set further restrictions. One property of dimension that will be necessary is the following: if $\mathfrak{p}$ and $\mathfrak{q}$ are prime ideals such that $\mathfrak{p}$ is properly contained in $\mathfrak{q}$ and such that there are no prime ideals properly between $\mathfrak{p}$ and $\mathfrak{q}$, then $\dim(A/\mathfrak{p}) = \dim(A/\mathfrak{q}) + 1$. This condition is satisfied, for example, for finitely generated algebras over a field and for many other important cases, but it is not true in general even for localizations of finitely generated algebras over a field. We discuss the question of dimension more thoroughly in Chapter 4 and give a precise statement of the class of rings that we consider. We will temporarily assume that prime ideals satisfy the above condition, as they do in many important cases.

Let X denote the set $\mathrm{Spec}(A)$ of prime ideals of A. For each nonnegative integer i we let $Z_i(X)$, or $Z_i(A)$, be the free Abelian group with basis consisting of all prime ideals $\mathfrak{p}$ such that the dimension of $A/\mathfrak{p}$ is i. We will on occasion refer to the dimension of $A/\mathfrak{p}$ as the dimension of $\mathfrak{p}$. The generator of $Z_i(X)$ corresponding to $\mathfrak{p}$ will be denoted $[A/\mathfrak{p}]$. The group $Z_i(X)$ is called the group of *cycles of X of dimension i*. The direct sum of $Z_i(X)$ over all i is called the group of cycles of A and is denoted $Z_*(X)$ or $Z_*(A)$.

As an example, we take $A = k[X, Y]$, a polynomial ring in two variables. Since $k[X, Y]$ has dimension 2, $Z_k(A)$ is zero for all k except for $k = 0, 1$, and 2. $Z_0(A)$ consists of the free Abelian group on the set of maximal ideals of A, and if k is algebraically closed it follows from the Hilbert nullstellensatz (see Matsumura [44, §5]) that this set corresponds to the set of points (a, b) in $k \times k$. $Z_1(A)$ has a basis consisting of primes $\mathfrak{p}$ with dimension 1, which correspond to irreducible curves. $Z_2(A)$ is free of rank 1 on the class of $[A]$, which corresponds to the 0 ideal.

Let M be an A-module; we wish to define a cycle associated to M. While it appears natural to take the set of associated or minimal primes, it turns out to be more useful to make a slightly different choice. Let the dimension of M be at most i. Then if $\mathfrak{p}$ is any prime ideal such that $\dim(A/\mathfrak{p}) = i$, $M_\mathfrak{p}$ is an $A_\mathfrak{p}$-module of finite length (possible zero). Denote the length of $M_\mathfrak{p}$ as an $A_\mathfrak{p}$-module by $\text{length}(M_\mathfrak{p})$. We let the cycle of dimension i associated to M, denoted $[M]_i$, be the sum

$$\sum_{\dim(A/\mathfrak{p})=i} \text{length}(M_\mathfrak{p})[A/\mathfrak{p}].$$

In Chapter 12 we will define a more complicated cycle associated with a module that has better naturality properties. In many important cases it will agree with the one defined here, but not in general.

We note that we could have summed only over those $\mathfrak{p}$ of dimension i in the support of M, since the coefficients of $[A/\mathfrak{p}]$ for $\mathfrak{p}$ not in the support of M are zero. Also note that if $M = A/\mathfrak{p}$ and i is the dimension of $A/\mathfrak{p}$, then $[A/\mathfrak{p}]_i$ is the same as the generator $[A/\mathfrak{p}]$ defined above. If the dimension of M is less than i, then $[M]_i = 0$.

We next define the Chow group, which is obtained by dividing the group $Z_*(\text{Spec}(A))$ of cycles by a certain equivalence relation, which we now describe. Let $\mathfrak{p}$ be a prime ideal of dimension $i + 1$, and let x be an element of A that is not in $\mathfrak{p}$. Then, since the support of $(A/\mathfrak{p})/x(A/\mathfrak{p})$ does not contain the unique minimal prime ideal in $\text{Spec}(A/\mathfrak{p})$, the dimension of $(A/\mathfrak{p})/x(A/\mathfrak{p})$ is at most i. We denote the cycle $[(A/\mathfrak{p})/x(A/\mathfrak{p})]_i$ by $\text{div}(\mathfrak{p}, x)$.

Definition 1.2.1 Rational equivalence *is the equivalence relation on* $Z_i(A)$ *generated by setting* $\text{div}(\mathfrak{p}, x) = 0$ *for all prime ideals* $\mathfrak{p}$ *of dimension* $i + 1$ *and all elements* $x \notin \mathfrak{p}$. *In other words, two cycles are rationally equivalent if their difference lies in the subgroup generated by the cycles of the form* $\text{div}(\mathfrak{p}, x)$.

The Chow *group of* A *is the direct sum of the groups* $A_i(A)$, *where* $A_i(A)$ *is the group of cycles* $Z_i(A)$ *modulo rational equivalence.*

The Chow group is denoted $A_*(A)$ or $A_*(\text{Spec}(A))$. Like the group of cycles, it comes with a natural grading by dimension.

As an example, we describe the Chow group of $k[X, Y]$, where k is an algebraically closed field. We have that $A_2(k[X, Y]) \cong \mathbb{Z}$ since there are no prime ideals of dimension 3 and thus no relations in dimension 2. Every prime ideal of dimension 1 is principal (since $k[X, Y]$ is a unique factorization domain), and hence every generator $[A/\mathfrak{p}]$ of $Z_1(A)$ is of the form $[A/fA] = \text{div}(0, f)$ where f generates $\mathfrak{p}$, so $A_1(A) = 0$. Every maximal ideal $\mathfrak{m}$ is generated by two elements $x - a, y - b$ with a and b in k. Then $[A/\mathfrak{m}] = \text{div}((x - a), y - b)$, so we also have $A_0(A) = 0$.

While the definition often makes it easy to check that an element of the Chow group is zero, it is usually more difficult to show that elements are not zero. One simple example is given by minimal prime ideals; since the cycle div(q, x) has nonzero coefficients only for nonminimal prime ideals, for any minimal prime ideal $\mathfrak{p}$, the class $[A/\mathfrak{p}]$ is not zero, and is in fact not torsion, in $A_*(A)$. We next give an extended example where an alternative construction is available, and which provides less trivial examples in which cycles can be shown not to vanish in $A_*(A)$.

Let A be an integrally closed domain of dimension d. We show that $A_{d-1}(A)$ can be expressed in terms of the divisor class group of A.

Let K be the quotient field of A. A nonzero finitely generated A-submodule of K is called a *fractional ideal*. A nonzero ideal of A is a special case of a fractional ideal, and if $\mathfrak{a}$ is a fractional ideal, since $\mathfrak{a}$ is finitely generated, there exists a nonzero element $x \in A$ such that $x\mathfrak{a}$ is an ideal in A. We call a fractional ideal $\mathfrak{a}$ *divisorial* if $K/\mathfrak{a}$ has no associated prime ideals of dimension less than $d - 1$. We note that a prime ideal has dimension $d - 1$ if and only if its height is 1.

If A is a Noetherian integrally closed domain and $\mathfrak{p}$ is a prime ideal of A of dimension $d - 1$, then the localization $A_\mathfrak{p}$ is a discrete valuation ring (see Atiyah and Macdonald [1, Chapter 9]). Thus, for any fractional ideal $\mathfrak{a}$ of A, the localization $\mathfrak{a}_\mathfrak{p}$ is a (possibly negative) power of the maximal ideal of $A_\mathfrak{p}$; we denote the power $v_\mathfrak{p}(\mathfrak{a})$. If $\mathfrak{a}$ and $\mathfrak{b}$ are fractional ideals, then the product $\mathfrak{a}\mathfrak{b}$ is also a fractional ideal, and we have

$$v_\mathfrak{p}(\mathfrak{a}\mathfrak{b}) = v_\mathfrak{p}(\mathfrak{a}) + v_\mathfrak{p}(\mathfrak{b})$$

for all $\mathfrak{p}$. If x is an element of K, we define $v_\mathfrak{p}(x)$ to be $v_\mathfrak{p}(Ax)$. Then $v_\mathfrak{p}(x) = 0$ for all but finitely many $\mathfrak{p}$, and $x \in A$ if and only if $v_\mathfrak{p}(x) \geq 0$ for all $\mathfrak{p}$.

We next show that the ring A itself, and in fact any principal fractional ideal, is divisorial. The proof uses the fact that a proper principal ideal cannot have height greater than one; this result will be proven in Chapter 2 (Theorem 2.3.8).

Theorem 1.2.1 *Let A be a Noetherian integrally closed domain. Then every principal fractional ideal is divisorial.*

Proof Let x/y be an element of the quotient field of A, and suppose that $K/(x/y)A$ has an associated prime ideal $\mathfrak{p}$ of height greater than one. Localizing at $\mathfrak{p}$, we may assume that A is local and that $\mathfrak{p}$ is its maximal ideal. Let w/z be an element of K whose annihilator modulo $(x/y)A$ is $\mathfrak{p}$. We then have

$$\mathfrak{p}(w/z) \subseteq (x/y)A \quad \text{but} \quad w/z \notin (x/y)A.$$

Multiplying these equations by y/x, we see that they are equivalent to

$$\mathfrak{p}(yw/zx) \subseteq A \quad \text{but} \quad yw/xz \notin A.$$

Hence it suffices to show that A itself is divisorial. We thus assume that $x/y = 1$.

Since $\mathfrak{p}$ is the maximal ideal of A and $\mathfrak{p}(w/z) \subseteq A$, we either have $\mathfrak{p}(w/z) = A$ or $\mathfrak{p}(w/z) \subseteq \mathfrak{p}$. If $\mathfrak{p}(w/z) \subseteq \mathfrak{p}$, then w/z is integral over A ([1, Proposition 2.4]), so $w/z \in A$ since A is integrally closed. Since we are assuming that $w/z \notin A$, we thus cannot have $\mathfrak{p}(w/z) \subseteq \mathfrak{p}$. Hence it remains to show that the alternative, that $\mathfrak{p}(w/z) = A$, is also impossible.

If $\mathfrak{p}(w/z) = A$, then there exists an $a \in \mathfrak{p}$ such that $a(w/z) = 1$. If b is any element of $\mathfrak{p}$, then $b(w/z) = c$ for some $c \in A$, so $b(w/z) = ca(w/z)$ and thus $b = ca$. But this argument shows that $\mathfrak{p}$ is a principal ideal generated by a, which contradicts the assumption that $\mathfrak{p}$ has height at least two. Thus, A is divisorial, and hence every principal fractional ideal is divisorial. $\qquad\square$

We next define a map ϕ from the set of divisorial ideals to the group of cycles $Z_{d-1}(A)$ by letting

$$\phi(\mathfrak{a}) = \sum v_{\mathfrak{p}}(\mathfrak{a})[A/\mathfrak{p}],$$

where the sum runs over all prime ideals of height one. Since $\mathfrak{a}_{\mathfrak{p}} = A_{\mathfrak{p}}$ for all but finitely many $\mathfrak{p}$ of height one, all but finitely many $v_{\mathfrak{p}}$ are zero and the sum is finite.

We next show that ϕ is a bijection. If $\mathfrak{a}$ and $\mathfrak{b}$ are divisorial ideals, then every associated prime of $\mathfrak{a}/\mathfrak{a} \cap \mathfrak{b}$ or $\mathfrak{b}/\mathfrak{a} \cap \mathfrak{b}$ has height 1, so $\mathfrak{a} = \mathfrak{b}$ if and only if $\mathfrak{a}_{\mathfrak{p}} = \mathfrak{b}_{\mathfrak{p}}$ for every prime ideal $\mathfrak{p}$ of height one; thus ϕ is injective. We claim that ϕ is also surjective. Suppose first that $\eta = \sum n_{\mathfrak{p}}[A/\mathfrak{p}]$ is an element of $Z_{d-1}(A)$ such that every $n_{\mathfrak{p}}$ is nonnegative. Let f be the natural map from A to the product of $A_{\mathfrak{p}}$ over those $\mathfrak{p}$ of height one for which $n_{\mathfrak{p}} \neq 0$, and let $\mathfrak{a}$ be the inverse image under f of the product of $\mathfrak{p}_{\mathfrak{p}}^{n_{\mathfrak{p}}}$. Then $\mathfrak{a}$ is a divisorial ideal and $\phi(\mathfrak{a}) = \eta$.

If some of the coefficients $n_{\mathfrak{p}}$ are negative, we first find an element $x \in A$ such that $v_{\mathfrak{p}}(x) \geq -n_{\mathfrak{p}}$ for all $\mathfrak{p}$; for example, we can choose x to be any nonzero element in the product of $\mathfrak{p}^{-n_{\mathfrak{p}}}$ over those $\mathfrak{p}$ for which $n_{\mathfrak{p}}$ is negative. Then $v_{\mathfrak{p}}(x) + n_{\mathfrak{p}} \geq 0$ for all $\mathfrak{p}$, so there exists an ideal $\mathfrak{a}$ of A such that $v_{\mathfrak{p}}(\mathfrak{a}) = v_{\mathfrak{p}}(x) + n_{\mathfrak{p}} \geq 0$ for all $\mathfrak{p}$ of height one. Thus, $\phi(x^{-1}\mathfrak{a}) = \sum n_{\mathfrak{p}}[A/\mathfrak{p}]$, so ϕ is surjective.

Since the map ϕ is a bijection, we can give the set of divisorial ideals the structure of an Abelian group by pulling back the structure on $Z_{d-1}(A)$. Since $v_{\mathfrak{p}}(\mathfrak{a}\mathfrak{b}) = v_{\mathfrak{p}}(\mathfrak{a}) + v_{\mathfrak{p}}(\mathfrak{b})$ for fractional ideals $\mathfrak{a}$ and $\mathfrak{b}$, this structure is essentially the same as multiplication of ideals; however, there is no reason why the product of divisorial ideals should be divisorial, so a little more work is required to define the group structure by multiplication (see the exercises for more details). However, the product of principal fractional ideals is principal and hence divisorial, and the set of principal fractional ideals forms a subgroup.

Let x/y be an element of K. Then $\phi((x/y)A) = \phi(xA) - \phi(yA)$. Furthermore, $\phi(xA)$ is the sum of $v_{\mathfrak{p}}(x)[A/\mathfrak{p}]$, and $v_{\mathfrak{p}}(x)$ is equal to the length of $A_{\mathfrak{p}}/xA_{\mathfrak{p}}$. Thus $\phi(xA) = \mathrm{div}((0), x)$. It follows that the image of the subgroup of principal ideals under ϕ is exactly the subgroup generated by cycles of the form $\mathrm{div}((0), x)$, which are exactly the cycles rationally equivalent to zero. Thus, if we let F denote the group of fractional ideals and P the subgroup of principal fractional ideals, then we have shown that ϕ induces an isomorphism

$$F/P \cong A_{d-1}(A).$$

The quotient group F/P is called the *divisor class group* of A.

It follows from this construction that if $\mathfrak{a}$ is any divisorial ideal of A that is not principal, then $\phi(\mathfrak{a})$ is not rationally equivalent to zero and so is not zero in $A_{d-1}(A)$. As a specific example, consider the ideal (X, Z) of the ring $A = k[[X, Y, Z, W]]/(XY - ZW)$, which is an integrally closed domain. The ideal (X, Z) is a prime ideal of height one and so is a divisorial ideal. However, (X, Z) is not principal, and hence it defines a nonzero element of the Chow group $A_{d-1}(A)$.

While the Chow group of X has some functorial properties, it is not a functor on the category of Noetherian rings. Instead, only certain classes of ring homomorphisms define maps of Chow groups. If A and B are rings, a map of rings f from A to B defines a map f^* from $\mathrm{Spec}(B)$ to $\mathrm{Spec}(A)$ that sends $\mathfrak{p}$ to $f^{-1}(\mathfrak{p})$. Thus, f will induce a map on cycles from $Z_*(B)$ to $Z_*(A)$. In general, however, this map does not preserve rational equivalence, so it does not induce a map on Chow groups.

One class of maps that does induce maps on Chow groups is the class of flat maps of constant relative dimension. Let k be a nonnegative integer and let $f : A \to B$ be a flat map of rings such that, for every prime ideal $\mathfrak{p}$ of A of dimension i, every minimal prime ideal of $B/\mathfrak{p}B$ has dimension $i + k$. If these conditions hold, we say that f is flat of relative dimension k. We then map $Z_i(A)$ to $Z_{i+k}(B)$ by sending $[A/\mathfrak{p}]$ to $[B/\mathfrak{p}B]_{i+k}$. Before proving that flat maps do induce maps on Chow groups, we prove a lemma that allows us to treat the result of dividing by not a zero-divisor on modules similarly to rational equivalence.

Proposition 1.2.2 *Let M be a finitely generated A-module of dimension at most $i + 1$. Let x be an element of A that is contained in no minimal prime* *ideal of dimension $i + 1$ in the support of M. Then*

$$[M/xM]_i - [_xM]_i = \sum_{\{\mathfrak{p}|\dim(\mathfrak{p})=i+1\}} \text{length}(M_\mathfrak{p}) \, \text{div}(\mathfrak{p}, x),$$

where $_xM$ is the set of elements of M annihilated by x. In particular, if x is not a zero-divisor on M, then $[M/xM]_i$ is rationally equivalent to zero.

Proof We first remark that the condition that x is in no minimal prime ideal in the support of M of dimension $i + 1$ implies that both M/xM and $_xM$ have dimension at most i.

We consider all modules of dimension at most $i + 1$ such that x is contained in no minimal prime of their support of dimension $i + 1$. We show that both sides of the equation we are trying to prove are additive on short exact sequences. Let

$$0 \to M' \to M \to M'' \to 0$$

be a short exact sequence of such modules. Since length is additive, it is clear that the right hand side of the equation is additive. Dividing by x, we obtain an exact sequence

$$0 \to_x M' \to_x M \to_x M'' \to M'/xM' \to M/xM \to M''/xM'' \to 0.$$

If $\mathfrak{p}$ is any prime ideal of dimension i, then localization at $\mathfrak{p}$ gives an exact sequence, so it now follows that the left-hand side is also additive.

Thus we can reduce to the case in which $M = A/\mathfrak{q}$ for a prime ideal $\mathfrak{q}$. If the dimension of $\mathfrak{q}$ is $i + 1$, then $x \notin \mathfrak{q}$, and the result follows from the definition of $\text{div}(\mathfrak{q}, x)$. If $M = A/\mathfrak{q}$ has dimension less than $i + 1$, then there are no prime ideals of dimension $i + 1$ in the support of M, and the expression on the right-hand side of the equation is zero. If $x \in \mathfrak{q}$, then $_xM = M/xM = M$, and we get zero on the left-hand side as well. If $x \notin \mathfrak{q}$ and $\dim(M) < i + 1$, the dimensions of M/xM and $_xM$ are less than i, so we again get zero on both sides of the equation. $\qquad\square$

Theorem 1.2.3 *If $f : A \to B$ is a flat map of relative dimension k, then the map from $Z_i(A)$ to $Z_{i+k}(B)$ that sends $[A/\mathfrak{p}]$ to $[B/\mathfrak{p}B]_{i+k}$ induces a map on Chow groups from $A_i(A)$ to $A_{i+k}(B)$.*

Proof Let $\mathfrak{p}$ be a prime ideal of A of dimension $i + 1$ and let x be an element of A that is not in $\mathfrak{p}$. We must show that the cycle $\text{div}(\mathfrak{p}, x)$ is mapped to a cycle that is rationally equivalent to zero in $Z_{i+k}(B)$. We have the short exact sequence

$$0 \to A/\mathfrak{p} \xrightarrow{x} A/\mathfrak{p} \to (A/\mathfrak{p})/x(A/\mathfrak{p}) \to 0$$

and $\operatorname{div}(\mathfrak{p}, x) = [(A/\mathfrak{p})/x(A/\mathfrak{p})]_i$. Since B is flat over A, tensoring with B gives the exact sequence

$$0 \to B/\mathfrak{p}B \xrightarrow{x} B/\mathfrak{p}B \to (B/\mathfrak{p}B)/x(B/\mathfrak{p}B) \to 0.$$

Since $B/\mathfrak{p}B$ has dimension at most $i + 1 + k$, it follows from Proposition 1.2.2 that $[(B/\mathfrak{p}B)/x(B/\mathfrak{p}B)]_{i+k}$ is rationally equivalent to zero. To complete the proof we must show that $[(B/\mathfrak{p}B)/x(B/\mathfrak{p}B)]_{i+k}$ is the image of $[(A/\mathfrak{p})/x(A/\mathfrak{p})]_i$.

Let

$$0 = M_0 \subset M_1 \subset \cdots \subset M_k = (A/\mathfrak{p})/x(A/\mathfrak{p})$$

be a filtration of $(A/\mathfrak{p})/x(A/\mathfrak{p})$ such that $M_{j+1}/M_j \cong A/\mathfrak{q}_j$ for prime ideals $\mathfrak{q}_j$ of A. Then $[(A/\mathfrak{p})/x(A/\mathfrak{p})]_i$ is the sum of $[A/\mathfrak{q}_j]_i$ over all $\mathfrak{q}_j$ of dimension i. Tensoring with B and again using flatness, we have a filtration of $(B/\mathfrak{p}B)/x(B/\mathfrak{p}B)$ with quotients of the form $A/\mathfrak{q}_j \otimes B$, and the associated cycle $[A/\mathfrak{q}_j \otimes B]_{i+k}$ is the image of $[A/\mathfrak{q}_j]_i$. If $\mathfrak{q}_j$ has dimension less than i, then all components of $[A/\mathfrak{q}_j \otimes B]_{i+k}$ have dimension less than $i + k$, since f is flat of relative dimension k. Thus, $[(B/\mathfrak{p}B)/x(B/\mathfrak{p}B)]_{i+k}$ is the sum of $[A/\mathfrak{q}_j \otimes B]_{i+k}$ for $\mathfrak{q}_j$ of dimension i, so $[(B/\mathfrak{p}B)/x(B/\mathfrak{p}B)]_{i+k}$ is the image of $\operatorname{div}(\mathfrak{p}, x)$ as required.   □

The map on Chow groups whose existence is proven in Theorem 1.2.3 is called flat pull-back by f and is denoted f^*. It is referred to as a pull-back because the map is considered to act from $\operatorname{Spec}(B)$ to $\operatorname{Spec}(A)$, and f^* of a cycle is the inverse image under this map.

Included in the class of flat maps of constant relative dimension are many of the usual constructions of commutative algebra. For example, if we define the dimension of a prime ideal in a localization of A to be its dimension in A, then localization is flat of relative dimension zero; note that, although the inverse image of a prime ideal may be empty, this is allowed by the definition. A polynomial ring in n variables defines a flat map of relative dimension n. We conclude this chapter with theorems on the behavior of the Chow group with respect to these operations.

Proposition 1.2.4 *Let S be a multiplicatively closed subset of A. For each prime ideal $\mathfrak{p}_S$ of A_S, define the dimension of $[A_S/\mathfrak{p}_S]$ to be the dimension of $[A/\mathfrak{p}]$. Let $Z_*(S, A)$ denote the subgroup of $Z_*(A)$ generated by those prime ideals of A that meet S. Then the inclusion of $Z_*(S, A)$ in $Z_*(A)$ induces an exact sequence*

$$Z_*(S, A) \to A_*(A) \to A_*(A_S) \to 0.$$

Proof The map from $A_*(A)$ to $A_*(A_S)$ is flat pull-back as defined above; our definition of dimension in $\text{Spec}(A_S)$ implies that $A_i(A)$ maps to $A_i(A_S)$ for each i. It is clear that every prime ideal that meets S goes to zero in $Z_*(A_S)$ so the composition is zero. Also, every prime ideal of A_S is extended from A, so the map on the right is surjective. The only nontrivial part is to show exactness at $A_*(A)$.

Let $\sum_j n_j[A/\mathfrak{p}_j]$ be a cycle in $Z_i(A)$ that goes to zero in $A_i(A_S)$. Then there exist prime ideals $\mathfrak{q}_k$ of A_S of dimension $i+1$ and elements x_k of A_S not in $(\mathfrak{q}_k)_S$ such that

$$\sum_j n_j[A_S/\mathfrak{p}_{jS}] = \sum_k \text{div}(\mathfrak{q}_k, x_k)$$

in $Z_i(A_S)$. Now the prime ideals $\mathfrak{q}_k$ are extended from prime ideals of A, which we also denote $\mathfrak{q}_k$. Furthermore, we may assume by multiplying by units in A_S that the elements x_k are in A. We now take the difference

$$\sum_j n_j[A/\mathfrak{p}_j] - \sum_k \text{div}(\mathfrak{q}_k, x_k),$$

which is a cycle in $Z_i(A)$. Since its image as a cycle in $Z_i(A_S)$ is zero, all of its components with nonzero coefficients must be prime ideals that do not survive in A_S, which means that they are in $Z_*(S, A)$. Thus, $\sum_j n_j[A/\mathfrak{p}_j]$ is rationally equivalent to a cycle in $Z_*(S, A)$, and the above sequence is exact. $\qquad\square$

A similar result is obtained when we restrict to an open subset of $\text{Spec}(A)$. We define a topology on $\text{Spec}(A)$, called the *Zariski topology*, by defining the closed sets to be sets of the form $V(\mathfrak{a}) = \{\mathfrak{p} \in \text{Spec}(A) \mid \mathfrak{a} \subset \mathfrak{p}\}$ for all ideals (or subsets) $\mathfrak{a}$. If U is an open subset of $X = \text{Spec}(A)$ we define the Chow group of U in the same way as that of X but take as generators only those prime ideals that are in U. We have a restriction map, which again is flat of relative dimension 0. On the other hand, if V is a closed subset, say $V = V(\mathfrak{a})$, then the inclusion of $\text{Spec}(A/\mathfrak{a})$ into $\text{Spec}(A)$ defines an inclusion on cycles that clearly passes to rational equivalence, so we have a map from $A_*(V)$ to $A_*(X)$. The proof of the following proposition is the same as that of Proposition 1.2.4.

Proposition 1.2.5 *Let U be an open subset of* $\text{Spec}(A)$. *We then have an exact sequence*

$$A_*(X - U) \to A_*(A) \to A_*(U) \to 0.$$

The final result of this chapter shows that the flat pull-back defined by adjoining an indeterminate is surjective. In fact, it is an isomorphism, but to define the inverse we need to consider intersection products that will not be defined in sufficient generality until Chapter 8. It suffices to prove that the map obtained by adjoining one variable T is surjective. The map from A to $A[T]$ is flat of relative dimension 1, so we have maps $A_i(A) \to A_{i+1}(A[T])$ for each i.

Theorem 1.2.6 *The map defined by flat pull-back of cycles from $A_i(A)$ to $A_{i+1}(A[T])$ is surjective for all i. In particular, $A_0(A[T]) = 0$.*

Proof If there exists a q such that $[A[T]/\mathfrak{q}]$ is not in the image of the Chow group of A, we choose one such that its intersection with A is maximal among all prime ideals of $A[T]$ with this property. The existence of such a q follows from the fact that A is Noetherian. Let the dimension of q be $i + 1$, and let p denote the intersection of q with A.

Let $\mathfrak{q}'$ be the extension of p to $A[T]$, and consider the localization at the multiplicatively closed set $S = A - \mathfrak{p}$. The ring $(A[T]/\mathfrak{q}')_S$ is a polynomial ring in one variable over the field $k(\mathfrak{p})$, so $(A[T]/\mathfrak{q}')_S$ is a principal ideal domain and the ideal $\mathfrak{q}_S$ is principal; by clearing denominators we may assume that $\mathfrak{q}_S$ is generated by a polynomial $f(T)$ with coefficients in A. We now take the cycle $[A[T]/\mathfrak{q}] - \mathrm{div}(\mathfrak{q}', f(T))$. Since $f(T)$ generates q in the localization of $A[T]/\mathfrak{q}'$ at $S = A - \mathfrak{p}$, the only prime ideal with nonzero coefficient in $\mathrm{div}(\mathfrak{q}', f(T))$ that contracts to p is q, and the coefficient of $[A[T]/\mathfrak{q}]$ in $\mathrm{div}(\mathfrak{q}', f(T))$ is 1. Thus, every prime ideal with nonzero coefficient in $[A[T]/\mathfrak{q}] - \mathrm{div}(\mathfrak{q}', f(T))$ must contract to a prime ideal of A that properly contains p. By the maximality of p, these prime ideals are in the image of $A_*(A)$, so $[A[T]/\mathfrak{q}]$ is rationally equivalent to a cycle in the image of $A_*(A)$, and thus q is in the image of $A_*(A)$ as well. $\square$

Exercises

1.1 Let p be a prime ideal of A. Show that the only associated prime of $A/\mathfrak{p}$ is p.

1.2 Show that every associated prime ideal of M is in the support of M.

1.3 Let S be a multiplicative set in A, and let f be the map from A to A_S. Show that for every ideal $\mathfrak{a}$ of A, $f^{-1}(\mathfrak{a}_S)$ is the set of $a \in A$ such that $sa \in \mathfrak{a}$ for some $s \in S$. Give an example of an ideal $\mathfrak{a}$ for which $f^{-1}(\mathfrak{a}_S) \neq \mathfrak{a}$.

1.4 Let p be a prime ideal that is minimal in the support of a module M. Show that the number of times $A/\mathfrak{p}$ occurs in any filtration of M of the type described in Theorem 1.1.4 is the length of $M_{\mathfrak{p}}$.

1.5 Let $A = k[[X, Y]]/(X^2, XY)$. Show that the associated prime ideals of A are (X) and (X, Y). Show that for every positive integer n there is a filtration as in Theorem 1.1.4 in which (X, Y) occurs n times.

1.6 Let A be the polynomial ring $k[X, Y]$, and let M be the ideal (X, Y) considered as an A-module. Show that any filtration of M as in Theorem 1.1.4 must include factors $A/\mathfrak{p}$ where $\mathfrak{p}$ is not associated to M.

1.7 Let M be a module of finite length, and let $\mathfrak{p}_1, \ldots, \mathfrak{p}_n$ be the prime ideals in the support of M. Show that $\mathfrak{p}_i$ is maximal for each i. Show that

$$M = M_{\mathfrak{p}_1} \oplus \cdots \oplus M_{\mathfrak{p}_n}.$$

1.8 Let A be a Dedekind domain (an integrally closed Noetherian domain of dimension 1). Show that every nonzero fractional ideal is divisorial. Show that the usual multiplication of ideals is the same as the multiplication defined in this chapter.

1.9 Show that the ring $k[X, Y, Z, W]/(XY - ZW)$ is an integrally closed domain.

1.10 Let A be an integrally closed domain of dimension d. For each nonzero fractional ideal $\mathfrak{a}$ of A, let $\bar{\mathfrak{a}}$ be the set of elements of K whose annihilator modulo $\mathfrak{a}$ is not contained in any prime ideal of height 1. Show that $\bar{\mathfrak{a}}$ is a divisorial fractional ideal. Show that the group structure on the set of divisorial ideals can be defined be letting the product of $\mathfrak{a}$ and $\mathfrak{b}$ be $\overline{\mathfrak{a}\mathfrak{b}}$.

1.11 Let $A = k[[X, Y, Z]]$ and $B = k[[X, Y, Z, W]]/(XY - ZW)$. Show that under the obvious map from A to B, (X, Z) is rational equivalent to zero but its image in B is not.

1.12 Show that the ideal $(X, Z)^n$ is a divisorial ideal in the ring $A = k[[X, Y, Z, W]]/(XY - ZW)$ and deduce that the class of $[A/(X, Z)]$ is not torsion in $A_2(A)$.

2

Graded Rings and Samuel Multiplicity

In this chapter we discuss graded and multigraded rings. The main examples for later constructions are graded rings defined by ideals in a ring, such as the Rees ring and associated graded ring of the ideal, but we also consider more general graded rings. We show that, under certain conditions, a graded ring defines a Hilbert polynomial; the degree and leading coefficient of this polynomial are important invariants of the ring. In the case of the associated graded ring of an ideal the degree is related to the dimension of the ring and the leading coefficient is used to define the Samuel multiplicity of the ideal. We also include a short account of dimension theory for local rings. In the last section we define generalizations of the concept of multiplicity of ideals, first to mixed multiplicities of a finite set of ideals and then to the case of submodules of a free module.

2.1 Graded Rings and Hilbert Polynomials

Definition 2.1.1 *Let G be the Abelian group $\mathbb{Z}^n$ for some integer $n \geq 0$. A graded ring is a ring, A, that has a decomposition as an Abelian group*

$$A = \oplus_{i \in G} A_i$$

such that $A_i A_j \subseteq A_{i+j}$ for all $i, j \in G$. A graded A-module is an A-module M that has a decomposition

$$M = \oplus_{i \in G} M_i$$

such that $A_i M_j \subseteq M_{i+j}$ for all $i, j \in G$. A ring homomorphism f from graded ring A to graded ring B is a homomorphism of graded rings if $f(A_i) \subseteq B_i$ for all $i \in G$; homomorphisms of graded modules are defined similarly.

18

An element of a graded module M is called *homogeneous* if it belongs to M_i for some i; i is the *degree* of the element. If m is an arbitrary element of M and we write $m = \sum m_i$ with $m_i \in M_i$, m_i is called the homogeneous component of m of degree i. A submodule N of a graded module M is a *graded submodule* if $N = \oplus_i (N \cap M_i)$, or, equivalently, if an element of M is in N if and only if every homogeneous component is in N. If N is a graded submodule of M, the quotient M/N is a graded module. A special case is that of a graded ideal in A. The annihilator of a homogeneous element is a graded ideal.

In the classical examples of graded rings and modules, the group G is the additive group of integers. However, in later chapters we will need to consider graded rings over the group $\mathbb{Z}^n$, so it will be useful to give the basic definitions in this generality. However, some of the results hold only for $\mathbb{Z}$-graded rings, and others will be proven first in this case and then extended to the case of $\mathbb{Z}^n$ to make the arguments simpler. The group G will often be called the index group of A and its elements will be referred to as indices.

For the most part, we will consider $\mathbb{Z}$-graded rings such that $A_i = 0$ for $i < 0$ and $\mathbb{Z}^n$-graded rings such that $A_i = 0$ for all $i = (i_1, \ldots i_n)$ such that $i_k < 0$ for some k, and we will assume that graded rings have this property unless stated otherwise. The summand A_0 is a subring of A, and, under our assumption for $\mathbb{Z}$-graded rings, the subgroup $\oplus_{i>0} A_i$ is an ideal. This ideal is called the *irrelevant* ideal of A.

If M is a graded module and $k \in \mathbb{Z}^n$, we denote by $M[k]$ the graded module obtained by shifting degrees by k; that is, we have $M[k]_i = M_{k+i}$ for all $i \in \mathbb{Z}^n$.

Proposition 2.1.1 *Let A be a $\mathbb{Z}^n$-graded ring, and let $\mathfrak{a}$ be a graded ideal of A. Then $\mathfrak{a}$ is prime if and only if it satisfies the condition that for all homogeneous elements a and b of A, if $ab \in \mathfrak{a}$, then $a \in \mathfrak{a}$ or $b \in \mathfrak{a}$.*

Proof Let ">" denote the lexicographic order on $\mathbb{Z}^n$, so that we have $(i_1, \ldots, i_n) > (j_1, \ldots, j_n)$ when $i_k > j_k$, where k is the lowest integer m for which $i_m \neq j_m$. Then if i and j are elements of $\mathbb{Z}^n$ with $i > j$, we have $i + k > j + k$ for all $k \in \mathbb{Z}^n$.

It is clear that the condition of the hypothesis is necessary for $\mathfrak{a}$ to be prime. Assume that it holds, and let a and b be arbitrary elements of A that are not in $\mathfrak{a}$. Let i and j be the largest indices under the lexicographic order such that the homogeneous components a_i and b_j are not in $\mathfrak{a}$. Then $a_i b_j$ is a homogeneous element of degree $i + j$, which is not in $\mathfrak{a}$. For every other product of homogeneous components $a_k b_m$ with $k + m = i + j$, we must have $k > i$ or $m > j$, so by the maximality of i and j, $a_k b_m$ must lie in $\mathfrak{a}$. Thus, the homogeneous component of ab of degree $i + j$ is not in $\mathfrak{a}$. Since $\mathfrak{a}$ is a graded ideal, we thus have $ab \notin \mathfrak{a}$, so $\mathfrak{a}$ is prime. $\square$

A further condition that we generally assume is that A is generated as an algebra over A_0 by elements of degree 1. Let e_i denote the element $(0, \ldots, 1, \ldots 0)$ of $\mathbb{Z}^n$, where the 1 is the ith position. We will use this notation throughout this chapter and also later in the book. Then this assumption states that A is generated over A_0 by $A_{e_1} \oplus \cdots \oplus A_{e_n}$. If $n = 1$, it states simply that A is generated over A_0 by A_1. We note that, since we are assuming that rings are Noetherian, it follows that A_0 is Noetherian and A_i is a finitely generated A_0-module for all i.

Example. Let R be a ring, and let $A = R[X_1, \ldots X_n]$ be a polynomial ring over R in n variables. Then A becomes a $\mathbb{Z}$-graded ring by letting A_i be the set of all homogeneous polynomials of degree i. We can also give $R[X_1, \ldots X_n]$ the structure of a $\mathbb{Z}^n$-graded ring by letting the part of degree $(i_1, \ldots i_n)$ be the free submodule of rank 1 generated by $X_1^{i_1} \cdots X_n^{i_n}$.

In general, under our assumptions, if A is $\mathbb{Z}$-graded and $a_1, a_2, \ldots, a_k$ form a set of generators of A_1 as an A_0-module, then there is a surjection of graded rings from $A_0[X_1, \ldots X_k]$ onto A obtained by mapping X_i to a_i. If A_0 is also Artinian, and thus a ring of finite length (for example, a field), then A_i, being finitely generated over A_0 by all monomials of degree i in the finite set of generators of A_1, is also a module of finite length. Furthermore, if M is a finitely generated graded A-module, then M_n must have finite length as an A_0-module for all n. We define the *Hilbert function* H_M of M by the formula

$$H_M(i) = \text{length}(M_i).$$

Similarly we may define the Hilbert function of a finitely generated $\mathbb{Z}^n$-graded module over a graded ring for which A_0 is Artinian; in this case i denotes an element of $\mathbb{Z}^n$.

Proposition 2.1.2 *If M is graded, then every associated prime ideal of M is graded.*

Proof It suffices to show that M has at least one graded associated prime $\mathfrak{p}$; we can then divide by a graded submodule isomorphic to $A/\mathfrak{p}$ and, following the method of Theorem 1.1.4, obtain a filtration of M with quotients that are isomorphic to $A/\mathfrak{p}$ for graded prime ideals $\mathfrak{p}$. Since the set of prime ideals thus obtained must contain all the associated primes, this will prove the result.

Let $\mathfrak{a}$ be a maximal annihilator of a homogeneous element. If a and b are homogeneous elements with $ab \in \mathfrak{a}$, then we may use the maximality of $\mathfrak{a}$ as in the proof of Proposition 1.1.1 to deduce that a or b is in $\mathfrak{p}$. Thus, $\mathfrak{p}$ is prime and is a graded associated prime of M. $\qquad\square$

The main result of this section is the existence of the Hilbert polynomial associated to a graded module over a graded ring. Let A be a $\mathbb{Z}^n$-graded ring such that A_0 is Artinian, and let M be a finitely generated graded A-module. Let H_M be the Hilbert function of M. The Hilbert polynomial is a polynomial P_M in n variables such that $P_M(i) = H_M(i)$ for sufficiently large i. We note that the additivity of the length function on short exact sequences implies that the Hilbert function is additive on short exact sequences of graded modules. Also, if N is a submodule or quotient module of M, then $H_N(i) \leq H_M(i)$ for all i. The Hilbert function also satisfies the property $H_{M[k]}(i) = H_M(i+k)$ for all i and k.

We first consider the case in which $n = 1$, so that A is graded over $\mathbb{Z}$. Suppose first that A is a polynomial ring $A_0[X_1, \ldots X_n]$ with the usual $\mathbb{Z}$-grading and that $M = A$. In this case the Hilbert function can be computed directly.

Proposition 2.1.3 *Let* $M = A_0[X_1, \ldots X_d]$ *considered as a* $\mathbb{Z}$-*graded module. Then*

$$H_M(i) = \binom{i+d-1}{d-1} \text{length}(A_0)$$

for all $i \geq 0$.

Proof Since M_i is a free A_0-module with the set of monomials of degree i as a basis, the theorem states that the number of monomials of degree i in d variables is given by the binomial coefficient $\binom{i+d-1}{d-1}$. There are various ways of counting the monomials; we give an inductive proof that generalizes to more general graded rings.

If $d = 1$, we have

$$\binom{i+d-1}{d-1} = \binom{i}{0} = 1,$$

which is indeed the number of monomials in one variable in degree i. Furthermore, if $i = 0$ and d is any positive integer, we have

$$\binom{i+d-1}{d-1} = \binom{d-1}{d-1} = 1,$$

which is the number of monomials of degree zero.

Assume that $d > 1$, and consider the short exact sequence

$$0 \rightarrow A_0[X_1, \ldots X_d][-1] \xrightarrow{X_d} A_0[X_1, \ldots X_d] \rightarrow A_0[X_1, \ldots X_{d-1}] \rightarrow 0.$$

Taking lengths in each degree, we have the formula

$$H_{A_0[X_1, \ldots X_d]}(i) = H_{A_0[X_1, \ldots X_d]}(i - 1) + H_{A_0[X_1, \ldots X_{d-1}]}(i). \tag{1}$$

On the other hand, we have the standard relation for binomial coefficients:

$$\binom{i + d - 1}{d - 1} = \binom{i + d - 2}{d - 1} + \binom{i + d - 2}{d - 2}. \tag{2}$$

By induction on i and d, the right-hand sides of Equations (1) and (2) are equal, and hence the left-hand sides are also equal, so we have

$$H_{A_0[X_1, \ldots X_d]}(i) = \binom{i + d - 1}{d - 1}.$$

$\square$

It follows from Proposition 2.1.3 that the Hilbert function of a polynomial ring in d variables is a polynomial in i of degree $d - 1$ for $i \geq 0$ and has a convenient expression as a binomial coefficient. It remains true for any finitely generated module over a graded ring that the Hilbert function is represented by a polynomial (under our usual assumptions) and that it can be written as a combination of binomial coefficients with integer coefficients. The property of this type of function that makes this work is the following.

Lemma 2.1.4 *Let $f(i)$ be a function defined for integers i greater than or equal to some fixed integer j, and suppose that*

$$f(i) - f(i - 1) = \sum_{k=0}^{m} c_k \binom{i + k}{k}$$

for some integers c_k and for all $i \geq j + 1$. Let r be the integer defined by the equation

$$r = f(j) - \left(\sum_{k=0}^{m} c_k \binom{j + k + 1}{k + 1} \right).$$

Then

$$f(i) = \sum_{k=0}^{m} c_k \binom{i + k + 1}{k + 1} + r$$

for all $i \geq j$.

Proof Let g be the function defined by

$$g(i) = c_k \binom{i+k+1}{k+1}$$

for all integers i. Then we have

$$g(i) - g(i-1) = c_k \binom{i+k+1}{k+1} - c_k \binom{i+k}{k+1} = c_k \binom{i+k}{k}.$$

Hence if we let $h(i) = \sum_{k=0}^{m} c_k \binom{i+k+1}{k+1} + r$, we have

$$h(i) - h(i-1) = f(i) - f(i-1)$$

for all $i \geq j+1$, so they will define the same function for all $i \geq j$ if and only if $h(j) = f(j)$. We have

$$h(j) = \sum_{k=0}^{m} c_k \binom{j+k+1}{k+1} + r$$

$$= \sum_{k=0}^{m} c_k \binom{j+k+1}{k+1} + \left(f(j) - \sum_{k=0}^{m} c_k \binom{j+k+1}{k+1} \right) = f(j)$$

so $f(i) = h(i)$ for all $i \geq j$ as was to be shown. $\qquad\square$

The next theorem proves the existence of the Hilbert polynomial, the polynomial that has the same values as the Hilbert function for large values of i. In addition, we show that the degree of the Hilbert polynomial is bounded by the number of generators of A over A_0.

Theorem 2.1.5 *Let A be a graded ring such that A_0 is Artinian and A is generated over A_0 by d elements of A_1. Let M be a finitely generated graded A-module. Then there is a polynomial P_M of degree at most $d-1$ such that $P_M(i) = H_M(i)$ for i sufficiently large. Moreover, the polynomial P_M can be expressed in the form*

$$P_M = \sum_{k=0}^{d-1} c_k \binom{i+k}{k}$$

with integer coefficients c_k.

Proof The proof is by induction on d. If $d = 0$, then $A = A_0$, which is Artinian, and every finitely generated A-module has finite length, so we have $M_i = 0$ for i large, and the theorem is true in this case.

If $d > 0$, suppose that A is generated by the elements $a_1, \ldots, a_d$ of degree 1. We divide by the ideal generated by a_1; the quotient $A/a_1 A$ is then generated by

$d - 1$ elements over A_0. Denoting $_{a_1} M$ the submodule of elements annihilated by a_1, we have a short exact sequence of graded modules

$$0 \to _{a_1} M[-1] \to M[-1] \overset{a_1}{\to} M \to M/a_1 M \to 0.$$

Now $_{a_1} M$ and $M/a_1 M$ are modules over the ring $A/a_1 A$, which, as we noted above, is a graded ring generated over A_0 by at most $d - 1$ generators of degree 1. Furthermore, the above short exact sequence gives the relation

$$H_M(i) - H_M(i - 1) = H_{M/a_1 M}(i) - H_{a_1 M}(i - 1)$$

for all i. By induction, the expression on the right can be written in the required form for i sufficiently large, so Lemma 2.1.4 implies that the function $H_M(i)$ may also be written in the required form for i sufficiently large.   □

If $n > 1$ the proof of the existence of Hilbert polynomials is somewhat more complicated, but it is essentially the same idea. We order $\mathbb{Z}^n$ by letting $i = (i_1, \ldots, i_n) \leq j = (j_1, \ldots, j_n)$ if $i_k \leq j_k$ for all k (this order is, of course, not the lexicographic order that we used before). To simplify the formulas, we introduce a notation for products of binomial coefficients. If $i = (i_1, \ldots, i_n)$ and $j = (j_1, \ldots, j_n)$ are elements of $\mathbb{Z}^n$, we let $\binom{i}{j}$ denote the product $\binom{i_1}{j_1} \cdots \binom{i_n}{j_n}$. Then the standard inductive equation for binomial coefficients implies that for each k we have

$$\binom{i + e_k}{j + e_k} = \binom{i}{j + e_k} + \binom{i}{j},$$

where $e_k = (0, \ldots, 0, 1, \ldots, 0)$ with 1 in the kth position.

The next lemma is a version of Lemma 2.1.4 for rings graded over $\mathbb{Z}^n$. In the notation for the sum we mean the sum over all elements of $\mathbb{Z}^n$ between the given bounds in the order defined above. If $i = (i_1, \ldots, i_n) \in \mathbb{Z}^n$, we denote the element $(i_1, \ldots, i_{n-1})$ of $\mathbb{Z}^{n-1}$ by $\tilde{i}$.

Lemma 2.1.6 *Let $f(i)$ be a function defined for $i \in \mathbb{Z}^n$ and suppose that*

$$f(i) - f(i - e_n) = \sum_{k=0}^{m} c_k \binom{i + k}{k}$$

for some integers c_k and for all $i \geq j + e_n$. Then

$$f(i) = \sum_{k=0}^{m} c_k \binom{i + k + e_n}{k + e_n} + f(i_0, \ldots, i_{n-1}, j_n)$$

$$- \sum_{k=0}^{m} c_k \binom{\tilde{i} + \tilde{k}}{\tilde{k}} \binom{j_n + k_n + 1}{k_n + 1} \tag{3}$$

for all $i \geq j$.

Proof The proof is essentially the same as that of Lemma 2.1.4. We note that this time the expression

$$f(i_0, \ldots, i_{n-1}, j_n) - \sum_{k=0}^{m} c_k \binom{\tilde{i} + \tilde{k}}{\tilde{k}} \binom{j_n + k_n + 1}{k_n + 1}$$

defines a function on $\mathbb{Z}^{n-1}$ rather than a constant. Let $h(i)$ be the function defined by the right-hand side of Equation (3). If we compute $h(i_1, \ldots, i_{n-1}, j_n)$, the first and third terms cancel, and we obtain $f(i_1, \ldots, i_{n-1}, j_n)$. Hence $h(i) = f(i)$ when $i_n = j_n$. As before, the hypothesis implies that

$$h(i + e_n) - h(i) = f(i + e_n) - f(i)$$

for all $i \geq j + e_n$. Hence the two functions agree for all $i \geq j$. $\qquad\square$

Theorem 2.1.7 *Let A be a $\mathbb{Z}^n$-graded ring and let M be a finitely generated graded module. Assume that A_0 is Artinian, that A is generated over A_0 by $A_{e_1} \oplus \cdots \oplus A_{e_n}$, and that A_{e_i} is generated as an A_0-module by d_i elements. Then there is a polynomial in n variables $P_M(i_1, \ldots i_n)$ and an element $j \in \mathbb{Z}^n$ such that the length of M_i is equal to $P_M(i)$ for all $i \geq j$. Moreover, letting $d' = (d_1 - 1, \ldots, d_n - 1)$, the polynomial P_M can be expressed in the form*

$$P_M = \sum_{k=0}^{d'} c_k \binom{i + k}{k}$$

with integer coefficients c_k.

Proof The proof is by induction, first on n and then on d_n. The case $n = 1$ is Theorem 2.1.5. If $n > 1$ and $d_n = 0$, then a finitely generated A-module is also finitely generated over the subring of elements of degrees of the form $(i_1, \ldots, i_{n-1}, 0)$, so the theorem holds in this case by induction on n.

Assume that $n > 1$ and $d_n \geq 1$, and let a be an element of A_{e_n} such that $A_{e_n}/a A_{e_n}$ is generated by at most $d_n - 1$ elements. As before, we obtain an exact sequence

$$0 \to_a M[-e_n] \to M[-e_n] \xrightarrow{a} M \to M/aM \to 0.$$

Let $d'' = (d_1 - 1, \ldots, d_{n-1} - 1, d_n - 2)$. By induction on d_n, the above exact sequence implies that we can write

$$H_M(i) - H_M(i - e_n) = \sum_{k=0}^{d''} c_k' \binom{i + k}{k}$$

for some integers c_k for sufficiently large i. Hence Lemma 2.1.6 implies that there exists a $j \in \mathbb{Z}^n$ such that we can write

$$H_M(i) = \sum_{k=0}^{d''} c_k' \binom{i+k+e_n}{k+e_n} + H_M(i_1, \ldots, i_{n-1}, j_n)$$

$$- \sum_{k=0}^{d''} c_k' \binom{\tilde{i}+\tilde{k}}{\tilde{k}} \binom{j_n+k_n+1}{k_n+1}.$$

The first and third terms of this expression are binomial coefficients, so they are in the correct form. The middle term is the Hilbert function of the finitely generated module consisting of elements of M of degrees $(i_1, \ldots, i_{n-1}, j_n)$ over the ring of elements of A of degrees of the form $(i_1, \ldots, i_{n-1}, 0)$, so it has the right form for large i by induction on n. Since the addition of e_n to k raises the degree of the last factor by one, we can thus conclude that we can write

$$H_M(i) = \sum_{k=0}^{d'} c_k \binom{i+k}{k}$$

for some integers c_k as stated in the theorem. $\qquad\square$

2.2 Rees Rings and Associated Graded Rings

In this section we discuss two examples of graded rings that will be used throughout the book.

Definition 2.2.1 *Let A be a ring (not necessarily graded), and let $\mathfrak{a}$ be an ideal of A. The* Rees ring *of $\mathfrak{a}$, denoted $R(\mathfrak{a})$, is the ring defined by*

$$R(\mathfrak{a}) = \oplus_{i \geq 0} \mathfrak{a}^i$$

where the multiplication from $\mathfrak{a}^i \times \mathfrak{a}^j$ to $\mathfrak{a}^{i+j}$ is induced by the multiplication in A. The associated graded ring *of $\mathfrak{a}$, denoted $G_\mathfrak{a}$, is the ring*

$$G_\mathfrak{a} = R(\mathfrak{a}) \otimes_A A/\mathfrak{a} = \oplus_{i \geq 0} \mathfrak{a}^i/\mathfrak{a}^{i+1}.$$

More generally, if M is an A-module, we can define corresponding graded modules $R(\mathfrak{a}, M)$ and $G_\mathfrak{a}(M)$ by letting $R(\mathfrak{a}, M) = \oplus_{i \geq 0} \mathfrak{a}^i M$ and letting $G_\mathfrak{a}(M) = \oplus_{i \geq 0} \mathfrak{a}^i M/\mathfrak{a}^{i+1} M$. We note that $R(\mathfrak{a})_0 = A$ and that $R(\mathfrak{a})$ is generated over $R(\mathfrak{a})_0$ by any set of generators of $\mathfrak{a}$ in degree 1. Similarly, $G_\mathfrak{a}$ has $A/\mathfrak{a}$ as the component of degree zero and again is generated by the images of any set of generators of $\mathfrak{a}$ in degree 1. In particular, if A is Noetherian, the Rees ring and associated graded ring of any ideal of A are also Noetherian. This fact has an important consequence, as follows.

Theorem 2.2.1 (The Artin-Rees Lemma) *Let M be a finitely generated module over a Noetherian ring A, and let N be a submodule of M. Let $\mathfrak{a}$ be an ideal of A. Then there exists an integer k such that*

$$\mathfrak{a}^{i+k} M \cap N = \mathfrak{a}^i (\mathfrak{a}^k M \cap N)$$

for all $i \geq 0$.

Proof We consider the graded module $R(\mathfrak{a}, M)$ over the Rees ring of A. The sum $\oplus(\mathfrak{a}^i M \cap N)$ is a graded submodule of $R(\mathfrak{a}, M)$, so, since $R(\mathfrak{a})$ is Noetherian, it must be finitely generated and hence have a finite set of homogeneous generators. If it has a set of generators in degree at most k, then k satisfies the conclusion of the theorem. $\qquad\square$

It is often convenient to represent $R(\mathfrak{a})$ as a subring of $A[T]$ by identifying it with the subring generated over A by $\mathfrak{a}T$. We thus have

$$R(\mathfrak{a}) = \{a_i T^i + \cdots + a_1 T + a_0 \mid a_i \in \mathfrak{a}^i \text{ for all } i\}.$$

If $\mathfrak{a}$ is an ideal of A such that $A/\mathfrak{a}$ has finite length, then $G(\mathfrak{a})$ satisfies the hypotheses of Theorem 2.1.5, and thus there exists a polynomial $P(i)$ such that $P(i) = \text{length}(\mathfrak{a}^i / \mathfrak{a}^{i+1})$ for large i. If M is an A-module such that $M/\mathfrak{a}M$ has finite length, then the graded module $G(\mathfrak{a}, M)$ also has a Hilbert polynomial; if $A/\mathfrak{a}$ does not have finite length, we let $\mathfrak{b}$ be the annihilator of M and replace A by $A/\mathfrak{b}$ and $\mathfrak{a}$ by the image of $\mathfrak{a}$ in $A/\mathfrak{b}$. This change does not affect M or $G(\mathfrak{a}, M)$, but now $A/\mathfrak{a}$ has finite length and we can again apply Theorem 2.1.5.

Using Lemma 2.1.4 together with the fact that

$$\text{length}(\mathfrak{a}^i M/\mathfrak{a}^{i+1} M) = \text{length}(M/\mathfrak{a}^{i+1} M) - \text{length}(M/\mathfrak{a}^i M)$$

we conclude that there exists a polynomial $P_\mathfrak{a}^M(i)$ such that $P_\mathfrak{a}^M(i) = \text{length}(M/\mathfrak{a}^i M)$ for sufficiently large i. This polynomial is called the *Hilbert-Samuel polynomial* of M with respect to $\mathfrak{a}$. In particular, if $A/\mathfrak{a}$ has finite length, then the polynomial $P_\mathfrak{a}(i)$ such that $P_\mathfrak{a}(i) = \text{length}(A/\mathfrak{a}^i)$ for large i is called the *Hilbert-Samuel polynomial of* $\mathfrak{a}$. The degree and leading coefficient of this polynomial are important invariants of the ideal and are discussed in later sections. If $\mathfrak{a}$ has a set of generators consisting of d elements, then the associated graded ring $G_\mathfrak{a}$ can be generated over the subring of elements of degree zero by d elements, so Theorem 2.1.5 implies that the degree of the Hilbert polynomial of the graded ring $G_\mathfrak{a}(M)$ is at most $d - 1$ for any finitely generated module M. It then follows that the degree of the Hilbert-Samuel polynomial of M is at most d.

2.3 Dimension Theory in Local Rings

We recall that a ring A is said to be local if it is a Noetherian ring with a unique maximal ideal $\mathfrak{m}$. In this section we review the basic theorems on the dimension of modules over local rings. The main point is that the Krull dimension of A as defined in terms of chains of prime ideals can also be defined in other ways, and that these equivalent definitions are very useful in applications.

We begin with some basic facts on modules of finite length.

Proposition 2.3.1 *Let M be a finitely generated module over a local ring A with maximal ideal $\mathfrak{m}$. The following are equivalent:*

 (i) *M is a module of finite length.*
 (ii) *$\mathfrak{m}^k M = 0$ for some k.*
 (iii) *The only prime ideal in the support of M is $\mathfrak{m}$.*

Proof Since A has only one simple module, namely $A/\mathfrak{m}$, a module of finite length k will have a composition series of length k with quotients $A/\mathfrak{m}$, so it will be annihilated by $\mathfrak{m}^k$. Thus, (i) implies (ii).

If $\mathfrak{p}$ is a minimal prime ideal in the support of M, then $\mathfrak{p}$ is an associated prime of M and M contains a submodule isomorphic to $A/\mathfrak{p}$. If $\mathfrak{p} \neq \mathfrak{m}$, then there is an element a of $\mathfrak{m}$ not in $\mathfrak{p}$, and a^k cannot annihilate M for any k. Hence (ii) implies (iii).

The implication (iii) implies (i) follows from the fact that there is a sequence of submodules with quotients $A/\mathfrak{p}$ for $\mathfrak{p}$ prime ideals in the support of M; if the only such prime ideal is $\mathfrak{m}$, then this filtration defines a composition series for M. $\square$

Proposition 2.3.2 *If M and N are finitely generated modules over a local ring A, then the support of $M \otimes_A N$ is the intersection of the supports of M and N.*

Proof This proposition follows from the isomorphism

$$(M \otimes_A N) \otimes A_\mathfrak{p} \cong (M \otimes_A A_\mathfrak{p}) \otimes_{A_\mathfrak{p}} (N \otimes_A A_\mathfrak{p}).$$

In fact, if $\mathfrak{p}$ is not in the support of M, then $M \otimes_A A_\mathfrak{p} = 0$, so

$$(M \otimes_A N) \otimes A_\mathfrak{p} \cong 0 \otimes_{A_\mathfrak{p}} (N \otimes_A A_\mathfrak{p}) = 0$$

and $\mathfrak{p}$ is not in the support of $M \otimes_A N$. Conversely, if $\mathfrak{p}$ is in the support of M and N, we may tensor both sides of the above equation over $A_\mathfrak{p}$ with $A_\mathfrak{p}/\mathfrak{p}_\mathfrak{p} = k(\mathfrak{p})$. By Nakayama's lemma, $M \otimes_A k(\mathfrak{p})$ and $N \otimes_A k(\mathfrak{p})$ are not zero, and hence

$$(M \otimes_A N) \otimes_A k(\mathfrak{p}) \cong (M \otimes_A k(\mathfrak{p})) \otimes_{k(\mathfrak{p})} (N \otimes_A k(\mathfrak{p})),$$

which is a tensor product of nonzero vector spaces over a field and so is not zero. Thus, $\mathfrak{p}$ is in the support of $M \otimes_A N$. □

Corollary 2.3.3 *If $\mathfrak{a}$ is an ideal of the local ring A, and M is a finitely generated A-module, then $M/\mathfrak{a}M$ is a module of finite length if and only if the only prime ideal in the support of M that contains $\mathfrak{a}$ is the maximal ideal.*

Proof It follows from Proposition 2.3.1 that $M/\mathfrak{a}M$ has finite length if and only if the only prime ideal in its support is $\mathfrak{m}$. Since $M/\mathfrak{a}M \cong M \otimes_A (A/\mathfrak{a})$, the Corollary follows from Proposition 2.3.2. □

We now return to the definition of dimension. Let $\dim(M)$ denote the Krull dimension of M as in Chapter 1.

Define $s(M)$ to be the minimum number k such that there exist $x_1, \ldots x_k$ in $\mathfrak{m}$ such that $M/(x_1, \ldots, x_k)M$ is a module of finite length. If $x_1, \ldots x_k$ generate $\mathfrak{m}$, then $M/(x_1, \ldots, x_k)M = M/\mathfrak{m}M$ has finite length, so $s(M)$ is at most equal to the number of elements in a minimal set of generators for $\mathfrak{m}$.

Let $d(M)$ be the degree of the Hilbert-Samuel polynomial of M with respect to an ideal $\mathfrak{a} \subseteq \mathfrak{m}$ such that $M/\mathfrak{a}M$ has finite length. It is not immediately clear that this definition is independent of the ideal chosen. To show that it does not depend on the ideal $\mathfrak{a}$, we first recall the definition, which states that $d(M)$ is the degree of the polynomial $P_\mathfrak{a}^M(i)$ such that $P_\mathfrak{a}^M(i) = \mathrm{length}(M/\mathfrak{a}^i M)$ for sufficiently large i. If $\mathfrak{b}$ is another ideal such that $M/\mathfrak{b}M$ has finite length, then there exists a k such that $\mathfrak{b}^k M \subset \mathfrak{a}M$, so that $P_\mathfrak{b}^M(ik) \geq P_\mathfrak{a}^M(i)$ for all sufficiently large i, and the degree of $P_\mathfrak{b}^M$ is at least as large as that of $P_\mathfrak{a}^M$. Since there is also an integer k' such that $\mathfrak{a}^{k'} M \subset \mathfrak{b}M$, the reverse inequality also holds, so the degrees of the polynomials are equal.

We prove that $\dim(M) = s(M) = d(M)$ by inductively proving a circle of inequalities as in [1, Chapter 11]. It follows from Proposition 2.3.1 that if any one of the three numbers is zero, then so are the other two. In fact, the first condition, that M itself has finite length, is true if and only if $s(M) = 0$; the second, that $\mathfrak{m}^n M = 0$ for some n, is true if and only if $d(M) = 0$; and the third states that the support of M has one element, which is true if and only if $\dim(M) = 0$.

It also follows from Theorem 2.1.5 of Section 2.1 that $d(M) \leq s(M)$, since by the definition of $s(M)$ there exists an ideal $\mathfrak{a}$ generated by at most $s(M)$ elements such that $M/\mathfrak{a}M$ has finite length. If we use the ideal $\mathfrak{a}$ to compute $d(M)$, we obtain that the degree of the Hilbert-Samuel polynomial of M with respect to $\mathfrak{a}$ is at most the number of generators of $\mathfrak{a}$, so that $d(M) \leq s(M)$.

Before proving the other two inequalities, we show that both $\dim(M)$ and $d(M)$ behave properly with respect to short exact sequences. Let

$$0 \to M' \to M \to M'' \to 0$$

be a short exact sequence of A-modules. Since the support of M is the union of the supports of M' and of M'' (Proposition 1.1.3), the Krull dimension of M is the larger of the Krull dimensions of M' and M''. For $d(M)$, the corresponding result follows the following Proposition.

Proposition 2.3.4 *Given a short exact sequence as above, and an ideal $\mathfrak{a}$ of A such that $M/\mathfrak{a}M$, $M'/\mathfrak{a}M'$, and $M''/\mathfrak{a}M''$ have finite length, we have*

(i) $d(M) = \max(d(M'), d(M''))$.
(ii) *If $d(M) = d$, and if a_d, a'_d, and a''_d are the coefficients of degree d of the Hilbert-Samuel polynomials of M, M', and M'', respectively, with respect to $\mathfrak{a}$, then*

$$a_d = a'_d + a''_d.$$

Proof For each i, we have a short exact sequence

$$0 \to (\mathfrak{a}^i M \cap M')/(\mathfrak{a}^{i+1} M \cap M') \to \mathfrak{a}^i M/\mathfrak{a}^{i+1} M \to \mathfrak{a}^i M''/\mathfrak{a}^{i+1} M'' \to 0.$$

Thus, we have a sum of Hilbert polynomials for the associated graded modules defined by the corresponding filtrations. The filtrations on M and M'' define the associated graded modules of M and M'' with respect to $\mathfrak{a}$, but the ith submodule in the filtration on M' is $\mathfrak{a}^i M \cap M'$ instead of $\mathfrak{a}^i M'$. However, by the Artin-Rees lemma there exists an integer k such that $\mathfrak{a}^{i+k} M \cap M' = \mathfrak{a}^i(\mathfrak{a}^k \cap M)$ for all $i \geq 0$. Let $P_\mathfrak{a}^{M'}$ be the Hilbert-Samuel polynomial of M' with respect to $\mathfrak{a}$, and let $Q_{M'}$ be the polynomial such that $Q_{M'}(i)$ is the length of $M'/M' \cap \mathfrak{a}^i M$ for large i. Since $\mathfrak{a}^i M' \subseteq M' \cap \mathfrak{a}^i M$ for all i, we have that $P_\mathfrak{a}^{M'}(i) \geq Q_{M'}(i)$ for large i. On the other hand, the fact that $\mathfrak{a}^{i+k} M \cap M' = \mathfrak{a}^i(\mathfrak{a}^k \cap M) \subseteq \mathfrak{a}^i M$ implies that $P_\mathfrak{a}^{M'}(i) \leq Q_{M'}(i + k)$ for large i. Hence $P_\mathfrak{a}^{M'}$ and $Q_{M'}$ have the same degree and leading coefficient. Thus, we deduce that we have additivity on the coefficients of highest degree for $P_\mathfrak{a}^{M'}$, $P_\mathfrak{a}^M$, and $P_\mathfrak{a}^{M''}$, and this proves the Proposition. $\qquad\square$

Proposition 2.3.5 *If M is a finitely generated module over a local ring A, we have $\dim(M) \leq d(M)$.*

Proof The proof is by induction on $d(M)$. By Proposition 2.3.4, we may reduce to the case in which $M = A/\mathfrak{p}$ for a prime ideal $\mathfrak{p}$. Furthermore, we may choose

$\mathfrak{p}$ so that $\dim(A/\mathfrak{p}) = \dim(M)$. Assume that $\dim(M) > d(M)$. There then exists a chain $\mathfrak{p} = \mathfrak{p}_0 \subset \mathfrak{p}_1 \subset \cdots \subset \mathfrak{p}_k$ of strict inclusions of prime ideals in the support of M with $k > d(M)$. If we choose $x \notin \mathfrak{p}$ but $x \in \mathfrak{p}_1$, we then have

$$\dim(M/xM) \geq k - 1 > d(M) - 1.$$

On the other hand, there is a short exact sequence

$$0 \to M \xrightarrow{x} M \to M/xM \to 0.$$

Using the second part of Proposition 2.3.4, we see that the degree of the Hilbert-Samuel polynomial of M/xM is strictly less than the degree of the Hilbert-Samuel polynomial of M. The inductive hypothesis, together with the previous inequality, now implies that

$$d(M) - 1 < \dim(M/xM) \leq d(M/xM) \leq d(M) - 1.$$

This is a contradiction, so we must have had $\dim(M) \leq d(M)$. $\qquad\square$

Proposition 2.3.6 *For a finitely generated module M over a local ring A, we have $s(M) \leq \dim(M)$.*

Proof This proposition is again proven by induction, this time on $\dim(M)$, which we now know is finite. Choose x in the maximal ideal of A but in no minimal prime ideal in the support of M. The dimension of M/xM is then at most $\dim(M) - 1$. On the other hand, the definition of $s(M)$ implies that $s(M/xM) \geq s(M) - 1$. Thus, the induction hypothesis implies that

$$s(M) - 1 \leq s(M/xM) \leq \dim(M/xM) \leq \dim(M) - 1$$

so that $s(M) \leq \dim(M)$. $\qquad\square$

We summarize these results as follows.

Theorem 2.3.7 *Let M be a finitely generated module over a local ring A. Then*

$$\dim(M) = s(M) = d(M).$$

We recall that the *height* of a prime ideal is the dimension of the localization $A_\mathfrak{p}$. The height of an arbitrary ideal is the minimum of the heights of minimal prime ideals containing it. For example, the ideal $\mathfrak{a} = (XY, XZ)$ in $k[X, Y, Z]$ has height one, although the prime ideal (Y, Z) is minimal over $\mathfrak{a}$ and has height two.

It follows from Theorem 2.3.7 that the height of any prime ideal $\mathfrak{p}$ in a Noetherian ring is finite and is bounded by the number of generators of the ideal. In fact, its height is bounded by the number of elements needed to generate an ideal over which $\mathfrak{p}$ is minimal. We state one special case of this result as a separate theorem.

Theorem 2.3.8 (Krull's Principal Ideal Theorem) *Let $\mathfrak{p}$ be a prime ideal that is minimal over a principal ideal. Then the height of $\mathfrak{p}$ is at most one.*

Definition 2.3.1 *A sequence $x_1, \ldots x_d$ of elements of $\mathfrak{m}$ such that d is minimal with $M/(x_1, \ldots x_d)M$ of finite length (i.e., d is the dimension of M) is a* system of parameters *for M. In particular, a sequence of $d = \dim(A)$ elements $x_1, \ldots x_d$ such that $A/(x_1, \ldots x_d)$ is Artinian is a system of parameters for A.*

We note an important consequence of the equality of the three definitions of dimension. If M is a module and x is an element in no minimal prime ideal in the support of M but is in the maximal ideal of A, then the difference in Krull dimension $\dim(M) - \dim(M/xM)$ is at least one, while the difference $s(M) - s(M/xM)$ is at most one. Thus, both differences are equal to one. Hence a sequence $x_1, \ldots x_d$ is a system of parameters for M if and only if x_1 is in no prime ideal $\mathfrak{p}$ in the support of M with $\dim(A/\mathfrak{p}) = \dim(M)$, x_2 is in no prime ideal $\mathfrak{p}$ in the support of $M/x_1 M$ with $\dim(A/\mathfrak{p}) = \dim(M/x_1 M)$, and so on.

2.4 The Samuel Multiplicity of an Ideal

Let A be a local ring and let M be a finitely generated module of dimension d. Let $\mathfrak{a}$ be an ideal of A contained in the maximal ideal such that $M/\mathfrak{a}M$ has finite length. As we have seen, this condition is equivalent to the condition that the maximal ideal $\mathfrak{m}$ of A is the only prime ideal in the support of M containing $\mathfrak{a}$. Let $P_\mathfrak{a}^M$ be the Hilbert-Samuel polynomial of M.

Definition 2.4.1 *The* Samuel multiplicity *or simply the* multiplicity, *of $\mathfrak{a}$ on M, denoted $e(\mathfrak{a}, M)$ is $d!$ times the coefficient of the degree d term of $P_\mathfrak{a}^M$. The multiplicity of $\mathfrak{a}$ is $e(\mathfrak{a}, A)$. The multiplicity of A is the multiplicity of the maximal ideal of A.*

The Samuel multiplicity of a module also has a simple expression as a limit.

Proposition 2.4.1 *Let M be a module of dimension d as above, and let $\mathfrak{a}$ be an ideal such that $M/\mathfrak{a}M$ has finite length. Then*

$$e(\mathfrak{a}, M) = d! \lim_{i \to \infty} \left(\frac{\text{length}(M/\mathfrak{a}^i M)}{i^d} \right).$$

Proof Let $P_\mathfrak{a}^M(i) = a_d i^d + a_{d-1} i^{d-1} + \cdots + a_0$ be the Hilbert-Samuel polynomial of M. Since $P_\mathfrak{a}^M(i) = \text{length}(M/\mathfrak{a}^i M)$ for large i, we have

$$d! \lim_{i \to \infty} \left(\frac{\text{length}(M/\mathfrak{a}^i M)}{i^d} \right) = d! \lim_{i \to \infty} \left(\frac{P_\mathfrak{a}^M(i)}{i^d} \right)$$

$$= d! \lim_{i \to \infty} \left(\frac{a_d i^d + a_{d-1} i^{d-1} + \cdots + a_0}{i^d} \right)$$

$$= d! \lim_{i \to \infty} (a_d + a_{d-1}/i + \cdots + a_0/i^d) = d! a_d = e(\mathfrak{a}, M).$$

$\square$

The origin of this concept of multiplicity comes from the multiplicity of a variety at a point. For example, a regular local ring has multiplicity one (we prove this below), and if A is the localization of $k[[X, Y]]/(XY)$ at the maximal ideal (X, Y), then A has multiplicity 2, which reflects the fact that the variety consists of two lines crossing at this point.

Before discussing multiplicities in general, we recall some facts about regular local rings. A local ring is *regular* if its maximal ideal can be generated by d elements, where d is the dimension of A.

Proposition 2.4.2 *Let A be a regular local ring of dimension d with maximal ideal $\mathfrak{m}$ and residue field k.*

(i) *The associated graded ring $G_\mathfrak{m}$ is a polynomial ring over k in d variables.*
(ii) *A is an integral domain.*
(iii) *The multiplicity of A is one.*

Proof Since A is regular, there exist d elements $x_1, \ldots, x_d$ that generate $\mathfrak{m}$. Map the polynomial ring $k[X_1, \ldots, X_d]$ onto $G_\mathfrak{m}$ by mapping X_i to x_i. The degree of the Hilbert polynomial of $G_\mathfrak{m}$ is $d - 1$, since it is one less than the degree of the Hilbert-Samuel polynomial of $\mathfrak{m}$, which in turn is equal to the dimension of A. The degree of the Hilbert polynomial of $k[X_1, \ldots, X_d]$ is also $d - 1$, and since $k[X_1, \ldots, X_d]$ is an integral domain, the degree of the Hilbert polynomial of any proper quotient of $k[X_1, \ldots, X_d]$ is at most $d - 2$.

Hence $G_{\mathfrak{m}}$ cannot be a proper quotient of $k[X_1, \ldots, X_d]$, so the above map is an isomorphism, proving (i).

The second statement of the proposition follows from the fact that $G_{\mathfrak{m}}$ is an integral domain. If a and b are nonzero elements of A, there exist m and n such that a is in $\mathfrak{m}^m$ but not in $\mathfrak{m}^{m+1}$ and b is in $\mathfrak{m}^n$ but not in $\mathfrak{m}^{n+1}$ (for a proof of this fact, see [1, Corollary 10.20]). Then the images of a and b in $\mathfrak{m}^m/\mathfrak{m}^{m+1}$ and $\mathfrak{m}^n/\mathfrak{m}^{n+1}$, respectively, are not zero, so since $G_{\mathfrak{m}}$ is an integral domain, their product in $\mathfrak{m}^{m+n}/\mathfrak{m}^{m+n+1}$ is not zero. Hence $ab \neq 0$ and A is an integral domain.

It follows from statement (i) that the Hilbert-Samuel of $\mathfrak{m}$ is given by the polynomial whose value at i is the number of monomials of degree less than i, so that

$$\text{length}(A/\mathfrak{m}^i) = \binom{i-1+d}{d}.$$

This polynomial has degree d and leading coefficient $1/d!$. Hence the multiplicity of A is $(d!)(1/d!) = 1$.   □

Under certain not very restrictive hypotheses, essentially assuming that the dimension of $A/\mathfrak{p}$ is the same for all minimal prime ideals $\mathfrak{p}$ of A, the condition that the multiplicity of a local ring A is equal to 1 is equivalent to regularity (Nagata [48, Theorem (40.6)]; see Huneke [19] for a related result).

We now return to arbitrary local rings. Let A be a local ring of dimension d, and let M be an A-module such that $M/\mathfrak{a}M$ has finite length. We let $c_d(\mathfrak{a}, M)$ denote $d!$ times the coefficient of the term of degree d in the Hilbert-Samuel polynomial $P_{\mathfrak{a}}^M$. It then follows from Proposition 2.3.4 that $e_d(\mathfrak{a}, M)$ is additive on short exact sequences and that it is zero for any module of dimension less than d.

It is not difficult to see that different ideals can have the same multiplicity, even when one is properly contained in the other. As an example, we let $A = k[[X, Y]]$ with ideals $\mathfrak{a} = (X^2, Y^2)$ and $\mathfrak{b} = (X^2, XY, Y^2)$. Then $\mathfrak{b}^n$ consists of the ideal generated by monomials of degree at least $2n$, while $\mathfrak{a}^n$ is generated by all such monomials $X^i Y^j$ where either i and j are even and $i + j \geq 2n$ or $i + j > 2n$. The quotient $\mathfrak{b}^n/\mathfrak{a}^n$ is generated by all monomials $X^i Y^j$ with i and j odd and $i + j = 2n$, and the number of such monomials is n. Hence the quotient $\mathfrak{a}^n/\mathfrak{b}^n$ has length n, and the Hilbert-Samuel polynomials of $\mathfrak{a}$ and $\mathfrak{b}$ differ by a polynomial of degree 1 so that the leading coefficients are equal. Hence $e(\mathfrak{a}) = e(\mathfrak{b})$. In the exercises we give more examples of multiplicities of ideals generated by monomials.

The fact that two ideals, one of which properly contains the other, can have the same multiplicity is related to the concept of integral extensions of ideals.

Definition 2.4.2 *Let $\mathfrak{a}$ be an ideal of a ring A, and let B be a ring that is an extension of A. An element $x \in B$ is* integral *over $\mathfrak{a}$ if there is an equation*

$$x^n + a_1 x^{n-1} + \cdots + a_n = 0$$

with $a_i \in \mathfrak{a}^i$ for all i.

If $\mathfrak{a} = A$, this condition states that x is integral over the ring A. We recall that the set of integral elements in an extension ring is a ring and that there are relations between the sets of prime ideals in an integral extension (see [1, Chapter 5] or [44, §9]).

Proposition 2.4.3 *Let $\mathfrak{a}$ be an ideal of A. Then the set of all elements that are integral over $\mathfrak{a}$ form an ideal of A.*

Proof We consider the Rees ring $R(\mathfrak{a})$ of $\mathfrak{a}$ as a subring of the polynomial ring $A[T]$. By the results quoted above, the set of elements of $A[T]$ that are integral over $R(\mathfrak{a})$ form a subring. An element of $A[T]$ of the form aT is integral over $R(\mathfrak{a})$ if and only if there exists an integer n and elements $b_1, \ldots, b_n \in R(\mathfrak{a})$ such that

$$(aT)^n + b_1 (aT)^{n-1} + \cdots + b_n = 0.$$

Taking the components of each term of this expression of degree n in the natural grading on $A[T]$, we obtain elements $a_0, \ldots a_{n-1}$ with $a_i \in \mathfrak{a}^i$ for each i such that

$$(aT)^n + (a_1 T)(aT)^{n-1} + \cdots + a_n T^n = 0.$$

Factoring out T^n, we see that $a^n + a_1 a^{n-1} + \cdots + a_n = 0$, so a is integral over $\mathfrak{a}$. Conversely, if a is integral over $\mathfrak{a}$ we can reverse these steps and conclude that aT is integral over $R(\mathfrak{a})$. Thus, aT is integral over $R(\mathfrak{a})$ if and only if a is integral over $\mathfrak{a}$.

Since the elements of $A[T]$ that are integral over $R(\mathfrak{a})$ form a subring of $A[T]$, in particular elements of the form aT that are integral over $R(\mathfrak{a})$ are closed under multiplication by elements of A. Thus, the set of elements of A that are integral over $\mathfrak{a}$ form an ideal of A. □

The condition that $\mathfrak{a}$ is integral over $\mathfrak{b}$ is also referred to by saying that $\mathfrak{b}$ is a *reduction* of $\mathfrak{a}$. The next proposition shows the relation between integral extensions and multiplicity; for a partial converse see Rees [52].

Proposition 2.4.4 *Let* $\mathfrak{a}$ *and* $\mathfrak{b}$ *be* $\mathfrak{m}$-*primary ideals of A such that* $\mathfrak{a} \supseteq \mathfrak{b}$. *If* $\mathfrak{a}$ *is integral over* $\mathfrak{b}$ *then the multiplicity of* $\mathfrak{a}$ *is equal to the multiplicity of* $\mathfrak{b}$.

Proof It suffices to prove this proposition in the case in which $\mathfrak{a}$ is obtained from $\mathfrak{b}$ by adjoining one element x that is integral over $\mathfrak{b}$. Suppose that x satisfies the equation

$$x^n + b_1 x^{n-1} + \cdots + b_n = 0$$

with $b_i \in \mathfrak{b}^i$ for all i. We claim that $\mathfrak{a}^m \subseteq \mathfrak{b}^{m-n}$ for all $m \geq n$. Since $\mathfrak{a}$ is generated by $\mathfrak{b}$ and x, $\mathfrak{a}^m$ is generated by the ideals $x^k \mathfrak{b}^{m-k}$ for all k between 0 and m. However, if $k \geq n$, we may multiply the above equation of integral dependence for x by x^{k-n} and conclude that $x^k \in x^{k-1}\mathfrak{b} + \cdots + x^{k-n}\mathfrak{b}^n$, so that $x^k \mathfrak{b}^{m-k} \subseteq x^{k-1}\mathfrak{b}^{m-k+1} + \cdots + x^{k-n}\mathfrak{b}^{m-k+n}$. Thus, $\mathfrak{a}^m$ is contained in the sum of ideals $x^k \mathfrak{b}^{m-k}$ for k between 0 and $n-1$. Hence $\mathfrak{a}^m \subseteq \mathfrak{b}^{m-n+1}$; since we also know that $\mathfrak{b}^m \subseteq \mathfrak{a}^m$, it follows that the Hilbert-Samuel polynomials of $\mathfrak{a}$ and $\mathfrak{b}$ have the same degree and leading coefficients, so the multiplicities of $\mathfrak{a}$ and $\mathfrak{b}$ are equal. $\qquad\qquad\square$

2.5 Mixed Multiplicities and Buchsbaum-Rim Multiplicities

There are several generalizations of the concept of Samuel multiplicity, and we describe two of them in this section. The first is mixed multiplicity, in which one ideal is replaced by a finite set of $\mathfrak{m}$-primary ideals. The second, Buchsbaum-Rim multiplicity, extends the definition to a submodule of a free module of rank greater than one, where the condition of being $\mathfrak{m}$-primary is replaced by the condition that the quotient has finite length.

Both extensions of the definition of multiplicity are based on generalizations of the construction of the Rees ring of an ideal. Let $\mathfrak{a}_1, \ldots, \mathfrak{a}_n$ be ideals of a local ring A. We define the Rees ring $R(\mathfrak{a}_1, \ldots, \mathfrak{a}_n)$ of $\mathfrak{a}_1, \ldots, \mathfrak{a}_n$ to be the direct sum of the products $\mathfrak{a}_1^{i_1} \cdots \mathfrak{a}_n^{i_n}$ over all $(i_1, \ldots, i_n)$ with $i_j \geq 0$ for all j. The Rees ring has a natural structure of a $\mathbb{Z}^n$-graded ring, and it can be considered as a subring of the graded polynomial ring $A[T_1, \ldots, T_n]$.

If M is a submodule of a free module F on k generators, we can consider the symmetric algebra on F, which is a polynomial ring over A in k variables (see Rees [53]). Denote these variables $T_1, \ldots, T_k$; we give the polynomial ring $A[T_1, \ldots, T_k]$ the usual $\mathbb{Z}$-grading. We may then consider M as a submodule of the component of the polynomial ring of degree 1. We define the Rees ring $R(M)$ of M to be the subring of $A[T_1, \ldots, T_k]$ generated over A by M.

The next two theorems show the existence of multiplicities in both of these situations. In neither case does a natural extension of the concept of an associated graded ring exist, so the methods of proof are slightly different from the case of the multiplicity of one ideal.

Theorem 2.5.1 *Let* $\mathfrak{a}_1, \ldots, \mathfrak{a}_n$ *be* $\mathfrak{m}$-*primary ideals of* A, *where* A *is a local ring of dimension* d. *Let* M *be a finitely generated* A-*module. Then there exists a polynomial* $P(i_1, \ldots i_n)$ *in* n *variables of degree at most* d *and integers* $m_1, \ldots, m_n$ *such that*

$$P(i_1, \ldots i_n) = \text{length}\left(M/\mathfrak{a}_1^{i_1}\mathfrak{a}_2^{i_2} \cdots \mathfrak{a}_n^{i_n} M\right)$$

when $i_j \geq m_j$ *for all* j. *Furthermore, the polynomial can be written*

$$\sum_k c_k \binom{i+k}{k}$$

for certain integers c_k *and indices* $k = (k_1, \ldots, k_n)$ *with* $k_i \geq 0$ *and* $k_1 + \cdots + k_n \leq d$.

Proof Let $R(\mathfrak{a}_i) = R(\mathfrak{a}_1, \ldots, \mathfrak{a}_n)$ be the Rees ring of the ideals $\mathfrak{a}_1, \ldots, \mathfrak{a}_n$, and let $R(\mathfrak{a}_i; M)$ denote the $R(\mathfrak{a}_i)$ module $\oplus \mathfrak{a}_1^{i_1}\mathfrak{a}_2^{i_2} \cdots \mathfrak{a}_n^{i_n} M$. Let N denote the tensor product $R(\mathfrak{a}_i; M) \otimes_A A/\mathfrak{a}_n$. The module N is a $\mathbb{Z}^n$-graded module whose component in each degree has finite length; hence by Theorem 2.1.7 there exists a polynomial Q in n variables such that

$$Q(i) = Q(i_1, \ldots, i_n) = \text{length}\left(\mathfrak{a}_1^{i_1} \cdots \mathfrak{a}_n^{i_n-1} M / \mathfrak{a}_1^{i_1} \cdots \mathfrak{a}_n^{i_n} M\right)$$

for sufficiently large i. Theorem 2.1.7 also states that Q can be written as a linear combination of binomial coefficients with integer coefficients. Let $H(i_1, \ldots, i_n) = \text{length}(M/\mathfrak{a}_1^{i_1}\mathfrak{a}_2^{i_2} \cdots \mathfrak{a}_n^{i_n} M)$. We then have

$$H(i) - H(i - e_n) = H(i_1, \ldots, i_{n-1}, i_n) - H(i_1, \ldots, i_{n-1}, i_n - e_n) = Q(i)$$

for sufficiently large i.

Let $j = (j_1, \ldots, j_n)$ be an element of $\mathbb{Z}^n$ such that this equality holds for all $i \geq j$. By induction on n applied to the module $\mathfrak{a}_n^{j_n} M$, there is a polynomial P' in $n - 1$ variables such that

$$P'(i_1, \ldots, i_{n-1}) = \text{length}\left(\mathfrak{a}_n^{j_n} M / \mathfrak{a}_1^{i_1} \cdots \mathfrak{a}_{n-1}^{i_{n-1}}\mathfrak{a}_n^{j_n} M\right)$$

for sufficiently large $(i_1, \ldots, i_{n-1})$. Adding a constant equal to the length of $M/\mathfrak{a}_n^{j_n} M$ to P', we obtain a polynomial P'' such that

$$P''(i_1, \ldots, i_{n-1}) = \text{length}\left(M/\mathfrak{a}_1^{i_1} \cdots \mathfrak{a}_{n-1}^{i_{n-1}}\mathfrak{a}_n^{j_n} M\right)$$

for sufficiently large $(i_1, \ldots, i_{n-1})$. It now follows from Lemma 2.1.6 that there exists a polynomial $P(i)$ such that $P(i) = H(i)$ for sufficiently large i and that $P(i)$ can be written in terms of binomial coefficients with integer coefficients.

We next show that P has degree d. Let $\mathfrak{a}$ and $\mathfrak{b}$ be $\mathfrak{m}$-primary ideals such that $\mathfrak{a} \subseteq \mathfrak{a}_i$ and $\mathfrak{b} \supseteq \mathfrak{a}_i$ for all i. Let $P_{\mathfrak{a}}^M$ and $P_{\mathfrak{b}}^M$ be the Hilbert-Samuel polynomials of M with respect to $\mathfrak{a}$ and $\mathfrak{b}$, respectively. We then have

$$P_{\mathfrak{a}}^M(i_1 + \cdots + i_k) \geq P(i_1, \ldots, i_k) \geq P_{\mathfrak{b}}^M(i_1 + \cdots + i_k)$$

for all sufficiently large $(i_1, \ldots, i_k)$. Since $P_{\mathfrak{a}}^M$ and $P_{\mathfrak{b}}^M$ have degree d, P must also have degree d.

Since $P(i)$ has degree d, the coefficients c_k in the expression of $P(i)$ in terms of binomial coefficients must be zero for all k with $k_1 + \cdots + k_n > d$. To see this, we note that

$$\binom{i+k}{k} = \binom{i_1 + k_1}{k_1}\binom{i_2 + k_2}{k_2} \cdots \binom{i_n + k_n}{k_n}$$

is a polynomial of degree $k_i + \cdots + k_n$ with exactly one term of this degree, namely $i_1^{k_1} i_2^{k_2} \cdots i_n^{k_n}$. It follows that these polynomials are linearly independent and that the degree of a linear combination of them is the maximum of the degrees of the terms with nonzero coefficients. Hence, since the degree of $P(i)$ is d, we have $c_k \neq 0$ only if $k_1 + \cdots + k_n \leq d$. $\qquad\square$

While the multiplicity of an ideal is simply a number, in the case of more than one ideal the entire homogeneous part of the polynomial of degree d is significant. The *mixed multiplicities* of the ideals are defined to be certain multiples of the coefficients of this polynomial. Let $P(i_1, \ldots, i_n)$ be the polynomial defined as in Theorem 2.5.1 for $M = A$, and let $k = (k_1, \ldots, k_n)$, where the k_i are nonnegative integers such that $k_1 + \cdots + k_n = d$. We define the mixed multiplicity $e_k(\mathfrak{a}_i)$ to be $k_1! \cdots k_n!$ times the coefficient of $i_1^{k_1} \cdots i_n^{k_n}$ in P.

As in the case of the multiplicity of a single ideal, the mixed multiplicities of a set of ideals are integers. To see this we use the fact the polynomial $P(i)$ can be written in the form

$$P(i) = \sum_k c_k \binom{i+k}{k}$$

with integers c_k and elements $k \in \mathbb{Z}^n$ with $k_1 + \cdots + k_n \leq d$. If $k_1 + \cdots + k_n = d$, then the only contribution in this sum to the coefficient of $i_1^{k_1} \cdots i_n^{k_n}$ comes from the term $c_k \binom{i+k}{k}$ and is equal to $c_k/(k_1! k_2! \cdots k_n!)$. Hence the mixed multiplicity $e_k(\mathfrak{a}_1, \ldots, \mathfrak{a}_n)$ is equal to c_k.

Mixed multiplicities were first introduced by Bhattacharya [7]. They were reintroduced later by Teissier [69] in a geometric context. Recently they have attracted a considerable amount of interest (see Teissier [70], Rees and Sharp [54], Verma [72, 73], Katz and Verma [32], and Swanson [67]).

We next define the Buchsbaum-Rim multiplicity of a module. Let M be a submodule of a free module F such that F/M is a module of finite length. If F has rank k, then the symmetric algebra $S(F)$ is a polynomial ring $A[T_1, \ldots, T_k]$, and we consider the Rees ring $R(M)$ as a subring of $S(F)$. We denote the graded components of $S(F)$ and $R(M)$ of degree i by $S_i(F)$ and $R_i(M)$, respectively.

As an example, we consider the case in which M is a direct sum of k $\mathfrak{m}$-primary ideals of A. In this case, $R(M)$ is the same as the Rees ring of these ideals, considered as a $\mathbb{Z}$-graded ring instead of a $\mathbb{Z}^k$-graded ring.

Theorem 2.5.2 *Let A be a local ring of dimension d, and let M be a submodule of a free module F of rank k such that $M \subseteq \mathfrak{m}F$ and F/M is a module of finite length. There exists a polynomial in one variable $P(i)$ of degree $d + k - 1$ such that*

$$P(i) = \text{length}(S_i(F)/R_i(M))$$

for large i.

Proof We first prove a special case of this theorem. Let $\mathfrak{a}$ be an $\mathfrak{m}$-primary ideal, let $F = A^k$, and let $M = \mathfrak{a} \oplus \cdots \oplus \mathfrak{a} = \mathfrak{a}F \subseteq F$. In this case $R_i(M) = \mathfrak{a}^i S_i(F)$, and $S_i(F)$ is a free A-module, so we have the formula

$$\text{length}(S_i(F)/R_i(F)) = \text{length}(A/\mathfrak{a}^i)(\text{rank}(S_i(F))).$$

Since the length of $A/\mathfrak{a}^i$ is a polynomial of degree d for large i and the rank of $S_i(F)$ is a polynomial of degree $k - 1$ for all $i \geq 0$, this proves that the length of $S_i(F)/R_i(M)$ is a polynomial of degree $d + k - 1$ in this case. In fact, if $P_\mathfrak{a}$ denotes the Hilbert-Samuel polynomial of $\mathfrak{a}$, then we have

$$\text{length}(S_i(F)/R_i(M)) = P_\mathfrak{a}(i)\binom{i + k - 1}{k - 1}$$

for large i.

We now prove the general case. Let $S = S(F)$, and denote the graded part of degree i of $R(M)$ by M^i. We consider the graded ring R whose component in degree i is $S_i \oplus S_{i-1}M \oplus S_{i-2}M^2 \oplus \cdots \oplus S_1 M^{i-1} \oplus M^i$. Then $R_0 = A$, and R is generated over R_0 by its component in degree 1, which is $S_1 \oplus M$, so in particular R is Noetherian. We next consider two graded ideals of R. Let $\mathfrak{b}$ denote the ideal consisting of all summands $S_r M^s$ for which $r > 0$; thus the component of $\mathfrak{b}$ in degree i is

$$\mathfrak{b}_i = S_i \oplus S_{i-1}M \oplus \cdots \oplus S_1 M^{i-1}.$$

Since $M \subseteq S_1$, we have a sequence of containments

$$S_{i-1}M \subseteq S_i, \ S_{i-2}M^2 \subseteq S_{i-1}M, \ldots, \ M^i \subseteq S_{i-1}M.$$

Let $\mathfrak{c}$ denote the ideal contained in $\mathfrak{b}$ defined by letting $\mathfrak{c}_i = S_{i-1}M \oplus S_{i-2}M^2 \oplus \cdots \oplus M^i$, where the summands are contained in the summands in the corresponding positions in the direct sum decomposition of $\mathfrak{a}$. Let N denote the quotient $\mathfrak{b}/\mathfrak{c}$. Then N is a graded R-module, and the component of N of degree i is

$$S_i/S_{i-1}M \oplus S_{i-1}M/S_{i-2}M^2 \oplus \cdots \oplus S_1 M^{i-1}/M^i.$$

Thus, the component in degree i has finite length, and this length is equal to the length of S_i/M^i. Since N is a finitely generated module over the graded ring R, the theory of Hilbert polynomials applies, and the Hilbert polynomial $P(i)$ of N has the required property that its value at i is the length of S_i/M^i for large i.

To show that $P(n)$ has the correct degree, we note that by assumption we have $M \subseteq \mathfrak{m}F$, and since F/M has finite length there exists an $\mathfrak{m}$-primary ideal $\mathfrak{a}$ such that $\mathfrak{a}F \subseteq M$. We have proven that the corresponding polynomials have degree $d + k - 1$ in both of these cases, and the value of $P(i)$ is in between these values, so $P(i)$ must also have degree $d + k - 1$. $\qquad \square$

Definition 2.5.1 *Let M be a submodule of a free module F of rank k. Let $P(i)$ be the polynomial as in Theorem 2.5.2, and let a_{d+k-1} be the leading coefficient of $P(i)$. Then $(d + k - 1)!a_{d+k-1}$ is called the* Buchsbaum-Rim multiplicity *of M.*

As before, this multiplicity is an integer. The Buchsbaum-Rim multiplicity was introduced by Buchsbaum and Rim [9] as a generalization of multiplicities defined by Euler characteristics; we will discuss this type of multiplicity in Chapter 5. More recent treatments can be found in Kirby [35] and Kirby and Rees [36]. A geometric application can be found in Kleiman [33].

Exercises

2.1 Let $A = \mathbb{Z}[X]/(X^2 - 1)$. Show that A can be given the structure of a ring graded over the group $\mathbb{Z}/2\mathbb{Z}$. Show that A has the property that if a and b are homogeneous elements of A and $ab = 0$, then $a = 0$ or $b = 0$, but A is not an integral domain.

2.2 Show that for every integer $d \geq 0$ and for every integer $n > 0$ there exists a graded ring of dimension $d + 1$ with the leading coefficient of the Hilbert polynomial equal to $n/(d - 1)!$.

2.3 Show that the set of monomials of degree n in d variables are in one-one correspondence with the set of subsets of $n + d - 1$ elements consisting of $d - 1$ elements as follows: map each subset $t_1 < \cdots < t_{d-1}$ of $\{1, \ldots, n + d - 1\}$ to the monomial $\prod_{i=1}^{d}(x_i^{t_i - t_{i-1} - 1})$, where $t_0 = 0$ and $t_d = n + d$. Show that this is indeed a one-one correspondence and deduce the formula for the Hilbert polynomial of the polynomial ring from it.

2.4 Show that the multiplicity of the ring $k[[X, Y]]/(XY)$ is equal to 2.

2.5 Let $A = k[[X, Y, Z]]/(XZ, YZ)$. Show that the multiplicity of A is 1 but that A is not regular.

2.6 Let A be a local ring, and let M be a module of finite length over A. Let F be a free module and let $f : F \to M$ be a surjective map with kernel K. Define the Buchsbaum-Rim multiplicity of M to be the Buchsbaum-Rim multiplicity of K. Show that this definition does not depend on the choice of F and f.

2.7 Let $\mathfrak{a}$ and $\mathfrak{b}$ be $\mathfrak{m}$-primary ideals of a local ring A. Give a formula for the Buchsbaum-Rim multiplicity of the module $\mathfrak{a} \oplus \mathfrak{b}$ in terms of the mixed multiplicities of $\mathfrak{a}$ and $\mathfrak{b}$.

2.8 Let $\mathfrak{a}$ be an $\mathfrak{m}$-primary ideal of $k[[X_1, \ldots X_n]]$ that is generated by monomials in $X_1, \ldots, X_n$. Consider the region S in the first quadrant of $\mathbb{R}^n$ consisting of points $(r_1, \ldots, r_n)$ with $(r_1, \ldots, r_n) \geq (i_1, \ldots, i_n)$ for some i_j with $X_1^{i_1} \cdots X_n^{i_n} \in \mathfrak{a}$. Show that

(a) The volume of the complement of S in the first quadrant of $\mathbb{R}^n$ is equal to the length of $A/\mathfrak{a}$.

(b) The volume of the convex hull of the complement of S in the first quadrant of $\mathbb{R}^n$ is equal to the multiplicity of $\mathfrak{a}$ divided by $n!$.

2.9 Let $\mathfrak{a}$ be an $\mathfrak{m}$-primary ideal. Show that all of the the mixed multiplicities of the set of r ideals $\mathfrak{a}_1, \mathfrak{a}_2, \ldots, \mathfrak{a}_r$ with $\mathfrak{a}_i = \mathfrak{a}$ for all i are equal to the multiplicity of $\mathfrak{a}$.

3

Complexes and Derived Functors

Although the use of homological methods in commutative algebra can be traced back to Hilbert, it is in more recent years that it has become a central part of the subject. One of the first uses of homological techniques was in taking free resolutions of quotients of polynomial rings by graded ideals, from which it is possible to compute the Hilbert function of the quotient. More recently, Serre introduced a homological definition of intersection multiplicities for what would now be called subschemes of a regular scheme, a development that contributed greatly to establishing the importance of homological methods. Most of these applications were originally to rings and ideals, but for various reasons it is preferable to consider the more general case of modules, and, finally, to consider the even more general case of complexes of modules, particularly complexes of the type that arise in free resolutions. In much of this book we consider complexes of modules to be the basic objects of study.

3.1 Derived Functors

In this section we give a brief account of some of the basic homological constructions that arise in the study of modules over commutative rings. As usual, we let A be a Noetherian ring, and we consider finitely generated A-modules. For more details and for proofs that we omit, we refer to a book on homological algebra such as Rotman [64] or a basic algebra book such as Lang [39].

A *complex* of A-modules is an indexed set E_i for $i \in \mathbb{Z}$ together with maps $d_i^E : E_i \to E_{i-1}$, called *boundary maps*, such that $d_i^E d_{i+1}^E = 0$ for all i. The complex will be denoted $E_\bullet$, and the boundary map will be denoted d_i when it is clear which complex is meant. We sometimes use notation with upper indices; in this case, the complex is denoted $E^\bullet$, and the maps d^i go from E^i to E^{i+1}. A map of complexes from $E_\bullet$ to $F_\bullet$ is defined to be a map f_i from E_i to F_i for each i commuting with the respective boundary maps, which means that

$f_{i-1}d_i^E = d_i^F f_i$ for all i. In this way the set of complexes, together with maps of complexes, becomes a category. In addition, the set of maps of complexes from $E_\bullet$ to $F_\bullet$ is an A-module, and the category of complexes is additive. If $F_\bullet$ is a complex, we denote by $F_\bullet[n]$ the complex $F_\bullet$ with degrees shifted by n and with boundary maps multiplied by $(-1)^n$; that is, $F_\bullet[n]_i = F_{n+i}$ and $d_i^{F[n]} = (-1)^n d_{i+n}$ for each i.

A complex is called *bounded above* if $E_i = 0$ for all i sufficiently large, *bounded below* if $E_i = 0$ for all i sufficiently small, and simply *bounded* if $E_i = 0$ for all but finitely many i. A complex in which every module is free is called a free complex, and similarly for complexes consisting of other classes of modules. Bounded free complexes will play an important role in later chapters.

The *homology* of a complex in degree i is defined to be the quotient $\mathrm{Ker}(d_i)/\mathrm{Im}(d_{i+1})$, and it is denoted $H_i(E_\bullet)$. The fact that the image of d_{i+1}^E is contained in the kernel of d_i^E follows from the condition that $d_i^E d_{i+1}^E = 0$. If $f_\bullet : E_\bullet \to F_\bullet$ is a map of complexes, the fact that it commutes with boundary maps implies that f_i maps $\mathrm{Ker}(d_i^E)$ to $\mathrm{Ker}(d_i^F)$ and maps $\mathrm{Im}(d_{i+1}^E)$ to $\mathrm{Im}(d_{i+1}^F)$ and thus induces a map from $H_i(E_\bullet)$ to $H_i(F_\bullet)$ for each i.

A sequence of complexes $E_\bullet \to F_\bullet \to G_\bullet$ is *exact* if the induced sequence of maps on modules is exact in each degree. As in the case of modules, a short exact sequence is a sequence

$$0 \to E_\bullet \to F_\bullet \to G_\bullet \to 0$$

where 0 denotes the complex which is zero in each degree. One of the basic techniques in homological algebra is to use the long exact sequence of homology modules derived from a short exact sequence of complexes. That is, given a short exact sequence as above, there is a long exact sequence

$$\cdots \to H_i(F_\bullet) \to H_i(G_\bullet) \to H_{i-1}(E_\bullet) \to H_{i-1}(F_\bullet) \to H_{i-1}(G_\bullet) \to \cdots$$

where the map from $H_i(G_\bullet)$ to $H_{i-1}(E_\bullet)$ is defined by lifting a representative of a homology class in G_i to F_i, mapping by the boundary map to F_{i-1}, and then lifting the result to E_{i-1}. We omit the verification of the numerous details that must be checked; they can be found in Lang [39] or worked out as an exercise.

The most common source of complexes in commutative algebra is free resolutions of modules and complexes derived from them. If M is a module, a free resolution is a complex $F_\bullet$ with $F_i = 0$ for $i < 0$ and a map from F_0 to M such that the sequence

$$\cdots \to F_i \to \cdots \to F_2 \to F_1 \to F_0 \to M \to 0$$

is exact. Free resolutions can always be constructed step by step, beginning with a map from a free module onto M and mapping a free module onto the kernel of the previous map at each stage. If A is Noetherian and M is finitely generated, all the free modules can be taken to be finitely generated. We use the notation $F_\bullet \to M$ to denote a free resolution. Similarly, an injective resolution $M \to I^\bullet$ is a complex $I^\bullet$ of injective modules with $I^i = 0$ for $i < 0$ together with a map $M \to I^0$ such that the resulting complex is exact.

If $\mathbf{F}$ is any additive functor on the category of modules and $E_\bullet$ is a complex, then $\mathbf{F}(E_\bullet)$ is also a complex; similarly, a map $f_\bullet : E_\bullet \to F_\bullet$ gives a map $\mathbf{F}(f_\bullet) :$ $\mathbf{F}(E_\bullet) \to \mathbf{F}(F_\bullet)$, and $\mathbf{F}$ defines a functor on the category of complexes.

If $f_\bullet$ is a map of complexes from $E_\bullet$ to $F_\bullet$, we say that $f_\bullet$ is *homotopic to zero* if there exist maps $s_i : E_i \to F_{i+1}$ such that $s_{i-1}d_i^E + d_{i+1}^F s_i = f_i$ for all i. If $f_\bullet$ and $g_\bullet$ are maps of complexes, we say that $f_\bullet$ and $g_\bullet$ are *homotopic* if $f_i - g_i$ is homotopic to zero. We say that $s_\bullet$ is a *homotopy* between $f_\bullet$ and $g_\bullet$.

Proposition 3.1.1 *If $f_\bullet$ is homotopic to $g_\bullet$, then $f_\bullet$ and $g_\bullet$ induce the same map on homology.*

Proof If x is in the kernel of d_i^E, then

$$f_i(x) - g_i(x) = s_{i-1}\left(d_i^E(x)\right) + d_{i+1}^F(s_i(x)) = d_{i+1}^F(s_i(x)),$$

which is in the image of d^F, so the images of $f_i(x)$ and $g_i(x)$ differ by a boundary and thus their images in $H_i(F_\bullet)$ are equal.   □

Proposition 3.1.2 *The set of maps from $E_\bullet$ to $F_\bullet$ that are homotopic to zero form a submodule of the module of maps of complexes. If $f_\bullet : E_\bullet \to F_\bullet$ is homotopic to zero, and if $g_\bullet : G_\bullet \to E_\bullet$ and $h_\bullet : F_\bullet \to H_\bullet$ are any maps of complexes, then $f_\bullet g_\bullet$ and $h_\bullet f_\bullet$ are homotopic to zero.*

Proof If $f_\bullet$ and $f'_\bullet$ are homotopic to zero with homotopies $s_\bullet$ and $s'_\bullet$, then $s_\bullet + s'_\bullet$ is a homotopy between $f_\bullet + f'_\bullet$ and zero, and $a s_\bullet$ is a homotopy between $a f_\bullet$ and zero for any $a \in A$. This proves the first assertion.

To prove the second statement, again let $s_\bullet$ be a homotopy between $f_\bullet$ and zero. Then the required homotopies for $f_\bullet g_\bullet$ and $h_\bullet f_\bullet$ are $s_\bullet g_\bullet$ and $h_\bullet s_\bullet$, respectively.   □

It follows from Proposition 3.1.2 that homotopy is an equivalence relation defined by congruence modulo a submodule, and that for any complexes $E_\bullet$ and $F_\bullet$ we can take the quotient of the module of maps of complexes modulo maps that are homotopic to zero. It also follows from Proposition 3.1.2 that composition is defined on this quotient. Thus, we may define a category of complexes together with maps modulo homotopy, which we call the *homotopy category*.

We say that a map $f_\bullet$ is a *quasi-isomorphism* if it induces an isomorphism on homology in every degree. The relation between complexes $F_\bullet$ and $G_\bullet$ defined by saying that $F_\bullet$ and $G_\bullet$ are related if there is a quasi-isomorphism from $F_\bullet$ to $G_\bullet$ is not symmetric, so it is not an equivalence relation. We say that two complexes $F_\bullet$ and $G_\bullet$ are *quasi-isomorphic* if they are equivalent in the equivalence relation generated by this relation. Quasi-isomorphic complexes have isomorphic homology, but the converse is not true; we give an example in the exercises at the end of Chapter 4.

Suppose that $f_\bullet$ and $g_\bullet$ are maps from $E_\bullet$ to $F_\bullet$ and from $F_\bullet$ to $E_\bullet$ such that $f_\bullet g_\bullet$ and $g_\bullet f_\bullet$ are homotopic to the identity maps; in other words, $f_\bullet$ and $g_\bullet$ are inverse isomorphisms in the homotopy category. Then it follows from Proposition 3.1.1 that they induce inverse isomorphisms on homology, so they are quasi-isomorphisms. In particular, this shows that the homology of a complex is defined in the homotopy category. The condition that $f_\bullet$ has a homotopy inverse is stronger than the condition that it be a quasi-isomorphism. If $f_\bullet$ has a homotopy inverse $g_\bullet$, then any additive functor $\mathbf{F}$ will send a homotopy from $f_\bullet g_\bullet$ to the identity to one from $\mathbf{F}(f_\bullet)\mathbf{F}(g_\bullet)$ to the identity, so an additive functor will preserve homotopy inverses and $\mathbf{F}$ defines a functor on the homotopy category. On the other hand, additive functors do not in general preserve quasi-isomorphisms. In fact, since a complex $F_\bullet$ is exact if and only if there is a quasi-isomorphism from $F_\bullet$ to the zero complex, an additive functor $\mathbf{F}$ that preserves quasi-isomorphisms must be exact.

We next wish to describe the construction of derived functors on the category of complexes. We first outline briefly the construction of the right derived functors of a left exact functor on the category of A-modules. The basic construction is as follows: let $\mathbf{F}$ be a functor that is left exact, so that if $0 \to M' \to M \to M''$ is exact, then $0 \to \mathbf{F}(M') \to \mathbf{F}(M) \to \mathbf{F}(M'')$ is also exact. Let M be an A-module. Let $M \to I^\bullet$ be an injective resolution of M. The right derived functors $R^i\mathbf{F}$ are then defined by the formula

$$R^i\mathbf{F}(M) = H^i(\mathbf{F}(I^\bullet)).$$

The independence of the definition on choice of resolution and the fact that it defines a functor come from the following proposition:

Proposition 3.1.3 *Let $E^\bullet$ and $F^\bullet$ be complexes such that $E^\bullet$ is exact and F^i is injective for all $i \geq n$. Then*

(i) *A map of complexes defined for $i < n$ can be completed to a map of complexes from $E^\bullet$ to $F^\bullet$.*

(ii) *Let $f^\bullet$ and $g^\bullet$ be maps of complexes from $E^\bullet$ to $F^\bullet$. Let s^i be a map from E^i to F^{i-1} for each $i < n$ such that the s^i satisfy the condition to define a homotopy between $f^\bullet$ and $g^\bullet$ for all $i < n - 1$. Then $s^\bullet$ can be continued to a homotopy between $f^\bullet$ and $g^\bullet$.*

Proof Both of these statements are proven by inductive constructions, and it suffices to carry out the construction for $i = n$ in such a way that the hypotheses remain satisfied for $n + 1$.

For statement (i): let $f_\bullet$ denote the partial homomorphism of complexes that is defined up through degree $n - 1$. We have the following diagram:

$$\begin{array}{ccccc}
E^{n-2} & \xrightarrow{d_E^{n-2}} & E^{n-1} & \xrightarrow{d_E^{n-1}} & E^n \\
f^{n-2} \downarrow & & f^{n-1} \downarrow & & \\
F^{n-2} & \xrightarrow{d_F^{n-2}} & F^{n-1} & \xrightarrow{d_F^{n-1}} & F^n
\end{array}.$$

Let α denote the composition $d_F^{n-1} f^{n-1}$ induced from E^{n-1} to F^n. Since the square on the left commutes and $d_F^{n-1} d_F^{n-2} = 0$, α factors through the cokernel of d_E^{n-2}. Since $E^\bullet$ is exact, this map from $\text{Coker}(d_E^{n-2})$ to E^n is one-one, and, since F^n is injective, the map extends to a map from E^n to F^n. This produces the required map from E^n to F^n, making the square on the right commute, and completes the inductive step.

For statement (ii): Since the maps s^i define a homotopy from $f^\bullet$ to $g^\bullet$ for $i < n - 1$, we have the diagram

$$\begin{array}{ccccccc}
E^{n-3} & \xrightarrow{d_E^{n-3}} & E^{n-2} & \xrightarrow{d_E^{n-2}} & E^{n-1} & \xrightarrow{d_E^{n-1}} & E^n \\
\downarrow & \overset{s^{n-2}}{\swarrow} & \downarrow & \overset{s^{n-1}}{\swarrow} & \downarrow & & \\
F^{n-3} & \xrightarrow{d_F^{n-3}} & F^{n-2} & \xrightarrow{d_F^{n-2}} & F^{n-1} & \xrightarrow{d_E^{n-1}F} & F^n
\end{array}$$

in which the vertical arrow in degree i is $f^i - g^i$. Since the maps s_i define a homotopy between $f^\bullet$ and $g^\bullet$ in degree $n - 2$ we have

$$s^{n-1} d_E^{n-2} + d_F^{n-3} s^{n-2} = f^{n-2} - g^{n-2}.$$

We compose the maps on both sides of this equality with d_F^{n-2} and, using the fact that $d_F^{n-2} d_F^{n-3} = 0$, we obtain

$$d_F^{n-2} s^{n-1} d_E^{n-2} = d_F^{n-2}(f^{n-2} - g^{n-2}) = (f^{n-1} - g^{n-1}) d_E^{n-2}.$$

Thus,

$$\left(d_F^{n-2} s^{n-1} - (f^{n-1} - g^{n-1})\right) d_E^{n-2} = 0$$

and, arguing as in part (i), we deduce that there exists a map t^n from E^n to F^{n-1} such that

$$t^n d_E^{n-1} = d_F^{n-2} s^{n-1} - (f^{n-1} - g^{n-1}).$$

Letting $s^n = -t^n$ and rearranging terms we have

$$s^n d_E^{n-1} + d_F^{n-2} s^{n-1} = (f^{n-1} - g^{n-1}).$$

Thus, we have constructed a map s^n satisfying the required properties. $\qquad\square$

From Proposition 3.1.3 it follows that a map from M to N induces a map of complexes on their injective resolutions, and that any two choices are homotopic. The functorial properties of $R^i \mathbf{F}$, as well as the fact that they are independent of the choice of resolution, then follow. Again we omit the details, which will follow from the more general construction below.

We remark that a dual version of the above proof shows that a partial map from a free complex into an exact complex can be extended to a map of complexes. This result is very useful in applications, and we state it as a separate proposition.

Proposition 3.1.4 *Let $E_\bullet$ and $F_\bullet$ be complexes such that E_i is projective and $F_\bullet$ is exact for all $i \geq n$. Then*

(i) *A map of complexes defined for $i < n$ can be completed to a map of complexes from $E_\bullet$ to $F_\bullet$.*

(ii) *Let $f_\bullet$ and $g_\bullet$ be maps of complexes from $E_\bullet$ to $F_\bullet$. Let s_i be a map from E_i to F_{i+1} for each $i < n$ such that the s_i satisfy the condition to define a homotopy between $f_\bullet$ and $g_\bullet$ for all $i < n - 1$. Then $s_\bullet$ can be continued to a homotopy between $f_\bullet$ and $g_\bullet$.*

In Propositions 3.1.3 and 3.1.4 the exactness of one of the two complexes is necessary for the inductive construction. If we do not know that the complex is exact but know only that the homology is annihilated by a given element of A we can still prove a weaker result. The next Proposition will be used in a later chapter.

Proposition 3.1.5 *Let $E_\bullet$ and $F_\bullet$ be complexes such that E_i is projective for all $i \geq n$, and let $f_\bullet$ be a map of complexes from $E_\bullet$ to $F_\bullet$ defined for $i \leq n$. Suppose that x is an element of A such that for $j = 0, \ldots, m - 1$ we have that $H_{n+j}(F_\bullet)$ is annihilated by x. Then the map $x^m f_\bullet$ can be extended to a map of complexes for $i \leq n + m$.*

Proof The proof uses the same kind of inductive construction as Propositions 3.1.3 and 3.1.4. We outline the first step. We have the diagram

$$
\begin{array}{ccccc}
E_{n+1} & \xrightarrow{d^E_{n+1}} & E_n & \xrightarrow{d^E_n} & E_{n-1} \\
 & & f_n \downarrow & & f_{n-1} \downarrow \\
F_{n+1} & \xrightarrow{d^F_{n+1}} & F_n & \xrightarrow{d^F_n} & F_{n-1}
\end{array} .
$$

To construct f_{n+1} we use the commutativity of the square on the right and the fact that $d^E_n d^E_{n+1} = 0$ to conclude that $d^F_n f_n d^E_{n+1} = 0$, so that $f_n d^E_{n+1}$ maps into the kernel of d^F_n. If $F_\bullet$ were exact we could use the fact that E_{n+1} is projective to lift $f_n d^E_{n+1}$ to F_{n+1} and define f_{n+1}. In this case, we know only that x annihilates the homology $H_n(F_\bullet)$, so we can lift $x f_n d^E_{n+1} = (x f_n) d^E_{n+1}$ to F_{n+1}. Thus, if we replace the original maps f_i by $x f_i$ for all $i \le n$, we can extend the resulting map of complexes to degree $n + 1$. Carrying out this process m times, we conclude that x^m times the original map can be extended to a map of complexes for $i \le m + n$. $\qquad\square$

We showed above how to define the right derived functors of a left exact functor. If **F** is a right exact functor, its left derived functors $L_i \mathbf{F}$ are defined by a similar process, taking a free (or projective) resolution, applying the functor **F**, and taking homology.

We now consider the more general case of derived functors of functors acting on complexes. As in the case of modules, there are two cases, that of left derived functors acting on complexes that are bounded below, and that of right derived functors acting on complexes that are bounded above. We carry out the construction only for left derived functors, since the case of right derived functors is essentially the same. The first step is to show the existence of free resolutions of complexes that are bounded below.

Proposition 3.1.6 *Let $E_\bullet$ be a complex that is bounded below. There exists a bounded below free complex $F_\bullet$ and a quasi-isomorphism $f_\bullet$ from $F_\bullet$ to $E_\bullet$.*

Proof As in the case for modules, one proceeds step by step. Since $E_\bullet$ is bounded below, we can let $F_i = 0$ for i sufficiently small.

Suppose that free modules F_i and maps from F_i to E_i have been constructed for $i < n$ such that f_i induces an isomorphism in homology for $i < n - 1$ and induces a surjection from the kernel of d^F_{n-1} onto the kernel of d^E_{n-1}. We will define F_n to be a direct sum $F' \oplus F''$ together with maps d^F_n and f_n as follows:

$$
\begin{array}{ccccc}
F' \oplus F'' & \xrightarrow{d^F_n} & F_{n-1} & \xrightarrow{d^F_{n-1}} & F_{n-2} \\
f_n \downarrow & & f_{n-1} \downarrow & & f_{n-2} \downarrow \\
E_n & \xrightarrow{d^E_n} & E_{n-1} & \xrightarrow{d^E_{n-1}} & E_{n-2}
\end{array} .
$$

We must define $F_n = F' \oplus F''$ and the maps $d_n^F = (d_n^F)' \oplus (d_n^F)''$ and $f_n = (f_n)' \oplus (f_n)''$ in such a way that the left square commutes, the composition of maps in the top row is zero, the map induced in homology by f_{n-1} is an isomorphism, and f_n induces a surjection from the kernel of d_n^F to the kernel of d_n^E.

Since f_{n-1} maps the kernel of d_{n-1}^F surjectively onto the kernel of d_{n-1}^E and the image of d_n^E is contained in the kernel of d_{n-1}^E, we can find a submodule M of the kernel of d_{n-1}^F such that $f_{n-1}(M) = \text{Im}(d_n^E)$. Choose F'' to be a free module together with a map $(d_n^F)''$ onto M, and let $(f_n)''$ be a lifting of $f_{n-1}(d_n^F)''$ to a map to E_n. Choose F' to be a free module together with a surjective map f_n' onto the kernel of d_n^E, and let $(d_n^F)' = 0$. The composition of maps in the top row is zero since both components of d_n^F map into the kernel of d_{n-1}^F, and the left square commutes; for $(d_n^F)''$ this follows since $(f_n)''$ was defined as a lifting, and for $(d_n^F)'$ since both compositions are zero. Furthermore, since the image of $(d_n^F)''$ maps onto the image of d_n^E under f_{n-1}, f_{n-1} now induces an isomorphism in homology. Since $(f_n)'$ maps onto the kernel of d_n^E, f_n does also, and the process can now be continued. $\qquad\square$

If $E_\bullet$ is a complex, a free (or projective) complex $F_\bullet$ together with a quasi-isomorphism from $F_\bullet$ to $E_\bullet$ will be called a *free* (or *projective*) *resolution* of $E_\bullet$. Similarly, a quasi-isomorphism from $E_\bullet$ to an injective complex will be called an injective resolution of $E_\bullet$.

We may now define the left derived functor $LF(E_\bullet)$ to be the complex $\mathbf{F}(F_\bullet)$ where $F_\bullet$ is a complex of free modules with a quasi-isomorphism to $E_\bullet$. This complex is not unique, but it is unique up to homotopy. To prove this fact and the functoriality of this process we need an analogue of Proposition 3.1.3. It is possible to give a direct proof, but it simplifies the argument considerably to introduce mapping cones of maps of complexes.

If $f_\bullet : E_\bullet \to F_\bullet$ is a map of complexes, its *mapping cone* is the complex $C_\bullet[f_\bullet]$, where $C_i[f_\bullet] = F_i \oplus E_{i-1}$ and $d^{C_i[f_\bullet]}(f, e) = (d_i^F(f) + f_{i-1}(e), -d_{i-1}^E(e))$. In matrix form, the boundary map on the mapping cone is written

$$\begin{pmatrix} d_i^F & f_{i-1} \\ 0 & -d_{i-1}^E \end{pmatrix}.$$

There is then a map of complexes from $F_\bullet$ to $C_\bullet[f_\bullet]$ and a map of complexes from $C_\bullet[f_\bullet]$ to $E_\bullet[-1]$ that form a short exact sequence

$$0 \to F_\bullet \to C_\bullet[f_\bullet] \to E_\bullet[-1] \to 0.$$

Note that since shifting degrees by -1 also multiplies the boundary maps by -1 the sign of the boundary map on $E_\bullet[-1]$ is correct. A direct verification shows that the boundary map induced by this short exact sequence from $H_i(E_\bullet[-1]) \cong H_{i-1}(E_\bullet)$ to $H_{i-1}(F_\bullet)$ is the map induced on homology by $f_\bullet$. It then follows from the long exact homology sequence that $f_\bullet$ is a quasi-isomorphism if and only if the mapping cone of $f_\bullet$ is exact.

Theorem 3.1.7 *Suppose that we have a diagram of maps of complexes*

$$
\begin{array}{c}
F_\bullet \\
\downarrow \rho_\bullet \\
E_\bullet \xrightarrow{\alpha_\bullet} G_\bullet
\end{array}
$$

where all three complexes are bounded below, $F_\bullet$ is a complex of free modules, and $\alpha_\bullet$ is a quasi-isomorphism. Then there is a map $\beta_\bullet$ from $F_\bullet \to E_\bullet$ such that $\alpha_\bullet \beta_\bullet$ is homotopic to $\rho_\bullet$. The map $\beta_\bullet$ is unique up to homotopy.

Proof Let $C_\bullet[\alpha_\bullet]$ denote the mapping cone of $\alpha_\bullet$, and let $\gamma_\bullet$ be the map from $G_\bullet$ into $C_\bullet[\alpha_\bullet]$ coming from the construction of the mapping cone. Let $\phi_\bullet$ be the composition $\gamma_\bullet \rho_\bullet$. Since $\alpha_\bullet$ is a quasi-isomorphism, the mapping cone $C_\bullet[\alpha_\bullet]$ is exact. Furthermore, $F_\bullet$ and $C_\bullet[\alpha_\bullet]$ are bounded below, so we can define a homotopy between $\phi_\bullet$ and the zero map in low degrees simply by letting $s_i = 0$, and $F_\bullet$ is a complex of free modules, so Proposition 3.1.4 implies that $\phi_\bullet$ is homotopic to zero. The remainder of the proof consists of unraveling this homotopy in terms of components to define maps between the original complexes that will give the result.

The homotopy described in the previous paragraph gives a map from F_i to $G_{i+1} \oplus E_i$ for each i; we denote this map in matrix form by $\binom{s_i}{\beta_i}$. Since $\phi_\bullet$ is defined to be $\rho_\bullet$ followed by embedding in the first component, the map ϕ_i in matrix form is $\binom{\rho_i}{0}$. The fact that we have a homotopy between the maps $\phi_\bullet$ and zero from $F_\bullet$ to $C_\bullet[\alpha_\bullet]$ is expressed by the matrix equation

$$
\begin{pmatrix} \rho_i \\ 0 \end{pmatrix} = \begin{pmatrix} d_{i+1}^G & \alpha_i \\ 0 & -d_i^E \end{pmatrix} \begin{pmatrix} s_i \\ \beta_i \end{pmatrix} + \begin{pmatrix} s_{i-1} \\ \beta_{i-1} \end{pmatrix} (d_i^F).
$$

This matrix equality produces two equalities:

(1) $\rho_i = d_{i+1}^G s_i + \alpha_i \beta_i + s_{i-1} d_i^F$.
(2) $-d_i^E \beta_i + \beta_{i-1} d_i^F = 0$.

The second equality states that the maps β_i define a map $\beta_\bullet$ of complexes, and the first equality states that $\alpha_\bullet \beta_\bullet$ is homotopic to $\rho_\bullet$. This proves the first statement of the theorem.

It remains to prove the uniqueness of $\beta_\bullet$ up to homotopy. By replacing two maps whose compositions with $\alpha_\bullet$ are homotopic to $\rho_\bullet$ by their difference, it suffices to show that if $\alpha_\bullet \beta_\bullet$ is homotopic to zero, then $\beta_\bullet$ is homotopic to zero.

Let $s_\bullet$ be a homotopy between $\alpha_\bullet\beta_\bullet$ and the zero map. As before, we embed $G_\bullet$ into the mapping cone $C_\bullet[\alpha_\bullet]$, which is exact. The facts that $s_\bullet$ is a homotopy between $\alpha_\bullet\beta_\bullet$ and zero and that $\beta_\bullet$ is a map of complexes imply that we have the following matrix equality for each i:

$$\begin{pmatrix} d_{i+1}^G & \alpha_i \\ 0 & -d_i^E \end{pmatrix}\begin{pmatrix} -s_i \\ \beta_i \end{pmatrix} = \begin{pmatrix} -s_{i-1} \\ \beta_{i-1} \end{pmatrix}(-d_i^F).$$

This equality implies that the map defined in degree i by the matrix $\begin{pmatrix} -s_i \\ \beta_i \end{pmatrix}$ is a map of complexes from $F_\bullet[-1]$ to $C_\bullet[\alpha_\bullet]$. Since $C_\bullet[\alpha_\bullet]$ is exact and bounded below and $F_\bullet$ is free, this map is homotopic to zero. If we follow a homotopy with the projection to $E_\bullet$ we obtain a map from F_i to E_{i+1} for each i, and a matrix computation shows that this map is a homotopy between $\beta_\bullet$ and the zero map. We leave the details, which are similar to the computation in part one, as an exercise. $\qquad\square$

Using Theorem 3.1.7 we may show the existence of derived functors. Let $\mathbf{F}$ be a left exact additive functor on the category of A-modules, and let $E_\bullet$ be a complex of A-modules that is bounded below. Let $F_\bullet$ be a complex of free modules together with a quasi-isomorphism $\alpha_\bullet$ from $F_\bullet$ to $E_\bullet$. We define the derived functor $L\mathbf{F}(E_\bullet)$ to be the complex $\mathbf{F}(F_\bullet)$.

Proposition 3.1.8

(i) *The complex $\mathbf{F}(F_\bullet)$ is uniquely defined up to homotopy equivalence by $E_\bullet$. Thus, this construction defines a unique complex $L\mathbf{F}(E_\bullet)$ in the homotopy category.*

(ii) *If $f_\bullet : E_\bullet \to E_\bullet'$ is a map of complexes, there is a naturally defined map of complexes from $L\mathbf{F}(E_\bullet)$ to $L\mathbf{F}(E_\bullet')$ that is unique up to homotopy.*

Proof Let $F_\bullet'$ be another complex of free modules, and suppose there is a quasi-isomorphism $\alpha_\bullet' : F_\bullet' \to E_\bullet$, so that we have a diagram

$$F_\bullet$$
$$\downarrow \alpha_\bullet$$
$$F_\bullet' \overset{\alpha_\bullet'}{\to} E_\bullet$$

where $\alpha_\bullet$ and $\alpha_\bullet'$ are quasi-isomorphisms. Theorem 3.1.7 implies first that there are maps $\beta_\bullet$ from $F_\bullet$ to $F_\bullet'$ and $\gamma_\bullet$ from $F_\bullet'$ to $F_\bullet$ such that $\alpha_\bullet\gamma_\bullet$ is homotopic to $\alpha_\bullet'$ and $\alpha_\bullet'\beta_\bullet$ is homotopic to $\alpha_\bullet$. Thus, we have that $\alpha_\bullet\gamma_\bullet\beta_\bullet$ is homotopic to $\alpha_\bullet$, so from the uniqueness of Theorem 3.1.7, $\gamma_\bullet\beta_\bullet$ is homotopic to the identity. Similarly, $\beta_\bullet\gamma_\bullet$ is homotopic to the identity, so the two maps are homotopy inverses. If we apply $\mathbf{F}$ to these maps, we obtain homotopy inverses between $\mathbf{F}(F_\bullet)$ and $\mathbf{F}(F_\bullet')$, proving the first part of the proposition.

To prove the second part, we let $\alpha'_\bullet : F'_\bullet \to E'_\bullet$ be a free resolution. Proposition 3.1.7 implies that there is a map $\beta_\bullet$ from $F_\bullet$ to $F'_\bullet$ such that $\alpha'_\bullet \beta_\bullet = f_\bullet \alpha_\bullet$, and that $\beta_\bullet$ is unique up to homotopy. Thus, we induce a map from $LF(E_\bullet)$ to $LF(E'_\bullet)$, and this map is also unique up to homotopy. The functoriality of this map comes from the functoriality up to homotopy of the constructions by which it is defined. □

Thus, we have defined left derived functors of an additive functor. Right derived functors are defined similarly, using injective rather than free resolutions. The result is a complex defined up to homotopy equivalence. Since a homotopy equivalence is a quasi-isomorphism, this implies that the homology of the resulting complex is a well-defined module in each degree.

The functors defined by Hom and the tensor product are among those whose derived functors are the most important. Let N be a fixed module. We consider the functor given by $\mathbf{F}(M) = \mathrm{Hom}_A(M, N)$, the A-module of homomorphisms from M to N. The derived functors are then defined by taking a free (or projective) resolution $F_\bullet \to M$; then the ith derived functor $\mathrm{Ext}^i_A(M, N)$ is defined by letting $\mathrm{Ext}^i_A(M, N) = H^i(\mathrm{Hom}(F_\bullet, N))$, where the maps on $\mathrm{Hom}(F_\bullet, N)$ are induced by those of $F_\bullet$. In this case the derived functor can also be computed by taking an injective resolution in the second variable, as will follow from the results of the next section.

3.2 Double Complexes

A *double complex* is a doubly indexed array of modules $E_{i,j}$, together with boundary maps $d^1_{i,j} : E_{i,j} \to E_{i-1,j}$ and $d^2_{i,j} : E_{i,j} \to E_{i,j-1}$ for each i, j satisfying

(i) $d^1_{i-1,j} d^1_{i,j} = 0$.
(ii) $d^2_{i,j-1} d^2_{i,j} = 0$.
(iii) $d^1_{i,j-1} d^2_{i,j} = d^2_{i-1,j} d^1_{i,j}$.

Double complexes arise in many computations in homological algebra. If we think of the double complex arranged in a matrix, the first two conditions say that each row and each column is a complex, and the third condition says that the set of maps from one row to the next forms a map of complexes, and similarly for the maps from one column to the next.

We use the notation $E_{\bullet\bullet}$ to denote a double complex. Associated to $E_{\bullet\bullet}$ is a simple complex, called the *total complex* of $E_{\bullet\bullet}$ and denoted $\mathrm{tot}(E_{\bullet\bullet})$. The modules that make up the total complex are defined as follows:

$$\mathrm{tot}(E_{\bullet\bullet})_k = \oplus_{i+j=k} E_{i,j}.$$

The map from $\text{tot}(E_{\bullet\bullet})_k$ to $\text{tot}(E_{\bullet\bullet})_{k-1}$ sends an element x in the summand $E_{i,j}$ of $\text{tot}(E_{\bullet\bullet})_k$ to $(\ldots, 0, d^1_{i,j}(x), (-1)^i d^2_{i,j}(x), 0, \ldots,)$. Conditions (i) through (iii) above imply that $\text{tot}(E_{\bullet\bullet})$ is a complex.

In the remainder of this section, we let $E_{\bullet\bullet}$ denote a double complex, and let $G_{\bullet}$ be its total complex. We assume furthermore that $E_{\bullet\bullet}$ has the property that for each integer k, we have $E_{i,j} = 0$ for all but finitely many i, j with $i + j = k$. Fix j, and consider the complex $E_{\bullet,j}$; this is the simple complex in the column indexed by j. Let $H^1_{i,j}$ denote the homology of this complex in degree i, j. Since the horizontal maps define maps of complexes from one column to the next, they induce maps on homology, and for each i we now have a complex that we denote $H^1_{i,\bullet}$. Denote the homology of the complex $H^1_{i,\bullet}$ in degree i, j by $H^{1,2}_{i,j}$. The main results of this complex relate the homology $H^{1,2}_{i,j}$ to the homology of the total complex. There are similar results, of course, for the homology $H^{2,1}_{i,j}$ taken in the reverse order.

We first define a filtration on the homology $H_k(G_{\bullet})$ of the total complex. Recall that G_k is the direct sum of $E_{i,j}$ over all i, j with $i + j = k$. We define $F^n(H_k(G_{\bullet}))$ to be the image in $H_k(G_{\bullet})$ of the intersection of $\text{Ker}(d^G_k)$ with the sum of those $E_{i,j}$ that satisfy the extra condition that $i \geq n$. In other words, $F^n(H_k(G_{\bullet}))$ is the submodule of the homology consisting of elements that can be represented by elements in G_k that have the property that all of their components in $E_{i,j}$ are zero for $i < n$. It is clear that $F^{n+1}(H_k(G_{\bullet})) \subseteq F^n(H_k(G_{\bullet}))$.

The aim of this section is to show that, with the filtration on $H_k(G_{\bullet})$ defined as above, the quotient $F^n(H_k(G_{\bullet}))/F^{n+1}(H_k(G_{\bullet}))$ of consecutive submodules in the filtration is isomorphic to a certain quotient of a submodule of $H^{1,2}_{i,j}(E_{\bullet\bullet})$ for $i = n$ and $i + j = k$.

The next step is to describe $H^{1,2}_{i,j}(E_{\bullet\bullet})$ in terms of elements. As above, we denote this module $H^{1,2}_{i,j}$. An element of $H^{1,2}_{i,j}$ is represented by an element of $H_i(E_{\bullet,j})$ that is mapped to zero in $H_i(E_{\bullet,j-1})$. The fact that it is an element of $H_i(E_{\bullet,j})$ says that it is represented by an element e_i of $E_{i,j}$ that goes to zero in $E_{i-1,j}$. The fact that its image in $H_i(E_{\bullet,j-1})$ is zero says that the element $d^2(e_i)$ in $E_{i,j-1}$ can be lifted to an element e_{i+1} in $E_{i+1,j-1}$. Thus, we have the following diagram.

$$
\begin{array}{ccccc}
& & E_{i+1,j-1} & & e_{i+1} \\
& & \downarrow & & \downarrow \\
E_{i,j} & \longrightarrow & E_{i,j-1} & & d^2_{i,j}(e_i) = d^1_{i+1,j-1}(e_{i+1}) \\
\downarrow & {\scriptstyle e_i} & & & \\
E_{i-1,j} & \downarrow & & & \\
& 0 & & &
\end{array}
$$

Hence an element of $H_{i,j}^{1,2}$ is represented by an element e_i of $E_{i,j}$ that fits into a diagram as above. It is zero in $H_{i,j}^{1,2}$ if and only if it can be lifted to an element of $H_{i,j+1}^1$. In terms of elements, this means that there is an $f_i \in E_{i,j+1}$ with $d_{i,j+1}^1(f_i) = 0$ and such that $e_i - d_{i,j+1}^2(f_i)$ can be lifted to an element f_{i+1} in $E_{i+1,j}$. In other words, it means that we can write

$$e_i = d_{i,j+1}^2(f_i) + d_{i+1,j}^1(f_{i+1})$$

where $d_{i,j+1}^1(f_i) = 0$.

In the next theorem we say that a module M is isomorphic to a subquotient of a module N if there exist submodules $P \subseteq Q$ of N such that $M \cong Q/P$.

Theorem 3.2.1 *With notation as above, the quotient of successive modules in the filtration $F^i(H_{i+j}(G_\bullet))/F^{i+1}(H_{i+j}(G_\bullet))$ is isomorphic to a subquotient of $H_{i,j}^{1,2}(E_{\bullet\bullet})$.*

Proof Let $k = i + j$. Let K denote the intersection of $\mathrm{Ker}(d_k^G)$ with the sum of those $E_{r,s}$ such that $r + s = k$ and $r \geq i$. By definition, $F^i(H_{i+j}(G_\bullet))$ is the image of K in $H_{i+j}(G_\bullet)$. An element of K can be written in the form $(0, \ldots, 0, e_i, e_{i+1}, e_{i+2}, \ldots)$, and the condition that it is in the kernel of d_k^G implies in particular that $d_{i,j}^1(e_i) = 0$ and $d_{i,j}^2(e_i) = (-1)^{i+1}d_{i+1,j-1}^1(e_{i+1})$. Hence we can define a map $\phi : K \to H_{i,j}^{1,2}$ by sending this element to e_i.

Let H be the submodule of K generated by the submodules $\mathrm{Im}(d_{k+1}^G) \cap K$ and $(\oplus_{r \geq i+1} E_{r,s}) \cap K$. Since the image of K in the homology of $G_\bullet$ is isomorphic to $K/(\mathrm{Im}(d_{k+1}^G) \cap K)$ and the image of $F^{i+1}(H_k(G_\bullet))$ in this quotient is generated by $(\oplus_{r \geq i+1} E_{r,s}) \cap K$, we have

$$F^i(H_{i+j}(G_\bullet))/F^{i+1}(H_{i+j}(G_\bullet)) \cong K/H. \tag{$*$}$$

Since we have a map ϕ from K to $H_{i,j}^{1,2}$, equation $(*)$ implies that to prove that $F^i(H_{i+j}(G_\bullet))/F^{i+1}(H_{i+j}(G_\bullet))$ is isomorphic to a subquotient of $H_{i,j}^{1,2}$, it suffices to show that ϕ induces an isomorphism from K/H onto $\phi(K)/\phi(H)$. Since this induced map is clearly surjective, it remains to show that it is injective, which means that if $\phi(k) = \phi(h)$ for $k \in K$ and $h \in H$, then $k \in H$. Replacing k by $k - h$, it thus suffices to show that if $\phi(k) = 0$, then $k \in H$.

Let $k = (0, \ldots, 0, e_i, e_{i+1}, e_{i+2}, \ldots)$, and suppose $\phi(k) = 0$. By the discussion before this theorem, we can write $e_i = d_{i,j+1}^2(f_i) + d_{i+1,j}^1(f_{i+1})$, where $d_{i,j+1}^1(f_i) = 0$. Hence we have

$$
\begin{aligned}
k &= (0, \ldots, 0, e_i, e_{i+1}, e_{i+2}, \ldots) \\
&= d_{k+1}^G(0 \ldots, 0, (-1)^i f_i, f_{i+1}, 0, \ldots) \\
&\quad + \left(0, \ldots, 0, 0, e_{i+1} - (-1)^{i+1} d_{i+1,j}^1(f_{i+1}), e_{i+2}, \ldots\right),
\end{aligned}
$$

so that $k \in \text{Im}(d_{k+1}^G) \cap K + ((\oplus_{r \geq i+1} E_{r,s}) \cap K = H$. This completes the proof of the theorem. $\qquad\square$

Applications of this theorem usually arise in situations where the homology modules $H_{i,j}^{1,2}$ can be computed easily. The subquotients of $H_{i,j}^{1,2}$ in the statement of the theorem can be determined by a sequence of maps between the $H_{i,j}^{1,2}$ that form a structure called a *spectral sequence*; see for example [64, Chapter 11]. For many purposes, however, the weaker statement made in this theorem suffices. We will give several applications of this result in later chapters. We note one special case in the next proposition.

Proposition 3.2.2 *Let $E_{\bullet\bullet}$ be a double complex such that for each integer k, the set of i, j with $i + j = k$ and $E_{i,j} \neq 0$ is finite. If $H_{i,j}^{1,2} = 0$ for all i, j, then the total complex $G_\bullet$ of $E_{\bullet\bullet}$ is exact.*

Proof The hypothesis that the set of $E_{i,j}$ with $i + j = k$ and $E_{i,j} \neq 0$ is finite implies that there exist integers m and n such that $F^m(H_k(G_\bullet)) = 0$ and $F^n(H_k(G_\bullet)) = H_k(G_\bullet)$. Thus, there is a finite filtration of $H_k(G_\bullet)$ with zero quotients. Hence $H_k(G_\bullet) = 0$. $\qquad\square$

A particular case of Proposition 3.2.2 is when all rows or all columns of $E_{\bullet\bullet}$ are exact. Another common situation is when all rows or columns except one are exact, or that the homology of the double complex can be reduced to the homology of one row or column in some more general way. The next proposition gives two examples.

Proposition 3.2.3 *Let $E_{\bullet\bullet}$ be a double complex, and let n be an integer such that $E_{i,j} = 0$ if $j < n$. Let $G_\bullet$ denote the total complex of $E_{\bullet\bullet}$.*

(i) *If all columns of $E_{\bullet\bullet}$ except the nth column are exact, then the homology of $G_\bullet$ is isomorphic to the homology of the nth column of $E_{\bullet\bullet}$.*

(ii) *If for each i the ith row of $E_{\bullet\bullet}$ is exact except at $E_{i,n}$, then the homology of $G_\bullet$ is isomorphic to the homology of the complex defined as the cokernel of $E_{\bullet,n+1} \to E_{\bullet,n}$.*

Proof The condition that $E_{i,j} = 0$ if $j < n$ implies that the complex $E_{\bullet,n}$ consisting of the nth column of $E_{\bullet\bullet}$ is a subcomplex of the total complex $G_\bullet$. The quotient $G_\bullet/E_{\bullet,n}$ is the total complex of the double complex obtained by removing the nth column from $E_{\bullet\bullet}$; we denote this complex $H_\bullet$. Every column of $H_\bullet$ is exact, so Proposition 3.2.2 implies that $H_\bullet$ is exact. Thus, the long exact homology sequence induced by the short exact sequence

$$0 \to E_{\bullet,n} \to G_\bullet \to H_\bullet \to 0$$

implies that the embedding of $E_{\bullet,n}$ in $G_\bullet$ induces an isomorphism in homology, which is statement (i).

Assume now that the ith row of $E_{\bullet\bullet}$ is exact except at $E_{i,n}$ for all i. Extend the double complex by adding the cokernel of $d_{i,n}^2$ in each row, and denote the new column $E_{\bullet,n-1}$. The map $d_{i,n-1}^1$ for each i is the map induced by $d_{i,n}^1$ on the respective cokernels. Denote the extended complex $F_{\bullet\bullet}$ and its total complex $K_\bullet$. The hypothesis states that each row of $F_{\bullet\bullet}$ is exact, which implies that $K_\bullet$ is exact by Proposition 3.2.2. Furthermore, as in the proof of part (i), we have a short exact sequence of complexes

$$0 \to E_{\bullet,n-1} \to K_\bullet \to G_\bullet \to 0,$$

and the associated long exact homology sequence shows that there is an isomorphism between $H_{i+1}(G_\bullet)$ and $H_i(E_{\bullet,n-1})$ for each i, as was to be shown.
□

One important application of Proposition 3.2.2 is to the derived functors of Hom and the tensor product. If $E_\bullet$ and $F_\bullet$ are bounded complexes, then $\mathrm{Hom}(E_\bullet, F_\bullet)$ and $E_\bullet \otimes F_\bullet$ are double complexes, and we use the notation $\mathrm{Hom}(E_\bullet, F_\bullet)$ and $E_\bullet \otimes F_\bullet$ also to denote the associated simple complexes. These define additive functors by fixing either variable and considering them as functors in the other variable. Thus, the theory of the previous section enables us to define derived functors, which we call Ext and Tor, respectively. However, there are a priori two different choices, one for each variable. The fact that these are the same follows from the next proposition, which we prove in the case of the Hom functor.

Proposition 3.2.4 *Let $F_\bullet$ be a projective complex that is bounded below, and suppose that $\alpha_\bullet : G_\bullet \to H_\bullet$ is a quasi-isomorphism of bounded above complexes. Then the induced map from $\mathrm{Hom}(F_\bullet, G_\bullet)$ to $\mathrm{Hom}(F_\bullet, H_\bullet)$ is a quasi-isomorphism.*

Proof Let $C_\bullet[\alpha_\bullet]$ be the mapping cone of $\alpha_\bullet$. The hypothesis that $\alpha_\bullet$ is a quasi-isomorphism implies that $C_\bullet[\alpha_\bullet]$ is exact. Consider the double complex $\mathrm{Hom}(F_\bullet, C_\bullet[\alpha_\bullet])$. For each i, $\mathrm{Hom}(F_i, C_\bullet[\alpha_\bullet])$ is exact since F_i is projective. Hence by Proposition 3.2.2 the associated simple complex is exact. However, it can be shown that $\mathrm{Hom}(F_\bullet, C_\bullet[\alpha_\bullet])$ is the mapping cone of $\mathrm{Hom}(F_\bullet, \alpha_\bullet)$ (see the exercises), so this latter map is a quasi-isomorphism.
□

To see that we can compute $\text{Ext}(G_\bullet, H_\bullet)$ with either a projective resolution of $G_\bullet$ or an injective resolution of $H_\bullet$, let $F_\bullet \to G_\bullet$ and $H_\bullet \to I_\bullet$ be projective and injective resolutions, respectively. Then Proposition 3.2.4 implies that we have quasi-isomorphisms between $\text{Hom}(F_\bullet, H_\bullet)$ and $\text{Hom}(F_\bullet, I_\bullet)$ and similarly between $\text{Hom}(F_\bullet, I_\bullet)$ and $\text{Hom}(G_\bullet, I_\bullet)$. Hence these are all quasi-isomorphic. Thus, we have a unique functor $\text{Ext}(G_\bullet, H_\bullet)$ and a unique functor $\text{Tor}(G_\bullet, H_\bullet)$ defined up to quasi-isomorphism.

We conclude this section with a technical property of the Ext and Tor functors for modules that is very useful.

Proposition 3.2.5 *Let M be an A-module and let x be an element of A. Then*

(i) *The map induced on $\text{Ext}^i(M, H_\bullet)$ by multiplication by x on M is multiplication by x on $\text{Ext}^i(M, H_\bullet)$.*

(ii) *If $xM = 0$ then $x\,\text{Ext}^i(M, H_\bullet) = 0$.*

Proof Multiplication by x on M induces multiplication by x on the module $\text{Hom}(M, N)$ for every module N, and hence, using the representation of $\text{Ext}^i(M, H_\bullet)$ as the homology of $\text{Hom}(M, I_\bullet)$ for an injective resolution $I_\bullet$, induces multiplication by x on $\text{Ext}^i(M, H_\bullet)$. And clearly if M is annihilated by x so are the modules $\text{Hom}(M, I_i)$ and hence also $\text{Ext}^i(M, H_\bullet)$ for each i. $\qquad\square$

Analogues of these theorems apply also to Tor and to Ext in the second variable.

3.3 The Koszul Complex

Let $x_1, \ldots, x_n$ be a sequence of elements of a ring A. The *Koszul complex* $K_\bullet(x_1, \ldots x_n)$ is defined as follows: let K_i be the exterior power $\Lambda^i(A^n)$. Then there is a map d_i from K_i to K_{i-1} that sends a basis element $e_{j_1} \wedge \cdots \wedge e_{j_i}$ to $\sum_{k=1}^i (-1)^{k+1} x_{j_k} e_{j_1} \wedge \cdots \wedge \widehat{e_{j_k}} \wedge \cdots \wedge e_{j_i}$. This expression is alternating in the e_{j_k}, so it does not depend on the order in which the e_{j_k} are written.

Proposition 3.3.1 *With K_i and d_i defined as above, we have $d_i d_{i+1} = 0$.*

Proof This proposition follows from a straightforward computation. We wish to show that if $e_{j_1} \wedge \cdots \wedge e_{j_{i+1}}$ is a generator of $\Lambda^{i+1}(A^n)$, then each component of $d_i d_{i+1}(e_{j_1} \wedge \cdots \wedge e_{j_{i+1}})$ is zero. Since the only nonzero coefficients of $d_{i+1}(e_{j_1} \wedge \cdots \wedge e_{j_{i+1}})$ correspond to basis elements obtained by removing one e_{j_k} and similarly for d_i, the only possible nonzero coefficients of $d_i d_{i+1}(e_{j_1} \wedge \cdots \wedge e_{j_{i+1}})$ correspond to basis elements obtained by removing two of the e_{j_k}.

Let $e_{j_1} \wedge \ldots \widehat{e_{j_k}} \wedge \ldots \wedge \widehat{e_{j_m}} \wedge \ldots \wedge e_{j_i}$ be such a basis element. A computation shows that its coefficient is $(-1)^{k+m} x_{j_k} x_{j_m} + (-1)^{k+m-1} x_{j_m} x_{j_k} = 0$. Thus, $d_i d_{i+1} = 0$. $\qquad\qquad\qquad\qquad\qquad\qquad\qquad\qquad\qquad\qquad\qquad\qquad\square$

The above definition is useful for computations, since it represents the Koszul complex as a complex of free modules with maps defined by specific matrices. For example, it follows immediately from this definition that the entries of the matrices defining the boundary maps are all $\pm x_i$ or 0. For some purposes, however, it is better to have a more intrinsic definition. To do this we define the Koszul complex defined by a map $\phi : F \to A$, where F is a free module. We let $K_\bullet$ be the complex with $K_i = \Lambda^i(F)$ and the map given by

$$d(f_1 \wedge \cdots \wedge f_i) = \sum_{j=1}^{i} (-1)^{j+1} (\phi(f_j) f_1 \wedge \cdots \wedge \hat{f}_j \wedge \cdots \wedge f_i).$$

This map is alternating so is well-defined, and the proof that it is a complex is the same as in Proposition 3.3.1. We denote this complex $K_\bullet(\phi)$. If F has basis $e_1, \ldots, e_n$ and the map ϕ takes e_i to x_i, then the boundary map in this definition takes the basis element $e_{j_1} \wedge \cdots \wedge e_{j_i}$ to

$$\sum_{k=1}^{i} (-1)^{k+1} \left(\phi(e_{j_k}) e_{j_1} \wedge \cdots \wedge \widehat{e_{j_k}} \wedge \cdots \wedge e_{j_i} \right)$$

$$= \sum_{k=1}^{i} (-1)^{k+1} x_{j_k} e_{j_1} \wedge \cdots \wedge \widehat{e_{j_k}} \wedge \cdots \wedge e_{j_i}$$

so this is the same complex as defined before.

If M is an A-module, we denote the complex $K_\bullet(x_1, \ldots, x_n) \otimes M$ by $K_\bullet(x_1, \ldots, x_n; M)$, and the homology of this complex will be denoted $H_i(x_1, \ldots, x_n; M)$. Since the Koszul complex is a complex of free modules, a short exact sequence of modules gives rise to a short exact sequence of associated Koszul complexes and thus to a long exact sequence of homology modules. We note that if $n = 1$, then the Koszul complex $K(x_1; M)$ has a copy of M in degree zero and another in degree 1, and the map is multiplication by x_1.

Proposition 3.3.2 *Suppose that $f : F \to G$ is a map of free modules, and suppose we have maps $\phi : F \to A$ and $\psi : G \to A$ such that $\psi f = \phi$. Then there is a map $K_\bullet(f) : K_\bullet(\phi) \to K_\bullet(\psi)$ induced on Koszul complexes. If f is an isomorphism, so is the induced map of Koszul complexes.*

Proof The map $K_\bullet(f)$ is simply the map induced by f on exterior powers by functoriality. We denote the boundary maps on $K_\bullet(\phi)$ and $K_\bullet(\psi)$ by d^ϕ and

d^ψ, respectively. Since $\psi f = \phi$, we have

$$K_{i-1}(f)(d^\phi(f_1 \wedge \cdots \wedge f_i))$$

$$= K_{i-1}(f)\left(\sum(-1)^{j+1}\phi(f_j)f_1 \wedge \cdots \wedge \hat{f_j} \wedge \cdots \wedge f_i\right)$$

$$= \sum(-1)^{j+1}\psi(f(f_j))f(f_1) \wedge \cdots \wedge \widehat{f(f_j)} \wedge \cdots \wedge f(f_i)$$

$$= d^\psi(f(f_1) \wedge \cdots \wedge f(f_i)) = d^\psi K_i(f)(f_1 \wedge \cdots \wedge f_i).$$

Thus, $K_\bullet(f)$ is a map of complexes. If f is an isomorphism, then $K_i(f)$ is an isomorphism for each i, so $K_\bullet(f)$ is an isomorphism of complexes. □

One simple special case of Proposition 3.3.2 is that the Koszul complexes defined by any permutation of the same elements are isomorphic; this fact can also be shown directly from the definition.

The importance of the Koszul complex comes partly from the fact, as proven below, that it provides a free resolution of the quotient module $A/(x_1, \ldots x_n)$ in many important situations and that it provides a criterion for a sequence of elements $x_1, \ldots, x_n$ to form a regular sequence (defined in Definition 3.3.1). However, perhaps more important is the fact that it can be built up from Koszul complexes on one element, a property that makes it possible to carry out inductive arguments.

Proposition 3.3.3 *The Koszul complex $K_\bullet(x_1, \ldots, x_n)$ has the following properties:*

(i) $K_\bullet(x_1, \ldots, x_n) \cong K_\bullet(x_1) \otimes K_\bullet(x_2, \ldots, x_n)$.
(ii) $K_\bullet(x_1, \ldots, x_n)$ *is the mapping cone of the map defined by multiplication by x_1 on $K_\bullet(x_2, \ldots, x_n)$.*
(iii) *There is a short exact sequence of complexes*

$$0 \to K_\bullet(x_2, \ldots x_n) \to K_\bullet(x_1, \ldots x_n) \to K_\bullet(x_2, \ldots x_n)[-1] \to 0.$$

Proof Since $K(x_1)$ has the form $\cdots 0 \to A \xrightarrow{x_1} A \to 0 \cdots$ where the nonzero components are in degrees zero and one, the double complex $K_\bullet(x_1) \otimes K_\bullet(x_2, \ldots, x_n)$ consists of copies of $K_\bullet(x_2, \ldots, x_n)$ in the rows indexed by zero and one, and the maps between them are multiplication by x_1. The component of the associated simple complex in degree k is

$$K_0(x_1) \otimes K_k(x_2, \ldots, x_n) \oplus K_1(x_1) \otimes K_{k-1}(x_2, \ldots, x_n)$$

$$\cong K_k(x_2, \ldots, x_n) \oplus K_{k-1}(x_2, \ldots, x_n).$$

This is precisely the module in degree k in the mapping cone $C_\bullet[x_1]$, and it is easy to compute that the boundary map is also the same. Hence the complexes in (i) and (ii) are the same. It remains to show that this complex is the Koszul complex $K_\bullet(x_1, \ldots, x_n)$. The isomorphism

$$\phi : K_k(x_1, \ldots, x_n) \to K_k(x_2, \ldots, x_n) \oplus K_{k-1}(x_2, \ldots, x_n)$$

is defined on basis elements by letting $\phi(e_{i_1} \wedge \cdots \wedge e_{i_k}) = (e_{i_1} \wedge \cdots \wedge e_{i_k}, 0)$ if none of the e_{i_j} are equal to e_1 and defining $\phi(e_1 \wedge e_{i_2} \wedge \cdots \wedge e_{i_k}) = (0, e_{i_2} \wedge \cdots \wedge e_{i_k})$. By the definition of the mapping cone, the boundary map sends an element (a, b) to $(d_k(a) + x_1 b, -d_{k-1}(b))$, and a straightforward computation shows that ϕ defines a map of complexes.

Statement (iii) is the short exact sequence of complexes associated to the mapping cone.   □

It follows from Proposition 3.3.3 that we have

$$K_\bullet(x_1, \ldots, x_n) \cong K_\bullet(x_1) \otimes K_\bullet(x_2) \otimes \cdots \otimes K_\bullet(x_n).$$

Since Koszul complexes defined by permutations of the same elements are isomorphic, we can also replace x_1 by any of the x_i in Proposition 3.3.3.

The short exact sequence of Koszul complexes of part (iii) of Proposition 3.3.3 produces a long exact sequence in homology, and the boundary map from $H_i(x_2, \ldots x_n)[-1] = H_{i-1}(x_2, \ldots x_n)$ to $H_{i-1}(x_2, \ldots x_n)$ in this long exact sequence is multiplication by x_1. These results also hold for $K_\bullet(x_1, \ldots x_n; M)$ for any module M.

One of the main uses of the Koszul complex is as a criterion for a sequence of elements to form a regular sequence.

Definition 3.3.1 *Let M be a nonzero module over a local ring A, and let $x_1, \ldots x_n$ be a sequence of elements in the maximal ideal of A. Then we say that $x_1, \ldots x_n$ form a* regular sequence on M *if, for each $i = 1, \ldots, n$, x_i is not a zero-divisor on $M/(x_1, \ldots x_{i-1})$.*

For $i = 1$, this definition says that x_1 is not a zero-divisor on M.

Theorem 3.3.4 *Let $x_1, \ldots x_n$ be a sequence of elements in the maximal ideal of a local ring A, and let M be a finitely generated A-module. Then the following are equivalent:*

(i) *$x_1, \ldots x_n$ form a regular sequence on M.*
(ii) *$H_i(x_1, \ldots x_n; M) = 0$ for $i > 0$.*
(iii) *$H_1(x_1, \ldots x_n; M) = 0$.*

Proof The proof is by induction on n. If $n = 1$, then the Koszul complex consists of one copy of M in degree 1 and one copy of M in degree zero, and d_1 is defined by multiplication by x_1 on M. Thus, its homology is zero in degree 1 if and only if x_1 is not a zero-divisor on M, and the homology in higher degrees vanishes automatically.

Suppose now that $n > 1$. We denote $K_\bullet(x_1, \ldots x_n; M)$ by $K_\bullet^n$ and $K_\bullet(x_1, \ldots x_{n-1}; M)$ by $K_\bullet^{n-1}$. From Proposition 3.3.3 we have a short exact sequence

$$0 \to K_\bullet^{n-1} \to K_\bullet^n \to K_\bullet^{n-1}[-1] \to 0.$$

This short exact sequence of complexes induces a long exact sequence:

$$\cdots H_i\left(K_\bullet^{n-1}\right) \to H_i\left(K_\bullet^n\right) \to H_{i-1}\left(K_\bullet^{n-1}\right) \xrightarrow{x_n} H_{i-1}\left(K_\bullet^{n-1}\right) \to H_{i-1}\left(K_\bullet^n\right) \cdots$$

Suppose first that $x_1, \ldots x_n$ is a regular sequence. Then $x_1, \ldots x_{n-1}$ is a regular sequence, so we know by induction that $H_i(K_\bullet^{n-1}) = 0$ for $i > 0$, and the long exact sequence implies that $H_i(K_\bullet^n) = 0$ for $i > 1$. For $i = 1$, we have the exact sequence

$$0 \to H_1\left(K_\bullet^n\right) \to M/(x_1, \ldots x_{n-1})M \xrightarrow{x_n} M/(x_1, \ldots x_{n-1})M \to \cdots$$

and the hypothesis that x_n is not a zero-divisor on $M/(x_1, \ldots x_{n-1})M$ implies that $H_1(K_\bullet^n) = 0$. Thus, we have shown that (1) implies (2).

It is clear that (2) implies (3). Assume that (3) holds, so that we have $H_1(K_\bullet(x_1, \ldots x_n; M)) = 0$. We now consider the following part of the above long exact sequence:

$$\cdots \to H_1\left(K_\bullet^{n-1}\right) \xrightarrow{x_n} H_1\left(K_\bullet^{n-1}\right) \to H_1\left(K_\bullet^n\right) \to M/(x_1, \ldots x_{n-1})M$$
$$\xrightarrow{x_n} M/(x_1, \ldots x_{n-1})M.$$

We are assuming that $H_1(K_\bullet^n) = 0$. Thus, multiplication by x_n on $H_1(K_\bullet^{n-1})$ is surjective, so by Nakayama's lemma $H_1(K_\bullet^{n-1}) = 0$. Hence we conclude by induction that $x_1, \ldots x_{n-1}$ is a regular sequence. From the second half of this sequence, again using that $H_1(K^n) = 0$, we see that the map from $M/(x_1, \ldots x_{n-1})M$ to $M/(x_1, \ldots x_{n-1})M$ given by multiplication by x_n is injective, so x_n is not a zero-divisor on $M/(x_1, \ldots x_{n-1})M$. Thus, $x_1, \ldots x_n$ is a regular sequence on M. $\qquad\square$

We remark that the exactness of the Koszul complex does not imply that the sequence is regular for nonlocal rings. An example is given in the exercises.

We prove one more fact about Koszul complexes, that they are self-dual.

Proposition 3.3.5 *Let* $K_\bullet = K_\bullet(x_1, \ldots, x_n)$. *Then* $\operatorname{Hom}(K_\bullet, A) \cong K_\bullet$.

Proof Let $e_1, \ldots, e_n$ be a basis for a free module F of rank n as in the definition of the Koszul complex. We use the fact that $\Lambda^n(F)$ is isomorphic to A and define an isomorphism of complexes from $K_\bullet$ to $\mathrm{Hom}(K_\bullet, \Lambda^n(F))$. Let $\alpha(i)$ denote the greatest integer in $i/2$ for all integers i. We define an isomorphism ϕ_i between K_i and $\mathrm{Hom}(K_{n-i}, \Lambda^n(F))$ by letting $\phi_i(e_{j_1} \wedge \cdots \wedge e_{j_i})$ be the map defined by

$$\phi_i\left(e_{j_1} \wedge \cdots \wedge e_{j_i}\right)\left(e_{k_1} \wedge \cdots \wedge e_{k_{n-i}}\right)$$
$$= (-1)^{\alpha(i)} e_{j_1} \wedge \cdots \wedge e_{j_i} \wedge e_{k_1} \wedge \cdots \wedge e_{k_{n-i}}.$$

We have to show that $\phi_\bullet$ commutes with the boundary maps, which says that for every $\eta \in K_i$ we have

$$\phi_{i-1}(d_i(\eta)) = \mathrm{Hom}(d_{n-i+1}, \Lambda^n(F))(\phi_i(\eta)).$$

Let $\eta = e_{j_1} \wedge \cdots \wedge e_{j_i}$. We compute the action of the maps defined by both sides of the above equation on a basis element $\omega = e_{k_1} \wedge \cdots \wedge e_{k_{n-i+1}}$ of K_{n-i+1}. If η and ω have more than one e_k in common, then every term of the expansion of $\phi_{i-1}(d_i(\eta))(\omega)$ and of $\mathrm{Hom}(d_{n-i+1}, \Lambda^n(F))(\phi_i(\eta))(\omega)$ is a wedge product of elements at least two of which are equal, so both sides are zero. The only other case is when η and ω have exactly one element in common, which we can take to be e_1. Thus, we let $\eta = e_1 \wedge e_{j_2} \wedge \cdots \wedge e_{j_i}$ and $\omega = e_1 \wedge e_{k_2} \wedge \cdots \wedge e_{k_{n-i+1}}$. In this case all terms but one vanish, and we are left with

$$\phi_{i-1}(d_i(\eta))(\omega) = (-1)^{\alpha(i-1)} x_1 e_{j_2} \wedge \cdots \wedge e_{j_i} \wedge e_1 \wedge e_{k_2} \wedge \cdots \wedge e_{k_{n-i+1}}$$

and

$$\mathrm{Hom}(d_{n-i+1}, \Lambda^n(F))(\phi_i(\eta))(\omega)$$
$$= (-1)^{\alpha(i)} x_1 e_1 \wedge e_{j_2} \wedge \cdots \wedge e_{j_i} \wedge e_{k_2} \wedge \cdots \wedge e_{k_{n-i+1}}.$$

These two expression differ by a factor of $(-1)^{i+1+\alpha(i)+\alpha(i-1)}$. A check of cases shows that $i + 1 + \alpha(i) + \alpha(i-1)$ is always even, so $\phi_\bullet$ is a map of complexes. $\qquad\square$

Exercises

3.1 Give an example to show that the relation "there exists a quasi-isomorphism from $F_\bullet$ to $G_\bullet$" is not symmetric.

3.2 Let $A = \mathbb{Z}$, let $F_\bullet$ be the complex $0 \to \mathbb{Z}/4\mathbb{Z} \xrightarrow{2} \mathbb{Z}/4\mathbb{Z} \to 0$.
   (a) Find a free resolution of $F_\bullet$.
   (b) Show that $F_\bullet$ is quasi-isomorphic to the complex $G_\bullet = 0 \to \mathbb{Z}/2\mathbb{Z} \xrightarrow{0} \mathbb{Z}/2\mathbb{Z} \to 0$ but that there is no quasi-isomorphism from $F_\bullet$ to $G_\bullet$ or from $G_\bullet$ to $F_\bullet$.

3.3 Let M be a module. Show that a projective resolution of M is the same as a projective complex $P_\bullet$ with $P_i = 0$ for $i < 0$ together with a quasi-isomorphism from $P_\bullet$ to M.

3.4 If $\alpha_\bullet : F_\bullet \to G_\bullet$ is a map of complexes, verify that the map from $H_i(F_\bullet)$ to $H_i(G_\bullet)$ induced by the long exact sequence of the mapping cone $C[\alpha_\bullet]$ is the map induced on homology by $\alpha_\bullet$.

3.5 Finish the proof of Theorem 3.1.7 showing that the map $\beta_\bullet$ defined there is unique up to homotopy.

3.6 Let $F_\bullet$ be a complex that is bounded below, let $\alpha_\bullet : G_\bullet \to H_\bullet$ be a map of bounded above complexes, and let $C_\bullet[\alpha_\bullet]$ be the mapping cone of $\alpha_\bullet$. Construct an isomorphism between $\operatorname{Hom}(F_\bullet, C_\bullet[\alpha_\bullet])$ and the mapping cone of $\operatorname{Hom}(F_\bullet, \alpha_\bullet)$.

3.7 Let M be a graded module over a $\mathbb{Z}$-graded ring A such that A_0 is Artinian. Suppose that M has a finite resolution by modules F_i, each of which is a finite sum of graded modules of the form $A[n_{ij}]$ for various integers n_{ij}, and the boundary maps preserve the grading. Give a formula for the Hilbert function of M in terms of the integers n_{ij} and the Hilbert function of A.

3.8 Let $\alpha_\bullet : F_\bullet \to G_\bullet$ be a map of complexes. Define a double complex $E_{\bullet\bullet}$ by letting $E_{1\bullet} = F_\bullet$, $E_{0\bullet} = G_\bullet$, $E_{i,j} = 0$ if i is not 0 or 1, and $d^1_{1,i} = \alpha_i$ for all i. Show that the total complex of $E_{\bullet\bullet}$ is the mapping cone of $\alpha_\bullet$.

3.9 Show that a free complex is not necessarily projective in the category of complexes. Show that a bounded complex is projective in the category of complexes if and only if it is a split exact complex of projective modules.

3.10 Let $E_\bullet$ and $F_\bullet$ be complexes, and let $\operatorname{Hom}(E_\bullet, F_\bullet)$ be the complex defined at the end of Section 2. Show that the homology of $\operatorname{Hom}(E_\bullet, F_\bullet)$ in degree zero is the set of homotopy classes of maps of complexes from $E_\bullet$ to $F_\bullet$.

3.11 Let $A = k[X, Y, Z]$ and consider the sequence $XY, X - 1, XZ$. Show that this sequence is regular but the permutation $XY, XZ, X - 1$ is not a regular sequence. (A regular sequence is defined as in the local case except that the condition that the x_i are in the maximal ideal is replaced by the condition that $M/(x_1, \ldots, x_n)M \neq 0$.) Show that the Koszul complex $K_\bullet(XY, XZ, X - 1)$ is exact.

4

Homological Properties of Rings and Modules

The topics discussed in this chapter, Cohen-Macaulay, Gorenstein, and regular rings and their related properties, were originally introduced by Macaulay [43], so they predate the introduction of homological algebra by several decades. However, there are now several convenient homological characterizations of these properties. In the first section we discuss the basic definitions of Cohen-Macaulay and Gorenstein rings; much of this material can be found in Matsumura [44, §17, 18]. In later sections we give various alternative homological interpretations. Section 2 is devoted to the concept of dimension that will be used throughout the remainder of the book.

4.1 Cohen-Macaulay and Gorenstein Rings

In this section we give the definition of depth and its relation to dimension, as well as standard definitions and basic properties of Cohen-Macaulay and Gorenstein rings.

Let A be a local ring with maximal ideal m, and let M be a nonzero finitely generated A-module. Let $k = A/\text{m}$.

Definition 4.1.1 *The supremum of integers k such that there exists a regular sequence of length k on M is called the* depth *of M.*

This definition applies in particular to the case in which $M = A$. The simplest example of a regular sequence is obtained by letting $A = k[[x_1, \ldots x_n]]$ and taking the sequence $x_1, \ldots x_n$. More generally, if A is a regular local ring of dimension d, and if $x_1, \ldots x_d$ form a basis for m/m^2, then $x_1, \ldots x_d$ form a regular sequence. To see this, we recall that the ring A is regular if its maximal ideal can be generated by d elements, and we have shown in the previous chapter that a regular local ring is an integral domain. If x is an element of m that is

64

not in $\mathfrak{m}^2$, then x is not a zero-divisor and A/xA has dimension $d - 1$ with its maximal ideal generated by $d - 1$ elements, so that A/xA is again regular, and the process can be continued. Thus, $x_1, \ldots x_d$ form a regular sequence, so a regular local ring of dimension d has depth at least equal to d.

If M has depth zero, then by definition every element of $\mathfrak{m}$ is a zero-divisor on M, and since the set of zero-divisors on M is the union of the associated primes of M, the maximal ideal of A must be an associated prime of M. Hence the depth of M is zero if and only if there is an element of M whose annihilator is $\mathfrak{m}$.

A priori the depth of a module M could be infinite. However, the next proposition implies that this cannot happen.

Proposition 4.1.1 *For every nonzero module M we have*

$$\text{depth}(M) \leq \dim(M).$$

Proof We prove this result by induction on the dimension of M. If M has dimension zero, it has finite length, so there exists a nonzero element of M annihilated by $\mathfrak{m}$. Thus, $\text{depth}(M) = 0$.

Suppose now that the dimension of M is greater than zero. If there exists an element x that is not a zero-divisor on M, then x is not in any associated prime ideal of M, so x cannot be contained in a minimal prime in the support of M, and we have the inequality $\dim(M/xM) \leq \dim(M) - 1$. Thus, by induction we have

$$\text{depth}(M/xM) \leq \dim(M/xM) \leq \dim(M) - 1.$$

Now suppose that $\text{depth}(M) \geq n$. We can then find x such that $\text{depth}(M/xM) \geq n - 1$. By induction, we have $\dim(M) - 1 \geq \text{depth}(M/xM) \geq n - 1$, so that $\dim(M) \geq n$. Thus, $\dim(M) \geq n$ for all integers n with $\text{depth}(M) \geq n$, so $\dim(M) \geq \text{depth}(M)$. $\qquad\qquad\square$

We have seen in Chapter 2 that the dimension of a module M goes down by exactly one when M is divided by an element in no minimal prime ideal in the support of M [or, more generally, in no minimal prime of dimension equal to $\dim(M)$]. It is clear from the definition that the depth of a module goes down by at least one when the module is divided by not a zero-divisor. It is also clear that for any module M of depth greater than zero there exists at least one such x that $\text{depth}(M/xM) = \text{depth}(M) - 1$. In fact, this formula holds for any element that is not a zero-divisor. To show this it is useful to have an alternative characterization of depth.

Theorem 4.1.2 *The depth of M is the minimum value of n such that* $\mathrm{Ext}^n_A(k, M) \neq 0$.

Proof We have shown above that the depth of M is zero if and only if there is a nonzero element of M annihilated by the maximal ideal of A; that is, if and only if $\mathrm{Hom}(k, M) \neq 0$. Thus, the theorem is true for depth zero.

We prove the general result by induction on the depth of M. Suppose that the depth of M is greater than zero, and denote the depth of M by d. Let x be an element of $\mathfrak{m}$ that is not a zero-divisor on M such that $\mathrm{depth}(M/xM) = d - 1$. We then have the short exact sequence

$$0 \to M \xrightarrow{x} M \to M/xM \to 0.$$

This short exact sequence gives rise to a long exact sequence of Ext modules of the form

$$\cdots \mathrm{Ext}^i(k, M) \xrightarrow{x} \mathrm{Ext}^i(k, M) \to \mathrm{Ext}^i(k, M/xM) \to \mathrm{Ext}^{i+1}(k, M)$$
$$\xrightarrow{x} \mathrm{Ext}^{i+1}(k, M) \to \cdots.$$

Now multiplication by x in this sequence is zero, since the module $\mathrm{Ext}^i(k, M)$ is annihilated by the maximal ideal of A for all i. Thus, if $\mathrm{Ext}^i(k, M/xM) = 0$ the following map in the sequence must be injective; since it is the zero map we may conclude that $\mathrm{Ext}^{i+1}(k, M) = 0$. By induction we know that $\mathrm{Ext}^i(k, M/xM) = 0$ for all $i < d - 1$, so from the above argument we can deduce that $\mathrm{Ext}^i(k, M) = 0$ for all $i \leq d - 1$. We also know by the induction hypothesis that $\mathrm{Ext}^{d-1}(k, M/xM) \neq 0$. The above exact sequence in degree d is thus

$$\cdots \to 0 \to \mathrm{Ext}^{d-1}(k, M/xM) \to \mathrm{Ext}^d(k, M) \xrightarrow{x} \mathrm{Ext}^d(k, M) \to \cdots.$$

Thus, $\mathrm{Ext}^d(k, M) \neq 0$, so d is precisely the lowest integer i for which $\mathrm{Ext}^i(k, M) \neq 0$. $\square$

Definition 4.1.2 *The module M is* Cohen-Macaulay *if the dimension of M is equal to the depth of M.*

Every module of dimension zero is Cohen-Macaulay. The ring A is defined to be Cohen-Macaulay if it is Cohen-Macaulay as a module. We have shown that the depth of a regular local ring of dimension d is at least d. Hence we have the following result.

Theorem 4.1.3 *A regular local ring is Cohen-Macaulay.*

Among the nice properties of Cohen-Macaulay modules, there are two that will be of major importance in the applications in this book. First, a Cohen-Macaulay module can have no embedded primes and its support must be catenary. A ring or a subset of a scheme is said to be *catenary* if all saturated chains of prime ideals between two given primes have the same length, where a chain of prime ideals is *saturated* if there are no prime ideals properly between any two consecutive primes in the chain. The second important property is that it often implies the vanishing of homology groups in complexes, as we show in Section 3.

It follows immediately from the definition that a Cohen-Macaulay module M of dimension greater than zero cannot have $\mathfrak{m}$ as an associated prime ideal. We show here that if d is the dimension of M and $\mathfrak{p}$ is an associated prime of M, then $A/\mathfrak{p}$ has dimension d. The main idea is contained in the following Lemma.

Lemma 4.1.4 *Let $\mathfrak{p}$ be an associated prime ideal of M, and assume that x is not a zero-divisor on M. Then there exists an associated prime ideal of M/xM that properly contains $\mathfrak{p}$.*

Proof Since $\mathfrak{p}$ is an associated prime ideal of M, there is an element z of M whose annihilator is exactly $\mathfrak{p}$. Suppose that z is a multiple of x, so that $z = xt$ for some t. Since x is not a zero-divisor, the annihilator of t is the same as that of xt, so the annihilator of t must be $\mathfrak{p}$ also. Thus, we may divide z by x and obtain another element whose annihilator is exactly $\mathfrak{p}$. Since A is Noetherian, the process of dividing by x cannot continue forever, so we can eventually find an element z with annihilator $\mathfrak{p}$ that is not a multiple of x.

These conditions imply that the image of z in M/xM is not zero, and that z is annihilated by x and $\mathfrak{p}$. Thus, there exists an associated prime ideal of M/xM that contains the ideal generated by x and $\mathfrak{p}$ and in particular that properly contains $\mathfrak{p}$. □

Theorem 4.1.5 *If M is a Cohen-Macaulay module of dimension d, and if $\mathfrak{p}$ is an associated prime ideal of M, then $\dim(A/\mathfrak{p}) = d$.*

Proof Let $x_1, \ldots x_d$ be a regular sequence on M, and suppose $\mathfrak{p}$ is an associated prime of M with $\dim(A/\mathfrak{p}) < d$. By Lemma 4.1.4, there is an associated prime ideal $\mathfrak{p}'$ of $M/x_1 M$ that properly contains $\mathfrak{p}$, so that $\dim(A/\mathfrak{p}') < d - 1$. Since $M/x_1 M$ is a Cohen-Macaulay module of dimension $d - 1$, the corollary thus follows by induction on the depth of M. □

One consequence of Corollary 1 is that every minimal prime ideal $\mathfrak{p}$ in the support of M satisfies the condition $\dim(A/\mathfrak{p}) = \dim(M)$, so that every component of the support of M has the same dimension. It also follows that M can have no embedded primes.

Proposition 4.1.6 *The following are equivalent:*

(i) *M is Cohen-Macaulay.*
(ii) *Every system of parameters for M is a regular sequence.*

Proof If even one system of parameters is a regular sequence, the depth is equal to the dimension, so it is clear that (2) implies (1).

Conversely, assume that M is Cohen-Macaulay, and let $x_1, \ldots x_d$ be a system of parameters for M. If x_1 is a zero-divisor on M, then it must be contained in an associated prime ideal $\mathfrak{p}$. Since x_1 is the first element in a system of parameters, it can be contained in no prime ideal $\mathfrak{q}$ such that $\dim(A/\mathfrak{q}) = d$, so we must have $\dim(A/\mathfrak{p}) < d$. But this contradicts Theorem 4.1.5. □

Proposition 4.1.7 *Let A be a Cohen-Macaulay ring and let $\mathfrak{p}$ be a prime ideal of A. Then*

$$\text{height}(\mathfrak{p}) + \dim(A/\mathfrak{p}) = \dim(A).$$

Proof The proof is by induction on the height of $\mathfrak{p}$. The proposition follows from Theorem 4.1.5 if the height of $\mathfrak{p}$ is zero. If the height of $\mathfrak{p}$ is $h \geq 1$, we can find an element x of $\mathfrak{p}$ in no minimal prime of A such that the image of $\mathfrak{p}$ in A/xA has height $h - 1$. Since A/xA is still Cohen-Macaulay, the proposition follows by induction. □

The above results show that the support of a Cohen-Macaulay ring or module is reasonably well behaved. We prove one more theorem in this direction.

Theorem 4.1.8 *Let $\mathfrak{q}$ and $\mathfrak{p}$ be prime ideals such that $\mathfrak{q}$ is properly contained in $\mathfrak{p}$ and there are no prime ideals properly between $\mathfrak{p}$ and $\mathfrak{q}$. Then*

$$\dim_{k(\mathfrak{p})}(\text{Ext}^{i+1}(k(\mathfrak{p}), M_\mathfrak{p})) \geq \dim_{k(\mathfrak{q})}(\text{Ext}^i(k(\mathfrak{q}), M_\mathfrak{q})).$$

Proof We first localize at $\mathfrak{p}$ and assume that $\mathfrak{p}$ is maximal. The condition that there are no prime ideals strictly between $\mathfrak{p}$ and $\mathfrak{q}$ then implies that the dimension of $A/\mathfrak{q}$ is one. Let x be an element of $\mathfrak{p}$ that is not in $\mathfrak{q}$. Since $\mathfrak{p}$ is minimal over $\mathfrak{q}$, $A/\mathfrak{q}/xA/\mathfrak{q}$ is a module of finite length. Thus, we have the short exact sequence

$$0 \to A/\mathfrak{q} \xrightarrow{x} A/\mathfrak{q} \to Q \to 0$$

where Q is a module over A of finite length. The long exact sequence associated to this short exact sequence gives an exact sequence

$$\text{Ext}^i(A/\mathfrak{q}, M) \xrightarrow{x} \text{Ext}^i(A/\mathfrak{q}, M) \to \text{Ext}^{i+1}(Q, M). \qquad (*)$$

The rank of $\text{Ext}^i(A/\mathfrak{q}, M)$ as a module over the integral domain $A/\mathfrak{q}$ is equal to the dimension of $\text{Ext}^i(k(\mathfrak{q}), M_\mathfrak{q})$ as a vector space over $k(\mathfrak{q})$; denote this integer r. Letting $\text{Ext}^i(A/\mathfrak{q}, M) = N$, and letting $_xN$ denote the submodule of N annihilated by x, it follows from Proposition 1.2.2 that

$$\text{length}(N/xN) \geq \text{length}(N/xN) - \text{length}(_xN)$$

$$= r(\text{length}(A/\mathfrak{q}/x(A/\mathfrak{q}))).$$

Hence the exact sequence $(*)$ implies that the length of $\text{Ext}^{i+1}(Q, M)$ is at least equal to $r(\text{length}(A/\mathfrak{q}/xA/\mathfrak{q})) = r(\text{length}(Q))$. Thus, we have

$$r(\text{length}(Q)) \leq \text{length}(\text{Ext}^{i+1}(Q, M)).$$

But, using the exact sequences arising from a composition series for Q, we also have

$$\text{length}(\text{Ext}^{i+1}(Q, M)) \leq \text{length}(\text{Ext}^{i+1}(A/\mathfrak{p}, M))\text{length}(Q).$$

Combining these two inequalities gives

$$\dim_{k(\mathfrak{q})}(\text{Ext}^i(k(\mathfrak{q}), M_\mathfrak{q})) = r \leq \text{length}(\text{Ext}^{i+1}(A/\mathfrak{p}, M))$$

$$= \dim_{k(\mathfrak{p})}(\text{Ext}^{i+1}(k(\mathfrak{p}), M_\mathfrak{p})),$$

which is the desired result. $\qquad \square$

Corollary 4.1.9 *If $\mathfrak{p}$ is an associated prime ideal of M and $\mathfrak{p} = \mathfrak{p}_0 \subset \cdots \subset \mathfrak{p}_i$ is a saturated chain of prime ideals such that $\mathfrak{p}_i$ is the maximal ideal of A, then* $\text{depth}(M) \leq i$.

Proof Since $\mathfrak{p}$ is an associated prime ideal of M, there is an embedding of $A/\mathfrak{p}$ into M and thus, after localizing at $\mathfrak{p}$, an embedding of $k(\mathfrak{p})$ into $M_\mathfrak{p}$. Thus, $\text{Hom}_{k(\mathfrak{p})}(k(\mathfrak{p}), M_\mathfrak{p})) \neq 0$, so by applying Theorem 4.1.8 i times we can conclude that $\text{Ext}^i(k(\mathfrak{p}_i), M_{\mathfrak{p}_i}) = \text{Ext}^i(k, M) \neq 0$. Thus, $\text{depth}(M) \leq i$. $\quad \square$

It follows from Corollary 4.1.9 that if $\mathfrak{p}_0 \subset \cdots \subset \mathfrak{p}_i = \mathfrak{m}$ is any saturated chain of prime ideals in the support of a Cohen-Macaulay module where $\mathfrak{p}_0$ is an associated prime ideal of M, then $i = \dim(M)$. Thus, in particular the support of M is catenary.

Theorem 4.1.10 *If A is Cohen-Macaulay and $\mathfrak{p}$ is a prime ideal of A, then $A_\mathfrak{p}$ is Cohen-Macaulay.*

Proof Let h be the height of $\mathfrak{p}$. Arguing as in the previous proposition, we can find $x_1, \ldots x_h$ such that $\mathfrak{p}$ is minimal over $(x_1, \ldots x_h)$ and such that $A/(x_1, \ldots x_h)$ has dimension $\dim(A) - h$. Thus, $x_1, \ldots x_h$ is a regular sequence on A. Since localization is exact, $x_1, \ldots x_h$ is a regular sequence on $A_\mathfrak{p}$, which has dimension h, so $A_\mathfrak{p}$ is Cohen-Macaulay. $\square$

A nonlocal Noetherian ring is defined to be Cohen-Macaulay if every localization at a prime ideal is Cohen-Macaulay. Theorem 4.1.10 shows that this definition agrees with the original one for local rings.

Definition 4.1.3 *A* complete intersection *is the quotient of a regular local ring by an ideal of height r generated by r elements.*

Proposition 4.1.11 *A complete intersection is Cohen-Macaulay.*

Proof We have seen that a regular local ring is Cohen-Macaulay. Let A be a regular local ring of dimension d, and let $\mathfrak{a}$ be an ideal of height r generated by $x_1, \ldots x_r$. Then this sequence can be completed to a system of parameters $x_1, \ldots x_r, x_{r+1}, \ldots x_d$ for A. By Proposition 4.1.6 this system of parameters is a regular sequence, so $x_{r+1}, \ldots x_d$ is a regular sequence on $A/\mathfrak{a}$, which has dimension $d - r$. Thus, $A/\mathfrak{a}$ is Cohen-Macaulay. $\square$

In particular, a complete intersection has no embedded prime ideals. This theorem in the case of a polynomial ring was the original "Unmixedness Theorem" of Macaulay [43, Section 48] which gave rise to this subject.

Definition 4.1.4 *A ring A is* Gorenstein *if it satisfies the following two conditions:*

(i) *A is Cohen-Macaulay.*
(ii) *If $x_1, \ldots x_d$ is a system of parameters for A, then the dimension of the k-module $\operatorname{Hom}_A(k, A/(x_1, \ldots x_d)A)$ is one.*

We note that $\operatorname{Hom}_A(k, M)$ can be identified with the submodule of elements of M annihilated by $\mathfrak{m}$. In general, if M is a Cohen-Macaulay module and $x_1, \ldots x_d$ is a system of parameters for M, we call the dimension of $\operatorname{Hom}(k, M/(x_1, \ldots x_d)M)$ the *type* of M. It is not clear that this definition is independent of the system of parameters chosen, but it follows from the next proposition that it is.

Proposition 4.1.12 *Let M be a Cohen-Macaulay module of dimension d, and let $x_1, \ldots x_d$ be a system of parameters for M. Then the dimension of $\operatorname{Hom}_A(k, M/(x_1, \ldots x_d))M$ is equal to the dimension of $\operatorname{Ext}_A^d(k, M)$ as a vector space over k.*

Proof If $d = 0$, this statement follows from the definition. Assume that it is true for dimension $d - 1$, and let $x_1, \ldots x_d$ be a system of parameters for M. The short exact sequence

$$0 \to M \xrightarrow{x_1} M \to M/x_1 M \to 0$$

gives rise to a long exact sequence on Ext modules as before, and we have the exact sequence

$$0 \to \operatorname{Ext}^{d-1}(k, M/x_1 M) \to \operatorname{Ext}^d(k, M) \xrightarrow{x_1} \operatorname{Ext}^d(k, M).$$

Since multiplication by x_1 is zero, we have an isomorphism

$$\operatorname{Ext}^{d-1}(k, M/x_1 M) \cong \operatorname{Ext}^d(k, M).$$

By induction, we know that the dimension of $\operatorname{Ext}^{d-1}(k, M/x_1 M)$ is equal to the dimension of $\operatorname{Hom}(k, M/(x_1, \ldots x_d)M)$, so this isomorphism implies that this dimension is equal to that of $\operatorname{Ext}^d(k, M)$ as was to be shown. $\square$

Hence the type of a module does not depend on the choice of regular sequence. If A is a regular local ring of dimension d, then we have seen that a system of parameters of A can be found that generates the maximal ideal of A. Since the dimension of $\operatorname{Hom}(k, k)$ is one, A is Gorenstein. Furthermore, if $x_1, \ldots x_r$ are elements of A that generate an ideal $\mathfrak{a}$ of height r, then, as we have shown above, the sequence can be extended to a system of parameters for A, so that the dimension of $\operatorname{Hom}_A(k, A/(x_1, \ldots x_d))$ will again be one by Proposition 4.1.12. Since $x_{r+1}, \ldots x_d$ form a system of parameters for $A/\mathfrak{a}$, this shows that $A/\mathfrak{a}$ is Gorenstein. Thus, in the following list of properties of local rings each condition is stronger than the next.

(i) Regular
(ii) Complete intersection
(iii) Gorenstein
(iv) Cohen-Macaulay

The exercises at the end of the chapter give examples of rings that satisfy each of these conditions but not the previous one in the list.

4.2 Dimension

In this section we make definite the definition of dimension we will be using throughout the remainder of the book, and we specify the class of rings we will be considering. As mentioned briefly in Chapter 1, the usual notion of Krull dimension does not have all the required properties in general, although in many cases it works correctly. The conditions we require are the following:

  (i) If $\mathfrak{p} \subset \mathfrak{q}$ are distinct prime ideals with no prime ideals properly between them, then $\dim(A/\mathfrak{p}) = \dim(A/\mathfrak{q}) + 1$.
 (ii) If S is a multiplicatively closed set and $\mathfrak{p}$ is a prime ideal of A that does not meet S, then $\dim(A/\mathfrak{p}) = \dim(A_S/\mathfrak{p}_S)$.
(iii) If we are considering only rings of finite type over a given field or homomorphic images of a given regular local ring, then we may take $\dim(A)$ to be the Krull dimension of A.

The first condition implies that we only allow catenary rings. In fact, if $\mathfrak{p}_0 \subset \mathfrak{p}_1 \cdots \subset \mathfrak{p}_n$ is a saturated chain of prime ideals, then the first condition implies that $\dim(A/\mathfrak{p}_0) = \dim(A/\mathfrak{p}_n) + n$, so every saturated chain of prime ideals between $\mathfrak{p}_0$ and $\mathfrak{p}_n$ must have length n. While noncatenary rings do not arise naturally in most applications, they do exist; see [48, Appendix].

If A is a finitely generated algebra over a field k, then the Krull dimension of $A/\mathfrak{p}$ is equal to the transcendence degree of the quotient field of $A/\mathfrak{p}$ over k for all $\mathfrak{p}$, and the fact that the required conditions are satisfied by Krull dimension follow from the properties of transcendence degree (see Atiyah and Macdonald [1] or Matsumura [44, §5]). However, even for localizations of finitely generated algebras over fields, the first condition need not hold for Krull dimension. For example, consider an integral domain A with two prime ideals $\mathfrak{p}$ of height one and $\mathfrak{q}$ of height two, respectively, such that $\mathfrak{p} \not\subseteq \mathfrak{q}$. Let S be the complement of $\mathfrak{p} \cup \mathfrak{q}$. Then the Krull dimension of the localization A_S is 2, and there are no prime ideals strictly between 0 and $\mathfrak{p}_S$, but the Krull dimension of $A_S/\mathfrak{p}_S$ is zero.

The condition we impose on rings is that they be finitely generated algebras over regular rings, where a ring is regular if the localization at every prime ideal is a regular local ring. In fact, we assume that there is a fixed regular ring R of finite Krull dimension such that all the rings under consideration are localizations of finitely generated rings over R. (For purposes of defining dimension it would actually suffice to require only that R be Cohen-Macaulay, but for later purposes we will want our rings to be homomorphic images of regular rings.) The definition of dimension is taken from Fulton [17, Chapter 20] and is the following:

Definition 4.2.1 *Let A be a ring that is a localization of a ring of finite type over the regular ring R. Let k be the Krull dimension of R. Let* $\mathfrak{p}$ *be a prime ideal of A, and let* $\mathfrak{q}$ *be the inverse image of* $\mathfrak{p}$ *in R. Then*

$$\dim(A/\mathfrak{p}) = \text{tr.deg.}(k(\mathfrak{p})/k(\mathfrak{q})) - \text{height}_R(\mathfrak{q}) + k.$$

We first note that if S is a multiplicative set in A and $\mathfrak{p}$ is a prime ideal that does not meet S, then neither the transcendence degree of $k(\mathfrak{p})$ over $k(\mathfrak{q})$ nor the height of $\mathfrak{q}$ in R changes under localization, so that the dimension remains the same. Thus, property (ii) holds.

To check property (iii), we first assume that R is a field. In that case the relations between transcendence degree and Krull dimension cited above, together with the fact that $k = 0$, imply that this definition is the same as the Krull dimension. If R is a regular local ring and we are considering only homomorphic images A of R, then for any $\mathfrak{p} \in \text{Spec}(A)$, if $\mathfrak{q}$ is the inverse image of $\mathfrak{p}$ in R, then $A/\mathfrak{p} = R/\mathfrak{q}$, so the transcendence degree of $k(\mathfrak{p})$ over $k(\mathfrak{q})$ is zero. Hence the dimension of $A/\mathfrak{p}$ according to the above definition is the Krull dimension of R minus the height of $\mathfrak{q}$, and, since R is Cohen-Macaulay, it follows from Proposition 4.1.7 that this definition again gives the Krull dimension of $A/\mathfrak{p} \cong R/\mathfrak{q}$.

To prove the first property, we first give a different characterization of the dimension.

Proposition 4.2.1 *Let R be a regular local ring, and let A be a localization of a homomorphic image of* $R[X_1, \ldots, X_n]$. *Let* $\mathfrak{p}$ *be a prime ideal of A, let* $\mathfrak{q}$ *be its inverse image in R, and let* $\bar{\mathfrak{p}}$ *be its inverse image in* $R[X_1, \ldots, X_n]$. *Then*

$$\text{tr.deg.}(k(\mathfrak{p})/k(\mathfrak{q})) - \text{height}(\mathfrak{q}) = n - \text{height}(\bar{\mathfrak{p}}).$$

Proof Let $\mathfrak{p}'$ be the prime ideal of $k(\mathfrak{q})[X_1, \ldots, X_n]$ corresponding to $\bar{\mathfrak{p}}$ after localization and dividing by $\mathfrak{q}$. Then, from the results on prime ideals in finitely generated algebras over a field quoted above, we have

$$\text{height}(\mathfrak{p}') + \text{tr.deg.}(k(\mathfrak{p})/k(\mathfrak{q})) = n.$$

On the other hand, the height of the ideal generated by $\mathfrak{q}$ in the ring $R_{\mathfrak{q}}[X_1, \ldots, X_n]$ is equal to the height of $\mathfrak{q}$ in R. Since $R_{\mathfrak{q}}[X_1, \ldots, X_n]$ is regular, it is Cohen-Macaulay, and Proposition 4.1.7 implies that we have

$$\text{height}(\mathfrak{q}R_{\mathfrak{q}}[X_1, \ldots, X_n]) + \text{height}(\mathfrak{p}') = \text{height}(\bar{\mathfrak{p}}).$$

Subtracting this equation from the earlier one gives the result. $\qquad\square$

The first property that we require for dimension follows immediately from Proposition 4.2.1. In fact, if $\mathfrak{p} \subset \mathfrak{p}'$ are two prime ideals with no prime ideals properly between them, then we can express them as prime ideals in a localization of a homomorphic image of $R[X_1, \ldots, X_n]$. Since $R[X_1, \ldots, X_n]$ is a regular ring, it then follows that $\mathrm{height}(\mathfrak{p}') = \mathrm{height}(\mathfrak{p}) + 1$, and then from Proposition 4.2.1 we have that $\dim(A/\mathfrak{p}) = \dim(A/\mathfrak{p}') + 1$.

As mentioned above, we will always assume that the local rings we use are homomorphic images of regular local rings. This includes all localizations of finitely generated algebras over a field or over the integers, and by the Cohen structure theorems (see Matsumura [44, §29]) it also includes all complete local rings. We will also occasionally assume that, given any two rings, we can find a regular ring that maps onto both of them, a property that also holds in these cases.

4.3 The Acyclicity Lemma

In this section we begin to discuss the relationship between depth and the vanishing of homology. One of the main consequences of this theory is that under certain conditions, a complex with homology of small dimension must actually be exact. This is a useful tool for proving that certain constructions give free resolutions.

Proposition 4.3.1 *Let*

$$0 \to M' \to M \to M'' \to 0$$

be a short exact sequence. If $\mathrm{depth}(M) > \mathrm{depth}(M'')$, *then*

$$\mathrm{depth}(M') = \mathrm{depth}(M'') + 1.$$

Proof This lemma follows from the definition of depth in terms of the vanishing of Ext modules and the long exact sequence for Ext. If $i < \mathrm{depth}(M'')$, then $\mathrm{Ext}^i(k, M'') = 0$, and the hypothesis that $\mathrm{depth}(M) > \mathrm{depth}(M'')$ implies that $i+1 < \mathrm{depth}(M)$, so we also have $\mathrm{Ext}^{i+1}(k, M) = 0$. Thus, the exact sequence

$$\cdots \to 0 = \mathrm{Ext}^i(k, M'') \to \mathrm{Ext}^{i+1}(k, M') \to \mathrm{Ext}^{i+1}(k, M) = 0$$

implies that $\mathrm{Ext}^{i+1}(k, M') = 0$. On the other hand, if $i = \mathrm{depth}(M'')$, then $\mathrm{Ext}^i(k, M'') \neq 0$, so the exact sequence

$$\cdots \to 0 = \mathrm{Ext}^i(k, M) \to \mathrm{Ext}^i(k, M'') \to \mathrm{Ext}^{i+1}(k, M')$$

shows that $\mathrm{Ext}^{i+1}(k, M') \neq 0$. Thus, the depth of M' is $i + 1$ as was to be shown. □

Using this result we can prove the following theorem of Peskine-Szpiro [50]:

Theorem 4.3.2 (The Acyclicity Lemma) *Let*

$$0 \to M_k \to \cdots \to M_0$$

be a complex of length k such that the depth of M_i is greater than or equal to i for $i = 0, \ldots, k$. If the homology $H_i(M_\bullet)$ has depth 0 for $i > 0$, then $H_i(M_\bullet) = 0$ for all $i > 0$.

Proof We prove this theorem by induction on k. If $k = 1$, then we have a complex of length 1 of the form $0 \to M_1 \to M_0$, and M_1 has depth at least one. The homology $H_1(M_\bullet)$ is a submodule of M_1 and has depth equal to 0 by hypothesis. Since a module of depth greater than zero can have no submodules of depth 0, the homology must vanish.

Now assume that $k > 1$. As above, the homology in degree k must be zero. Consider the short exact sequence

$$0 \to M_k \to M_{k-1} \to Q \to 0$$

where Q is the cokernel of the map from M_k to M_{k-1}. By Proposition 4.3.1, the depth of Q is at least $k - 1$. We now consider the complex

$$0 \to Q \to M_{k-2} \to \cdots \to M_0.$$

Since the depth of Q is at least $k - 1$, this complex satisfies the depth condition of the hypothesis. Furthermore, its homology is the same as the homology of the original complex in each degree. Hence the theorem follows by induction. $\square$

As mentioned above, the main application of this theorem is to show that certain complexes of free modules are free resolutions. We give a simple example in the case of determinantal ideals in the exercises.

4.4 Modules of Finite Projective Dimension

We say that a module M has *finite projective dimension* if it has a finite resolution

$$0 \to P_k \to \cdots \to P_0 \to M$$

in which each P_i is a projective module. Since we are generally considering local rings, we will usually take the modules P_i to be free. The smallest integer k for which there exists a projective resolution of length k is called the *projective dimension* of M. If no such k exists, the projective dimension of M is said to be infinite. Modules of finite projective dimension have special properties, and there are still many open questions concerning them.

If A is a local ring, then we define a complex of free modules to be *minimal* if all the matrices defining the maps in the complex have entries in the maximal ideal of A. For the rest of this section we assume that A is a local ring with maximal ideal $\mathfrak{m}$ and residue field k. Similarly, a minimal free resolution of a module or a complex is a complex that is a free resolution and is minimal in this sense. A minimal free resolution of a finitely generated module can be constructed by taking a minimal set of generators for the kernel at each step. In fact, if $\phi : F_0 \to M$ is a map of a free module onto M such that a basis for F_0 is mapped to a minimal set of generators for M, then ϕ induces an isomorphism from $F_0 \otimes A/\mathfrak{m}$ to $M \otimes A/\mathfrak{m}$, so every element of the kernel of ϕ lies in $\mathfrak{m}F_0$. Thus, if $\phi_1 : F_1 \to F_0$ is a map from a second free module onto the kernel of ϕ, the matrix that defines ϕ_1 with respect to any basis will have coefficients in $\mathfrak{m}$. Continuing in this fashion, we obtain a minimal free resolution of M.

If $F_\bullet \to M$ is a minimal free resolution, then $\mathrm{Tor}_i(k, M)$ can be computed by taking the homology of $F_\bullet \otimes k$, and the minimality implies that all maps in the complex $F_\bullet \otimes k$ are zero. Hence the ranks of F_i are equal to the dimension of $\mathrm{Tor}_i(k, M)$, so in particular they are independent of the minimal resolution. These numbers are called the *Betti numbers* of M. A minimal free resolution is unique up to isomorphism of complexes; this result follows from the following Proposition:

Proposition 4.4.1 *If $E_\bullet$ and $F_\bullet$ are minimal complexes that are bounded below, and if $f_\bullet$ is a quasi-isomorphism between $E_\bullet$ and $F_\bullet$, then $f_\bullet$ is an isomorphism of complexes.*

Proof We prove by induction that $f_i : E_i \to F_i$ is an isomorphism for all i. Suppose that f_i is an isomorphism for all $i \leq n$; these maps then induce an isomorphism from $\mathrm{Ker}(d_n^E)$ to $\mathrm{Ker}(d_n^F)$. We thus have a diagram

$$0 \to \mathrm{Im}\big(d_{n+1}^E\big) \to \mathrm{Ker}\big(d_n^E\big) \to H_n(E_\bullet) \to 0$$
$$\downarrow \qquad\qquad \downarrow \qquad\qquad \downarrow$$
$$0 \to \mathrm{Im}\big(d_{n+1}^F\big) \to \mathrm{Ker}\big(d_n^F\big) \to H_n(F_\bullet) \to 0$$

in which the second and third vertical maps are isomorphisms. In fact, we have just shown that the middle map is an isomorphism, and the map on the right is also since $f_\bullet$ is assumed to be a quasi-isomorphism. Thus, the induced map from $\mathrm{Im}(d_{n+1}^E)$ to $\mathrm{Im}(d_{n+1}^F)$ is an isomorphism.

To conclude the proof that $f_{n+1} : E_{n+1} \to F_{n+1}$ is an isomorphism we consider the diagram

$$0 \to H_{n+1}(E_\bullet) \to E_{n+1}/\text{Im}\big(d_{n+2}^E\big) \to \text{Im}\big(d_{n+1}^E\big) \to 0$$
$$\downarrow \qquad\qquad \downarrow \qquad\qquad \downarrow$$
$$0 \to H_{n+1}(F_\bullet) \to F_{n+1}/\text{Im}\big(d_{n+2}^F\big) \to \text{Im}\big(d_{n+1}^F\big) \to 0.$$

We have just shown that the vertical map on the right is an isomorphism, and the vertical map on the left is an isomorphism since $f_\bullet$ is a quasi-isomorphism. Hence the map in the middle is an isomorphism. Since $E_\bullet$ and $F_\bullet$ are minimal complexes, the image of d_{n+2}^E is contained in $\mathfrak{m} E_{n+1}$ and similarly for $F_\bullet$. Hence f_{n+1} is an isomorphism modulo $\mathfrak{m}$. Since E_{n+1} and F_{n+1} are free modules, this implies that f_{n+1} is an isomorphism. □

While Proposition 4.4.1 is somewhat simpler for resolutions of modules, it is useful to have the result in this more general form. If $M_\bullet$ is a complex that is bounded below and has homology of finite length, we recall that by Proposition 3.1.6 there exists a quasi-isomorphism from a bounded below free complex $F_\bullet$ to $M_\bullet$ and that we call $F_\bullet \to M_\bullet$ a free resolution of $M_\bullet$. As above, $F_\bullet$ is called minimal if the boundary maps are zero modulo $\mathfrak{m}$. If $E_\bullet \to M_\bullet$ is a second minimal free resolution, then Theorem 3.1.7 implies that there exists a quasi-isomorphism $f_\bullet$ from $E_\bullet$ to $F_\bullet$, and Proposition 4.4.1 then implies that $f_\bullet$ is an isomorphism of complexes. Thus, a minimal free resolution of a complex is unique up to isomorphism of complexes.

To prove the existence of minimal resolutions of complexes, it is possible to modify the construction of Proposition 3.1.6 so that the result is minimal. However, it is easier to proceed as follows: If $F_\bullet$ is a resolution that is not minimal, and if the map $d_i : F_i \to F_{i-1}$ is not zero modulo $\mathfrak{m}$, then a matrix defining d_i has an entry that is a unit in A. By change of basis we can assume this matrix is of the form $\big(\begin{smallmatrix} 1 & 0 \\ 0 & M \end{smallmatrix}\big)$, where M is a matrix of appropriate size. It follows that the complex can be split into a complex of the form $\cdots \to 0 \to A \xrightarrow{1} A \to 0 \to \cdots$ and a complex $F_\bullet'$ in which F_i' and F_{i-1}' have ranks 1 less than the ranks of F_i and F_{i-1} and that is quasi-isomorphic to $F_\bullet$. This process can be continued until a minimal complex is obtained. Putting this together with Proposition 4.4.1, we obtain the following.

Proposition 4.4.2 *A complex $E_\bullet$ of finitely generated modules that is bounded below has a unique minimal free resolution. If $F_\bullet$ is any free resolution of $E_\bullet$, then $F_\bullet$ is a direct sum of a split exact complex and the minimal free resolution of $E_\bullet$.*

Proof Both statements follow from the above discussion and Proposition 4.4.1.
                                                                        □

One consequence of Proposition 4.4.2 is that if a module M has finite projective dimension, then the projective dimension is the length of the minimal
free resolution of M.

The simplest example of a module of finite projective dimension is, of course,
a free module. A less trivial example is obtained from a regular sequence. If
$x_1, \ldots, x_k$ is a regular sequence on A, then the Koszul complex on $x_1, \ldots x_k$ is a
free resolution of $A/(x_1, \ldots x_k)$, so $A/(x_1, \ldots x_k)$ is a module of projective dimension k. In particular, there exist modules of projective dimension k for all k
between 0 and the depth of A. We see next that the depth of A is in fact the upper
bound for the projective dimension of modules of finite projective dimension.

Theorem 4.4.3 (Auslander-Buchsbaum) *If M is a finitely generated module
of finite projective dimension over a local ring A, then*

$$\mathrm{depth}(M) + \mathrm{proj.dim.}(M) = \mathrm{depth}(A).$$

Proof We first show that if the depth of A is zero, then every module of finite
projective dimension is free. Let M be such a module, and let $F_\bullet \to M$ be a
minimal free resolution of M. If k is the projective dimension of M, and if
$k > 0$, then we have the exact sequence

$$0 \to F_k \to F_{k-1}.$$

Now this map is defined by a matrix with coefficients in $\mathfrak{m}$ by the minimality
of the resolution. Since the depth of A is zero, there is an element in A that is
annihilated by the maximal ideal of A, so there is an element in F_k annihilated
by the maximal ideal, and such a map cannot be injective. Thus, $k = 0$, so M
must be free.

Assume now that the depth of the ring A is greater than 0. If the A-module
M has depth zero, we map a free module F onto M and let K be the kernel, so
that we have a short exact sequence

$$0 \to K \to F \to M \to 0.$$

Then $\mathrm{depth}(K) = \mathrm{depth}(M) + 1$ by Proposition 4.3.1, and we have that
$\mathrm{proj.dim.}(K) = \mathrm{proj.dim.}(M) - 1$, since this exact sequence can be taken as the
first step in a free resolution. If the equality holds for K, then it holds for M, so
in this way we may reduce to the case in which the depth of M is at least one.
Since the union of the sets of prime ideals associated to A and M is finite, we can
then find an element x of $\mathfrak{m}$ that is not a zero-divisor on either A or M. The depth

of M/xM is one less than the depth of M, and the depth of A/xA is one less than the depth of A. Thus, to complete the proof it suffices to show that the projective dimension of M/xM over A/xA is equal to the projective dimension of M over A.

Let $F_\bullet \to M$ be a minimal free resolution of M over A. Then the tensor product $F_\bullet \otimes_A A/xA$ is a complex of free A/xA–modules mapping onto M/xM, and we claim that this complex is exact in degrees greater than zero, so that it is a minimal free resolution. The homology of this complex in degree i is $\mathrm{Tor}_i(M, A/xA)$. Now $\mathrm{Tor}_i(M, A/xA)$ can also be computed by taking a free resolution of A/xA and tensoring with M. Since x is not a zero-divisor on A, the Koszul complex on x is a free resolution of A/xA, and since x is not a zero-divisor on M, this complex remains exact in degree 1 after tensoring with M; thus, it is exact in all degrees greater than zero. Hence $F_\bullet \otimes_A A/xA$ is a minimal free resolution of M/xM over A/xA, so the projective dimension of M/xM over A/xA is equal to the projective dimension of M over A as required. $\square$

The Auslander-Buchsbaum theorem implies that the homology of a bounded free complex cannot vanish if the complex is too long. Combined with the Acyclicity Lemma, it determines the degree of the highest nonvanishing homology module for a bounded free complex with homology of finite length.

Proposition 4.4.4 *Let $F_\bullet = 0 \to F_k \to \cdots \to F_0 \to 0$ be a minimal bounded free complex over a local ring A such that $F_k \neq 0$. Let d be the depth of A. Then*

(i) *There exists an integer i with $k - d \leq i \leq k$ such that $H_i(F_\bullet) \neq 0$.*

(ii) *If $F_\bullet$ has homology of finite length, then the largest integer such that $H_i(F_\bullet) \neq 0$ is exactly $k - d$.*

Proof Suppose that $H_i(F_\bullet) = 0$ for all $i \geq k - d$. Let M be the cokernel of the map from F_{k-d} to F_{k-d-1}. If $M \neq 0$, then the complex $F_\bullet$ truncated at $k - d - 1$ is a minimal free resolution of M of length $d + 1 > \mathrm{depth}(A)$, contradicting the Auslander-Buchsbaum theorem. It is possible that $M = 0$, but in this case, since $F_\bullet$ is minimal, we have $F_{k-d-1} = 0$. Thus, if i is the smallest integer greater than $k - d - 1$ such that $F_i \neq 0$ (since we are assuming that $F_k \neq 0$ such an i must exist), the minimality of $F_\bullet$ implies that $H_i(F_\bullet) \neq 0$. This proves (i).

If the homology of $F_\bullet$ has dimension 0, then the Acyclicity Lemma implies that $H_i(F_\bullet) = 0$ for all $i > k - d$, so that $k - d$ must be the largest i such that $H_i(F_\bullet) \neq 0$, proving (ii). $\square$

Theorem 4.4.4 applies in particular to the Koszul complex $K_\bullet$ on a system of parameters of A. The concept of projective dimension provides a useful criterion for a ring to be regular.

Theorem 4.4.5 (Auslander-Buchsbaum-Serre) *Let A be a local ring. The following are equivalent:*

  (i) *A is regular.*
 (ii) *The residue field k has finite projective dimension.*
(iii) *Every finitely generated module has finite projective dimension.*

Proof If A is regular, then a minimal set of generators for the maximal ideal is a regular sequence, so the Koszul complex is a free resolution of k, which thus has finite projective dimension. Thus, (i) implies (ii). If M is any finitely generated module, let $F_\bullet$ be a minimal free resolution of M; then the dimension of $\mathrm{Tor}_i^A(k, M)$ is the rank of F_i for all i; if k has finite projective dimension, then $\mathrm{Tor}_i^A(k, M) = 0$ for i large, so $F_i = 0$ and M has finite projective dimension. Thus, (ii) implies (iii). It is clear that (iii) implies (ii), so the only part left to prove is that if k has finite projective dimension, then A is regular.

Let $F_\bullet$ be a minimal free resolution of k, and let $K_\bullet$ be the Koszul complex on a minimal set of generators for the maximal ideal of A. Since $K_\bullet$ is a complex of free modules and $F_\bullet$ is exact, the identity map lifts to a map $f_\bullet$ from $K_\bullet$ to $F_\bullet$. We will show by induction that f_i is an injective map modulo $\mathfrak{m}$; that is, that the map induced by f_i from $K_i \otimes (A/\mathfrak{m})$ to $F_i \otimes (A/\mathfrak{m})$ is injective, or, equivalently, that a minimal set of generators of K_i is mapped to part of a minimal set of generators of F_i, so that $f_i(K_i)$ is a direct summand of F_i.

The map f_i is an isomorphism for i equal to 0 and 1, so we may assume that a minimal set of generators of K_i is mapped to part of a minimal set of generators of F_i for $i < n$ and prove that this property holds for $i = n$. We have

$$
\begin{array}{ccccc}
K_n & \to & K_{n-1} & \to & K_{n-2} \\
\downarrow & & \downarrow & & \downarrow \\
F_n & \to & F_{n-1} & \to & F_{n-2}
\end{array}
$$

where $f_\bullet$ is given by the vertical maps. Suppose that there exists an element x in K_n that is not in $\mathfrak{m}K_n$ but such that $f_n(x)$ is in $\mathfrak{m}F_n$. We claim that $d_n^K(x)$ is not in $\mathfrak{m}^2 K_{n-1}$. To see this, we note that a basis element $e_{j_1} \wedge \cdots \wedge e_{j_n}$ of K_n is mapped by d_n^K to an alternating sum of elements of the form $x_{j_k} e_{j_1} \wedge \cdots \wedge \widehat{e_{j_k}} \wedge \cdots \wedge e_{j_n}$, where the x_r form a minimal set of generators of the maximal ideal of A. Now the elements $x_{j_k} e_{j_1} \wedge \cdots \wedge \widehat{e_{j_k}} \wedge \cdots \wedge e_{j_n}$ are all distinct, so the facts that the x_r are linearly independent in $\mathfrak{m}/\mathfrak{m}^2$ and that the $e_{j_1} \wedge \cdots \wedge e_{j_{n-1}}$ form a basis for K_{n-1} imply that these elements are linearly independent in $\mathfrak{m}K_{n-1}/\mathfrak{m}^2K_{n-1}$ over $A/\mathfrak{m}$. Hence, since $x \notin \mathfrak{m}K_n$, we have $d_n^K(x) \notin \mathfrak{m}^2K_{n-1}$.

Our induction assumption implies that $f_{n-1}(K_{n-1})$ is a direct summand of F_{n-1}, so it follows from the above that $f_{n-1}d_n^K(x)$ is not in $\mathfrak{m}^2F_{n-1}$. Thus, $d_n^F f_n(x) \notin \mathfrak{m}^2F_{n-1}$. But $F_\bullet$ is minimal, so that if we had $f_n(x) \in \mathfrak{m}F_n$, we would have $d_n^F f_n(x) \in \mathfrak{m}^2F_{n-1}$. This contradiction implies that f_n imbeds K_n as a direct summand of F_n as claimed.

Thus, the Koszul complex embeds into $F_\bullet$. But the maximum length of a minimal free resolution of a finitely generated module is the depth of A, which is at most equal to its dimension d. Thus, the length of the Koszul complex on a minimal set of generators of $\mathfrak{m}$ is d, so $\mathfrak{m}$ is generated by d elements and A is regular. $\square$

We remark that the fact that all finitely generated modules have finite projective dimension is equivalent to the fact that all (not necessarily finitely generated) modules have finite projective dimension, but we will not need this fact. Second, the main part of the proof was to show that if $x_1, \dots, x_n$ form part of a minimal set of generators of $\mathfrak{m}$, then the Koszul complex on $x_1, \dots, x_n$ embed into a free resolution of $A/(x_1, \dots, x_n)$ in the sense that the map in each degree embeds K_i into F_i as a direct summand, a result that is true for any local ring, whether regular or not. It is conjectured that this result holds also for any system of parameters of a local ring A, whether they are linearly independent in $\mathfrak{m}/\mathfrak{m}^2$ or not; we discuss this question further in Chapter 6.

We end this section with one of the most striking consequences of the Auslander-Buchsbaum-Serre theorem.

Corollary 4.4.6 *If A is a regular local ring, then the localization $A_\mathfrak{p}$ is regular for any prime ideal $\mathfrak{p}$.*

Proof Let $F_\bullet$ be a finite free resolution of $A/\mathfrak{p}$ over A. Then the localization $F_\bullet \otimes A_\mathfrak{p}$ is a finite free resolution of the residue field of $A_\mathfrak{p}$ over $A_\mathfrak{p}$. Thus, $A_\mathfrak{p}$ is regular. $\square$

4.5 Injective Modules over Noetherian Rings

While free modules are easy to describe over any ring and projective modules are often still not too difficult, the class of injective modules is considerably more complicated. However, over Noetherian rings there is a structure theorem that makes it possible to classify them completely.

We recall that a module I is *injective* if, whenever $M \subset N$ and we have a map from M to I, the map can be extended from a map from N to I. This criterion can be checked on ideals contained in A and, hence, for Noetherian rings, on finitely generated modules. An equivalent condition is that whenever I is a submodule of a module M, the imbedding splits and I is a direct summand of M.

We refer to [64, Chapter 3] for the general properties of injective modules, and the fact that every module can be embedded into an injective module. In this section we discuss special properties of injective modules over commutative Noetherian rings.

Proposition 4.5.1 *If A is Noetherian, a direct sum of injective modules is injective.*

Proof Let I_α be a set of injective modules. It suffices to check that any module homomorphism from an ideal $\mathfrak{a}$ of A to the direct sum of the I_α can be extended to a homomorphism on A. Since A is Noetherian, $\mathfrak{a}$ is finitely generated.

Let ϕ be a map from $\mathfrak{a}$ to $\oplus_\alpha I_\alpha$. Since $\mathfrak{a}$ is finitely generated, the image of ϕ is contained in a finite direct sum of the I_α. We may extend the projection onto each component of this finite sum to a map on A, since each I_α is injective. Since a finite direct sum is also a product, this set of maps defines a map from A to $\oplus_\alpha I_\alpha$ that extends the original map on $\mathfrak{a}$. $\square$

Definition 4.5.1 *The* injective hull *of a module M is an injective module $E(M)$ containing M such that every nonzero submodule meets M nontrivially.*

We show next that the injective hull of a module exists and is unique up to isomorphism. The crucial idea is that of an essential extension. We say that an extension $M \subset N$ is *essential* if every nonzero submodule of N meets M nontrivially. If $M \subset N$ is any extension, then Zorn's Lemma applies to essential extensions of M contained in N, and there is a maximal essential extension of M contained in N.

Proposition 4.5.2 *Let $M \subset I$, where I is injective. If N is a maximal essential extension of M in I, then $I = N \oplus P$ for some submodule P of I.*

Proof Let P be a maximal submodule of I such that $N \cap P = 0$. We wish to prove that $N + P = I$. Let ϕ be the composition $N \to I \to I/P$. Since $N \cap P = 0$, ϕ is an injective map; we wish to show that ϕ is also surjective. Since I is an injective module and ϕ is one-one, the imbedding of N into I

can be extended to I/P; in other words, there is a homomorphism ψ from I/P to I such that $\psi\phi$ is the identity on N. Let Q be the image of ψ. Since Q contains N, and N is a maximal essential extension, if $N \neq Q$, there is a nonzero submodule Q' of Q that meets N trivially. Let P'/P be the inverse image of Q' under ψ. Then P' is a nontrivial extension of P that meets N trivially. This contradicts the maximality of P, so we must have had $N = Q$. Thus, the map from N to I/P is surjective, so $I = N \oplus P$. $\square$

The existence of injective hulls follows from Proposition 4.5.2, since we can imbed M into an injective module I, and a maximal essential extension of M in I is a direct summand of I and hence is injective. The injective hull of M is also unique up to isomorphism, since, if $E(M)$ and $E'(M)$ are two injective hulls of M, the identity map on M can be extended to a map $E(M) \to E'(M)$ that must be one-one since otherwise its kernel would meet M nontrivially. The injectivity of $E(M)$ implies that this map splits; since $E'(M)$ is essential it must thus be surjective and hence an isomorphism.

Proposition 4.5.3 *Let M be an A-module.*

(i) *The associated prime ideals of $E(M)$ are the same as the associated prime ideals of M.*

(ii) *The support of $E(M)$ is equal to the support of M.*

Proof Since M is a submodule of $E(M)$, every associated prime of M is also associated to $E(M)$. Conversely, let $\mathfrak{p}$ be an associated prime of $E(M)$. Since $E(M)$ is an essential extension of M, a submodule of $E(M)$ isomorphic to $A/\mathfrak{p}$ must have a nonzero intersection with M. If $m \neq 0$ is in this intersection, its annihilator is $\mathfrak{p}$, so $\mathfrak{p}$ is associated to M. The second statement of the proposition now follows from the fact that the minimal prime ideals in the support of a module are associated primes of the module. $\square$

For each prime ideal $\mathfrak{p}$ of A, let $E(A/\mathfrak{p})$ be the injective null of $A/\mathfrak{p}$. This module is indecomposable, since any two nonzero submodules meet $A/\mathfrak{p}$ nontrivially, and any two nonzero submodules of $A/\mathfrak{p}$ have a nonzero intersection. Now if I is any nonzero injective module, we may find an associated prime ideal of I, so that we have a module isomorphic to $A/\mathfrak{p}$ contained in I. Since I is injective, we may extend this map to give a map from $E(A/\mathfrak{p})$ to I that must have zero kernel since a nonzero kernel would have to meet $A/\mathfrak{p}$ nontrivially. This map splits, so we have a direct sum. Thus, we have shown that every injective module over a Noetherian ring A has a direct summand of the form $E(A/\mathfrak{p})$.

Before proving that every injective module can be uniquely represented as a sum of modules of the form $E(A/\mathfrak{p})$, we prove some facts about these modules.

Proposition 4.5.4 *Let $\mathfrak{p}$ be a prime ideal of A. Then*

(i) *The only associated prime ideal of $E(A/\mathfrak{p})$ is $\mathfrak{p}$.*
(ii) *$E(A/\mathfrak{p})$ is an $A_{\mathfrak{p}}$-module.*
(iii) *$\mathrm{Hom}(k(\mathfrak{p}), E(A/\mathfrak{p})) \cong k(\mathfrak{p})$.*
(iv) *If $\mathfrak{q}$ is a prime ideal of A that is not equal to $\mathfrak{p}$, then we have $\mathrm{Hom}(k(\mathfrak{q}), E(A/\mathfrak{p})) = 0$.*

Proof Statement (i) follows immediately from Proposition 4.5.3. We note that it follows that if $a \in \mathfrak{p}$ and $x \in E(A/\mathfrak{p})$, then there exists an n such that $a^n x = 0$.

To show that $E(A/\mathfrak{p})$ is an $A_{\mathfrak{p}}$-module, we have to show that if $s \notin \mathfrak{p}$, then multiplication by s on $E(A/\mathfrak{p})$ is bijective. Since the only associated prime of $E(A/\mathfrak{p})$ is $\mathfrak{p}$, multiplication by s is one-one. On the other hand, if x is a nonzero element of $E(A/\mathfrak{p})$, let $\mathfrak{a}$ be the annihilator of x. Since $A/\mathfrak{a}$ is isomorphic to a submodule of $E(A/\mathfrak{p})$, its only associated prime is $\mathfrak{p}$. Thus, multiplication by s is one-one on $A/\mathfrak{a}$ and defines an embedding of $A/\mathfrak{a}$ into itself. Hence we can define a map ϕ from $s(A/\mathfrak{a})$ to $E(A/\mathfrak{p})$ that sends s to x. Since $E(A/\mathfrak{p})$ is injective, the map ϕ can be extended to $A/\mathfrak{a}$, which proves that there is a y with $sy = x$ and that multiplication by s is surjective.

To prove (iii) we can localize and assume that A is local with maximal ideal $\mathfrak{p}$. In this case, $A/\mathfrak{p}$ is a field, so since the image of every nonzero map of $A/\mathfrak{p}$ into $E(A/\mathfrak{p})$ has to meet $A/\mathfrak{p}$ nontrivially, it has to be contained in $A/\mathfrak{p}$. Hence $\mathrm{Hom}(A/\mathfrak{p}, E(A/\mathfrak{p})) = \mathrm{Hom}(A/\mathfrak{p}, A/\mathfrak{p}) \cong A/\mathfrak{p}$.

Let $\mathfrak{q}$ be a prime ideal other than $\mathfrak{p}$. If there is an element $s \in \mathfrak{q}$ that is not in $\mathfrak{p}$, then s annihilates $k(\mathfrak{q})$ while from part (i) of the proposition there are no nonzero elements of $E(A/\mathfrak{p})$ annihilated by s, so $\mathrm{Hom}(k(\mathfrak{q}), E(A/\mathfrak{p})) = 0$. If there is an element s in $\mathfrak{p}$ that is not in $\mathfrak{q}$, then s is not a zero-divisor on $k(\mathfrak{q})$ while every element of $E(A/\mathfrak{p})$ is annihilated by a power of s, so $\mathrm{Hom}(k(\mathfrak{q}), E(A/\mathfrak{p})) = 0$ in this case also. $\qquad\square$

Proposition 4.5.5 *Every injective module I is a direct sum of modules isomorphic to $E(A/\mathfrak{p})$ for various prime ideals $\mathfrak{p}$. The number of summands isomorphic to $E(A/\mathfrak{p})$ for a given $\mathfrak{p}$ is uniquely determined by I.*

Proof Apply Zorn's Lemma to the set of direct sums of modules of the form $E(A/\mathfrak{p})$ contained in I. Let J be a maximal direct sum of copies of $E(A/\mathfrak{p})$

for various prime ideals $\mathfrak{p}$. Since a direct sum of injective modules is injective, J is injective, so the embedding of J into I must split, and we have $I = J \oplus K$ for some submodule K of I. But a direct summand of an injective module is injective, so K is injective. If K is not zero, it has a summand of the form $E(A/\mathfrak{p})$ and the direct sum can be extended. Hence $K = 0$ and I itself is a sum of submodules of the form $E(A/\mathfrak{p})$.

The uniqueness of the number of times each factor $E(A/\mathfrak{p})$ occurs in the direct sum follows from Proposition 4.5.4, which implies that this number is the dimension of $\mathrm{Hom}(k(\mathfrak{p}), I)$ over $k(\mathfrak{p})$. $\qquad\square$

The decomposition theorem for injective modules and the existence of injective hulls make it possible to prove analogues for injective resolutions of many of the theorems we have proved for free resolutions in previous sections. In particular, we can define minimal injective resolutions of modules and show that they are unique as well as defining invariants of the module similar to the Betti numbers for minimal free resolutions. We recall that a free complex $F_\bullet$ is called minimal if all the maps in the complex $F_\bullet \otimes k$ are zero and that, if $F_\bullet$ is minimal, then the rank of F_i is equal to the dimension of $\mathrm{Tor}_i(k, F_\bullet)$ for all i. We define a complex $I^\bullet$ of injective modules to be minimal if, for every prime ideal $\mathfrak{p}$, every map in the complex $\mathrm{Hom}(k(\mathfrak{p}), (I^\bullet)_\mathfrak{p})$ is zero. If $I^\bullet$ is minimal, then for all prime ideals $\mathfrak{p}$ and all integers i, since every map in the complex $\mathrm{Hom}(k(\mathfrak{p}), (I^\bullet)_\mathfrak{p})$ is zero, it follows from Proposition 4.5.5 that the number of summands of I^i isomorphic to $E(A/\mathfrak{p})$ is equal to the dimension of $\mathrm{Ext}^i_{A_\mathfrak{p}}(k(\mathfrak{p}), (I^\bullet)_\mathfrak{p})$.

We now restrict to the case in which $I^\bullet$ is an injective resolution of a finitely generated module M; in this case if $I^\bullet$ is minimal it is called a minimal injective resolution of M. The number of times $E(A/\mathfrak{p})$ occurs as a summand in a minimal injective resolution is equal to the dimension of $\mathrm{Ext}^i_{A_\mathfrak{p}}(k(\mathfrak{p}), M_\mathfrak{p})$ and so is finite. These numbers are called the *Bass numbers* (see Bass [4]) of the module and are denoted $\mu_i(\mathfrak{p}, M)$. There is one Bass number for each i and each prime ideal $\mathfrak{p}$. The statement of Theorem 4.1.8 can be expressed in terms of Bass numbers by saying that if $\mathfrak{q}$ and $\mathfrak{p}$ are prime ideals such that $\mathfrak{q}$ is properly contained in $\mathfrak{p}$, and there are no prime ideals properly between $\mathfrak{p}$ and $\mathfrak{q}$, then $\mu_{i+1}(\mathfrak{p}, M) \geq \mu_i(\mathfrak{q}, M)$.

Proposition 4.5.6 *Every module has a minimal injective resolution.*

Proof Let M be a module. Construct an injective resolution of M as follows: in the first step, imbed M into its injective hull $E(M)$. If the resolution has been constructed up through degree n, construct I^{n+1} by imbedding the cokernel of

d_I^{n-1} into its injective hull (in this notation d^{-2} is the map from 0 to M and d^{-1} is the imbedding of M in $E(M)$). We claim that an injective resolution constructed in this way is minimal.

Let $\mathfrak{p}$ be a prime ideal of A. Suppose that for some $i \geq 0$ the map from $\mathrm{Hom}(k(\mathfrak{p}), (I^i)_\mathfrak{p})$ to $\mathrm{Hom}(k(\mathfrak{p}), (I^{i+1})_\mathfrak{p})$ is not zero. Let Q^i denote the cokernel of d_I^{i-2}, so that I^i is the injective hull of Q^i. We then have that $d^i(Q^i) = 0$. Let $\phi : k(\mathfrak{p}) \to (I^i)_\mathfrak{p}$ be a map whose composition with $(I^i)_\mathfrak{p} \to (I^{i+1})_\mathfrak{p}$ is not zero. Since $k(\mathfrak{p})$ is a field, ϕd_I^i is one-one, and hence the image of ϕ does not meet Q^i. Multiplying by an element $s \notin \mathfrak{p}$, we obtain a nonzero submodule of I^i that meets Q^i trivially, contradicting the assumption that I^i is the injective hull Q^i. Thus, we must have that the map from $\mathrm{Hom}(k(\mathfrak{p}), (I^i)_\mathfrak{p})$ to $\mathrm{Hom}(k(\mathfrak{p}), (I^{i+1})_\mathfrak{p})$ is zero, so $I^\bullet$ is minimal. $\qquad\square$

4.6 Dualizing Complexes

The main results of this section give alternative characterizations of Cohen-Macaulay and Gorenstein rings in terms of dualizing complexes. In addition, we prove a result on the homology of bounded free complexes.

Definition 4.6.1 *Let A be a Noetherian local ring of dimension d. A dualizing complex for A is a complex*

$$0 \to D^0 \to \cdots \to D^d \to 0$$

such that

(i) *The homology of $D^\bullet$ is finitely generated.*
(ii) *Each module D^i is a sum of $E(A/\mathfrak{p})$ for $A/\mathfrak{p}$ with $\dim(A/\mathfrak{p}) = d - i$, each occurring exactly once.*

An alternative version of condition (ii) is that for every prime ideal $\mathfrak{p}$, $\mathrm{Ext}^i(k(\mathfrak{p}), D_\mathfrak{p}^\bullet)$ is equal to $k(\mathfrak{p})$ if $i = \dim(A) - \dim(A/\mathfrak{p})$ and is zero otherwise. We note that condition (ii) implies in particular that $D^\bullet$ is a complex of injective modules.

Proposition 4.6.1 *Let A be a regular local ring. Then a minimal injective resolution of A is a dualizing complex for A.*

Proof Let $I^\bullet$ be a minimal injective resolution of A, and let $\mathfrak{p}$ be a prime ideal of A. We have to show that the dimension of $\mathrm{Ext}^i(k(\mathfrak{p}), A_\mathfrak{p})$ over $k(\mathfrak{p})$ is one, where $\dim(A/\mathfrak{p}) = \dim(A) - i$. Since the localization of A at $\mathfrak{p}$ is also regular by Theorem 4.4.5, we can assume that $\mathfrak{p}$ is the maximal ideal of A. Let

$k = A/\mathfrak{p}$. Then the Koszul complex $K_\bullet$ on a regular system of parameters is a free resolution of k, and, letting d be the dimension of A, we have

$$\text{Ext}^d(k, A) \cong \text{Coker}(\text{Hom}(K_{d-1}, A) \to \text{Hom}(K_d, A)),$$

and this cokernel is isomorphic to k. □

The above proposition makes it possible to construct a dualizing complex for every local ring A that is a homomorphic image of a regular local ring R. We let $I^\bullet$ be the resolution over R, and let $D^\bullet = \text{Hom}_R(A, I^\bullet)$. Then $D^\bullet$ is a complex of A-modules, and since its homology is $\text{Ext}_R^i(A, R)$, its homology is finitely generated. In fact, if M is any finitely generated A-module, then the homology $H^i(\text{Hom}_R(M, D^\bullet))$ is $\text{Ext}_R^i(M, R)$ and is finitely generated. The fact that $D^\bullet$ is a dualizing complex for A follows from the following lemma.

Lemma 4.6.2 *Let $A = R/\mathfrak{a}$, and let $\mathfrak{p}$ be a prime ideal of R that contains $\mathfrak{a}$, so that $\mathfrak{p}$ can also be considered as a prime ideal of A. Then*

$$\text{Hom}_R(A, E_R(A/\mathfrak{p})) \cong E_A(A/\mathfrak{p}).$$

Proof In the statement of the lemma we are letting $E_R(A/\mathfrak{p})$ denote the injective hull of $A/\mathfrak{p}$ as an R-module and letting $E_A(A/\mathfrak{p})$ denote the injective hull of $A/\mathfrak{p}$ as an A-module. We note that $\text{Hom}_R(A, E_R(A/\mathfrak{p}))$ is simply the set of elements of $E_R(A/\mathfrak{p})$ annihilated by $\mathfrak{a}$. If $M \subseteq N$ are A-modules and ϕ is a map of A-modules from M to $\text{Hom}_R(A, E_R(A/\mathfrak{p}))$, then ϕ is also a map of R-modules from M to $E_R(A/\mathfrak{p})$ and so can be extended to a map ψ on N. Since N is an A-module, it is annihilated by $\mathfrak{a}$, so the image of ψ is contained in $\text{Hom}_R(A, E_R(A/\mathfrak{p}))$. Hence any map of A-modules from the submodule M to $\text{Hom}_R(A, E_R(A/\mathfrak{p}))$ can be extended to a map of A-modules on N, so $\text{Hom}_R(A, E_R(A/\mathfrak{p}))$ is injective.

Since $\text{Hom}_R(A, E_R(A/\mathfrak{p}))$ is a submodule of $E_R(A/\mathfrak{p})$ that contains $A/\mathfrak{p}$, it is an essential extension of $A/\mathfrak{p}$. Thus, $\text{Hom}_R(A, E_R(A/\mathfrak{p})) = E_A(A/\mathfrak{p})$. □

A dualizing complex can be used to determine the dimension and depth of a module M.

Proposition 4.6.3 *Let A be a local ring of dimension d, and let $D^\bullet$ be a dualizing complex for A. Let M be a finitely generated module. Then*

(i) *The lowest integer i for which $\text{Ext}^i(M, D^\bullet) \neq 0$ is $d - \dim(M)$.*
(ii) *The highest integer i for which $\text{Ext}^i(M, D^\bullet) \neq 0$ is $d - \text{depth}(M)$.*
(iii) *$D^\bullet$ is exact except in degree zero if and only if A is Cohen-Macaulay.*

Proof Let $k = d - \dim(M)$. Then if $j < k$, it follows from Proposition 4.5.4 that every associated prime ideal of D^j has dimension greater than $d - k = \dim(M)$, so $\mathrm{Hom}(M, D^j) = 0$, and thus $\mathrm{Ext}^j(M, D^\bullet) = 0$. Let $\mathfrak{p}$ be an associated prime ideal in the support of M of dimension equal to $\dim(M)$. Then there is a submodule of M isomorphic to $A/\mathfrak{p}$, and $E(A/\mathfrak{p})$ occurs as a summand of D^k. The imbedding of $A/\mathfrak{p}$ into $E(A/\mathfrak{p})$ can be extended to a nonzero map on M. Thus, $\mathrm{Hom}(M, D^k) \neq 0$, and since we have shown that $\mathrm{Hom}(M, D^{k-1}) = 0$, we have $\mathrm{Ext}^k(M, D^\bullet) \neq 0$. This proves (i).

The second statement is proven by induction on the depth of M. We note first that since $D^i = 0$ for $i > d$, we have $\mathrm{Ext}^i(M, D^\bullet) = 0$ for $i > d$. If the depth of M is zero, then M has a submodule isomorphic to $A/\mathfrak{m} = k$, and from part (i) we know that $\mathrm{Ext}^d(k, D^\bullet) \neq 0$. From the short exact sequence

$$0 \to k \to M \to M' \to 0$$

we derive an exact sequence

$$\mathrm{Ext}^d(M, D^\bullet) \to \mathrm{Ext}^d(k, D^\bullet) \to \mathrm{Ext}^{d+1}(M', D^\bullet) = 0,$$

so $\mathrm{Ext}^d(M, D^\bullet) \neq 0$.

Assume now that the depth of M is at least one, and let x be an element of the maximal ideal of A which is not a zero-divisor on M. We have the short exact sequence

$$0 \to M \xrightarrow{x} M \to M/xM \to 0,$$

and the depth of M/xM is one less than the depth of M. The associated long exact sequence gives the exact sequence

$$\mathrm{Ext}^i(M, D^\bullet) \xrightarrow{x} \mathrm{Ext}^i(M, D^\bullet) \to \mathrm{Ext}^{i+1}(M/xM, D^\bullet).$$

If $\mathrm{Ext}^{i+1}(M/xM, D^\bullet) = 0$, then, since $\mathrm{Ext}^i(M, D^\bullet)$ is finitely generated and $x \in \mathfrak{m}$, $\mathrm{Ext}^i(M, D^\bullet) = 0$. Thus, if i is the highest integer such that $\mathrm{Ext}^{i+1}(M/xM, D^\bullet) \neq 0$, we must have $\mathrm{Ext}^{i+1}(M, D^\bullet) = 0$. It now follows from the exact sequence

$$\mathrm{Ext}^i(M, D^\bullet) \to \mathrm{Ext}^{i+1}(M/xM, D^\bullet) \to \mathrm{Ext}^{i+1}(M, D^\bullet) = 0$$

that $\mathrm{Ext}^i(M, D^\bullet) \neq 0$. Thus, statement (ii) follows by induction.

Statement (iii) follows immediately from statements (i) and (ii) applied to $M = A$. $\qquad\square$

It follows from Proposition 4.6.3 that a module M is Cohen-Macaulay if and only if there is exactly one i for which $\mathrm{Ext}^i(M, D^\bullet) \neq 0$.

There is also a characterization of Gorenstein rings using a dualizing complex. We prove one implication.

Proposition 4.6.4 *If the dualizing complex of A is a minimal injective resolu-tion of A, then A is Gorenstein.*

Proof If $D^\bullet$ is a minimal free resolution of A, then A is Cohen-Macaulay by Proposition 4.6.3, and $\mathrm{Ext}(k, A) \cong \mathrm{Ext}^d(k, D^\bullet) = k$, so by Proposition 4.1.12 A is Gorenstein. $\square$

Further characterizations of Gorenstein rings can be found in Bass [4].

Proposition 4.6.5 *Let M be a module of finite length, and let $E = E(k)$. Then*

(i) *The length of* $\mathrm{Hom}(M, E)$ *is equal to the length of M.*
(ii) *We have $M \cong \mathrm{Hom}(\mathrm{Hom}(M, E), E)$.*
(iii) *The annihilator of* $\mathrm{Hom}(M, E)$ *is equal to the annihilator of M.*

Proof If $M = k$, then every homomorphism from k to $E(k)$ maps k into k, so $\mathrm{Hom}(k, E) = \mathrm{Hom}(k, k) \cong k$, so the result is true for k. Since E is injective, $\mathrm{Hom}(-, E)$ is exact, so the first statement is true for all M by induction on the length of M.

The second statement is also clearly true for $M = k$. There is a natural map from M to $\mathrm{Hom}(\mathrm{Hom}(M, E), E)$ that takes m to evaluation at m, that is, to the map that takes ϕ to $\phi(m)$. Again we use induction on the length of M. There is a short exact sequence

$$0 \to k \to M \to M' \to 0$$

where the length on M' is one less than the length of M. Since E is injective, we have a diagram with exact rows, where for every module N we denote $\mathrm{Hom}(\mathrm{Hom}(N, E), E)$ by N^{**}:

$$\begin{array}{ccccccccc} 0 \to & k & \to & M & \to & M' & \to 0 \\ & \downarrow & & \downarrow & & \downarrow & \\ 0 \to & k^{**} & \to & M^{**} & \to & M'^{**} & \to 0 \end{array}$$

By induction, the outer maps are isomorphisms, so the center map is also an isomorphism.

Since the annihilator of $\mathrm{Hom}(M, N)$ contains the annihilator of M for any module N, the third statement follows from statement (ii). $\square$

The duality for modules of finite length in Proposition 4.6.5 can be extended to a duality for bounded complexes with finitely generated homology. We refer to Roberts [55] for details.

It follows from the Acyclicity Lemma that if

$$F_\bullet = 0 \to F_d \to \cdots \to F_0 \to 0$$

is a complex of free modules of length d with homology of finite length, and
if A is Cohen-Macaulay, then the homology of $F_\bullet$ is zero in degrees greater
than zero. The next theorem gives a generalization of this result to non-Cohen-
Macaulay rings. Instead of saying that the homology is zero, it says that the
homology is annihilated by a fixed nonzero ideal, independent of the complex.
If A is Cohen-Macaulay, it follows from Proposition 4.6.3 that all of these ideals
are equal to A, showing again that the homology vanishes in this case.

Theorem 4.6.6 *Let*

$$F_\bullet = 0 \to F_k \to \cdots \to F_0 \to 0$$

*be a bounded free complex with homology of finite length. Let $\mathfrak{a}_i$ be the anni-
hilator of $H^i(D^\bullet)$, where $D^\bullet$ is a dualizing complex. Then $\mathfrak{a}_d\mathfrak{a}_{d-1}\cdots\mathfrak{a}_{d-k+i}$
annihilates $H_i(F_\bullet)$.*

Proof We consider the double complex $\mathrm{Hom}(F_\bullet, D^\bullet)$. The picture looks like

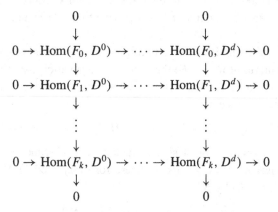

By Theorem 3.2.1 there are two filtrations of the homology of the associated
total complex of $\mathrm{Hom}(F_\bullet, D^\bullet)$ whose quotients are subquotients of the iterated homology modules taken in the two possible orders. We compute these
filtrations.

We first take homology along the column $\mathrm{Hom}(F_\bullet, D^j)$ for fixed j. For
$j < d$, D^j is a sum of injective modules that are sums of $E(A/\mathfrak{p})$, which are
$A_\mathfrak{p}$-modules by Proposition 4.5.4, for prime ideals $\mathfrak{p}$ other than the maximal
ideal. Since $F_\bullet$ has homology of finite length, the homology of $\mathrm{Hom}(F_\bullet, D^j)$
is thus zero if $j < d$. For $j = d$, the complex in column j is $\mathrm{Hom}(F_\bullet, E(k))$,
and since $E(k)$ is injective its homology in degree $i + d$ is $\mathrm{Hom}(H_i(F_\bullet), E(k))$.
By Proposition 4.6.5, $\mathrm{Hom}(H_i(F_\bullet), E(k))$ has the same annihilator as $H_i(F_\bullet)$.
We have shown that every column except the dth column is exact, so by

Proposition 3.2.3, the homology $\text{Hom}(H_i(F_\bullet), E(k))$ is isomorphic to the homology of the total complex in degree $i + d$. Putting these facts together, we conclude that the annihilator of the total complex in degree $i + d$ is equal to the annihilator of $H_i(F_\bullet)$.

We now use the other filtration to show that the homology of the total complex in degree $i + d$ is annihilated by $\mathfrak{a}_d \mathfrak{a}_{d-1} \cdots \mathfrak{a}_{d-k+i}$. If we take homology of $\text{Hom}(F_m, D^\bullet)$ for fixed m, since F_m is free we obtain $\text{Hom}(F_m, H^j(D^\bullet))$ in degree $m + j$, which has annihilator $\mathfrak{a}_j$. There is therefore a filtration of the homology of the total complex with quotients annihilated by $\mathfrak{a}_j$, and it follows that the homology of the total complex is annihilated by the product of those $\mathfrak{a}_j$ that occur. It remains to determine which $\mathfrak{a}_j$ occur in the filtration of the homology of the total complex in degree $i + d$. The integers j that occur are those for which there exists an integer m with $m + j = i + d$, $F_m \neq 0$, and $D^j \neq 0$. The highest such integer is $j = d$, and the lowest occurs when $m = k$, so that $j = i + d - k$. Thus, the product $\mathfrak{a}_d \mathfrak{a}_{d-1} \cdots \mathfrak{a}_{d-k+i}$ annihilates the homology of the total complex, and hence it also annihilates $H_i(F_\bullet)$. $\quad\square$

If the complex $F_\bullet$ has length exactly d, then Theorem 4.6.6 implies that there is a nonzero ideal of A independent of $F_\bullet$ that annihilates the homology of $F_\bullet$ in positive degrees. In fact, if $i > 0$, then $\mathfrak{a}_d \mathfrak{a}_{d-1} \cdots \mathfrak{a}_{d-k+i} = \mathfrak{a}_d \mathfrak{a}_{d-1} \cdots \mathfrak{a}_i$ is a product of ideals $\mathfrak{a}_j$ where for each j we have $\dim(A/\mathfrak{a}_j) < \dim(A)$. Hence these ideals cannot be contained in a prime ideal $\mathfrak{p}$ with $\dim(A/\mathfrak{p}) = \dim(A)$, so their product is nonzero.

If A is Cohen-Macaulay, then the homology of $D^\bullet$ is zero in degrees greater than zero, and all of these ideals $\mathfrak{a}_j$ are equal to A. We thus have another proof that the homology of $F_\bullet$ is zero in positive degrees for a Cohen-Macaulay ring.

Exercises

4.1 Give an example of a complete intersection that is not regular and an example of a Cohen-Macaulay ring that is not Gorenstein.

4.2 Let R be a regular local ring, and let $A = R/\mathfrak{a}$, where $r_1, \ldots r_k$ is a minimal set of generators for $\mathfrak{a}$ and r_i is in the square of the maximal ideal of R for all i. Let $K_\bullet$ be the Koszul complex over A on a minimal set of generators of the maximal ideal $\mathfrak{m}$ of A. Show that $H_1(K_\bullet)$ is a vector space over $A/\mathfrak{m}$ of dimension k. Show that this result can be used to give a definition of complete intersection that does not require a specific map from a regular local ring onto A.

4.3 Show that $k[[X, Y]]/(X^2, XY)$ is a ring of dimension 1 that is not Cohen-Macaulay.

4.4 Prove that the subring of the ring $k[[X, Y]]$ consisting of power series with only terms of even degree is Gorenstein.

4.5 Show that the subring of the polynomial ring $k[X, Y]$ generated over k by the elements X^4, X^3Y, XY^3, Y^4 is a non-Cohen-Macaulay integral domain.

4.6 The ring $k[X, Y, Z]/(X^2 - Y^2, Y^2 - Z^2, XY, XZ, YZ)$ is a Gorenstein ring that is not a complete intersection. This example, as well as the previous one, are due to Macaulay [43].

4.7 Let $A = k[X, Y]$, and let $F_\bullet$ be the complex

$$\cdots \to 0 \to A \oplus A \overset{\phi}{\to} A \to 0$$

where ϕ is the map defined by the matrix $(X \ Y)$. Let $G_\bullet$ be the complex

$$\cdots \to 0 \to A \overset{0}{\to} A/(X, Y) \to 0.$$

Show that $F_\bullet$ and $G_\bullet$ have isomorphic homology but are not quasi-isomorphic.

4.8 Let $T = \left(\begin{smallmatrix} a & b & c \\ d & e & f \end{smallmatrix} \right)$ be a matrix of elements in a Cohen-Macaulay ring A such that the ideal of 2 by 2 minors of T has height 2. Let D_{23}, D_{13}, D_{12} be the minors of this matrix, and let $\mathfrak{a}$ denote the ideal they generate. Use the Acyclicity Lemma to show that the sequence

$$0 \to A^2 \overset{\left(\begin{smallmatrix} a & d \\ b & e \\ c & f \end{smallmatrix} \right)}{\longrightarrow} A^3 \overset{(D_{23}, -D_{13}, D_{12})}{\longrightarrow} A \to A/\mathfrak{a} \to 0$$

is a free resolution of $A/\mathfrak{a}$.

4.9 Show that if a complex $I^\bullet$ of injective modules is not minimal, then there exists a prime ideal $\mathfrak{p}$ of A such that $I^\bullet$ has a direct summand of the form $\cdots 0 \to E(A/\mathfrak{p}) \overset{\phi}{\to} E(A/\mathfrak{p}) \to 0 \to \cdots$ where ϕ is an isomorphism.

4.10 Show that if $I^\bullet$ and $J^\bullet$ are minimal injective complexes and if $f_\bullet : I^\bullet \to J^\bullet$ is a quasi-isomorphism, then $f_\bullet$ is an isomorphism of complexes.

5

Intersection Multiplicities

In this chapter we discuss the definition of intersection multiplicities, a topic that we will take up again in the last chapter. The basic idea is to give a definition of the multiplicity of intersection of two subschemes of a scheme in such a way that it agrees with the intuitive notion of counting the number of intersections at a point and so that certain reasonable conditions will be satisfied. One condition is that the result should be independent of rational equivalence. A second is that the intersection of two smooth subschemes that meet transversally at a point should be one; algebraically, this condition states that if $x_1, \ldots x_d$ form a minimal set of generators of the maximal ideal of a regular local ring, and if the subschemes are defined by the ideals $x_1, \ldots x_k$ and $x_{k+1}, \ldots x_d$, respectively, their intersection multiplicity should be one. Another condition, which we will discuss in a later chapter, is that Bézout's theorem must hold; that is, in projective space the total number of intersections of two subschemes should be the product of the degrees of the subschemes.

Another version of this problem is to attempt to define an intersection product on the Chow group to create a ring structure on the Chow group. There are a couple of differences between this point of view and that described in the previous paragraph. Suppose that we are considering intersection multiplicities defined by cycles in $A_*(\mathrm{Spec}(A))$, where A is a local ring of dimension 2. We would like to define the intersection multiplicity of two cycles of dimension 1 that meet only at the closed point of $\mathrm{Spec}(A)$, and to come out with a number. On the other hand, if we were to take a product in the Chow group, the answer would be an element of the Chow group of dimension 0, and the component of dimension zero of the Chow group of a local ring is usually zero. The correct interpretation is that the intersection multiplicity is an element of the Chow group of $\mathrm{Spec}(A/\mathfrak{m})$ whose image in the Chow group of $\mathrm{Spec}(A)$ is the product defined by the ring structure. From this point of view intersection multiplicity carries more information. On the other hand, we often prove results

93

94   *Multiplicities*

only in the case in which the subschemes meet in the proper codimension, while
to have a product on the Chow group the product must be defined for all rational
equivalence classes of cycles whether they intersect properly or not.

At present there is no definition of intersection multiplicities that has been
proposed for general subschemes of general schemes or even for schemes of
finite type over a field. In this chapter we define them in two cases. First,
we define the mutiplicity for intersection with a scheme defined by a principal ideal (i.e., a divisor) in Spec(A). We complete this definition in a later
chapter, where we define intersection multiplicities for divisors in projective
schemes. In the last section we give Serre's definition for ideals in regular local
rings. The main part of this chapter is devoted to relations between these definitions and with Samuel multiplicities coming from properties of the Koszul
complex.

5.1 Intersection with Divisors

Let A be a Noetherian ring, and let x be an element of A. In this section we
define an operation on the Chow group of A that represents intersection with the
subscheme Spec(A/xA). More generally, if V is any closed subset of Spec(A),
and if $W = \text{Spec}(A/xA)$, we define a map from $A_i(V)$ to $A_{i-1}(V \cap W)$ for
each i. We let (x) denote this map, which we refer to as "intersection with the
divisor (x)." We first define the intersection of (x) with a cycle; it suffices to
consider the case where $\eta = [A/\mathfrak{p}]$ and extend by additivity. We recall that, if
M is an A-module of dimension at most d, then $[M]_d$ denotes the cycle

$$\sum_{\dim(A/\mathfrak{p})=d} \text{length}(M_\mathfrak{p})([A/\mathfrak{p}]).$$

If $A/\mathfrak{p}$ has dimension d, and if x is not in $\mathfrak{p}$, we define the intersection to
be the cycle $[A/\mathfrak{p}/x(A/\mathfrak{p})]_{d-1}$; this is the cycle that we denoted div($\mathfrak{p}, x$) in
Chapter 1. If x is in $\mathfrak{p}$, we define the intersection to be zero. For any cycle η
we denote the intersection of (x) with η by $(x) \cap \eta$.

We still must show that this definition preserves rational equivalence in
$Z_*(V)$; we do so in Chapter 8 after generalizing to intersection with
divisors in projective schemes. Another important property is that the operations of intersection with two divisors will commute modulo rational equivalence. If x and y are elements of A, and $\mathfrak{p}$ is a prime ideal of A such that
$\dim((A/\mathfrak{p})/(x, y)(A/\mathfrak{p})) = \dim(A/\mathfrak{p}) - 2$, then it will follow from the results
of the next section that we have an equality of cycles

$$(x) \cap (y) \cap [A/\mathfrak{p}] = (y) \cap (x) \cap [A/\mathfrak{p}].$$

In the more difficult case, where the codimension of the ideal generated by x and y in $A/\mathfrak{p}$ is one, these cycles may not be equal. Equality modulo rational equivalence will be proven in Chapter 8.

5.2 The Euler Characteristic of the Koszul Complex

The Koszul complex on a set of elements has many special properties that have important consequences for intersection multiplicities. If $F_\bullet$ is a bounded complex with homology of finite length, we define the *Euler characteristic* of $F_\bullet$ to be $\sum(-1)^i \operatorname{length}(H_i(F_\bullet))$. The Euler characteristic of $F_\bullet$ is denoted $\chi(F_\bullet)$. One of the main elementary properties of the Euler characteristic is its additivity on short exact sequences of complexes.

Proposition 5.2.1 *If*

$$0 \to E_\bullet \to F_\bullet \to G_\bullet \to 0$$

is a short exact sequence of bounded complexes with homology of finite length, then $\chi(F_\bullet) = \chi(E_\bullet) + \chi(G_\bullet)$.

Proof This formula is a consequence of the long exact homology sequence and the fact that length is additive on exact sequences. □

Let A be a local domain of dimension d, and let $x_1, \ldots, x_d$ be a system of parameters for A. The main result of this section is to show that three multiplicities defined by the system of parameters are equal. The first is the Samuel multiplicity of the ideal generated by $(x_1, \ldots, x_d)$, the second is $(x_1) \cap \cdots \cap (x_d) \cap [A]$ as defined in the previous section, and the third is the Euler characteristic of the Koszul complex $K_\bullet(x_1, \ldots, x_d)$. We also show similar results for modules, and show that the Euler characteristic of the complex $K_\bullet(x_1, \ldots, x_d; M)$ is zero if the dimension of M is less than d. Throughout this section we assume that A is a local ring, that M is a finitely generated A-module, and that $x_1, \ldots x_k$ is a sequence of elements of A such that $M/(x_1, \ldots, x_k)M$ has finite length. We denote the Euler characteristic of $K_\bullet(x_1, \ldots, x_k; M)$ by $\chi(x_1, \ldots, x_k; M)$. It follows from Proposition 5.2.1 that if

$$0 \to M' \to M \to M'' \to 0$$

is a short exact sequence of modules satisfying the above conditions, then $\chi(x_1, \ldots, x_k; M) = \chi(x_1, \ldots, x_k; M') + \chi(x_1, \ldots, x_k; M'')$.

For graded rings the Euler characteristic of the Koszul complex can be computed in terms of Hilbert polynomials. We have defined the Hilbert polynomial of a graded module M to be the polynomial P_M such that $P_M(n) = \text{length}(M_n)$ for large values of n. In the next proposition we use a slightly different Hilbert polynomial, a polynomial P such that $P(n) = \text{length}(\oplus_{k \leq n} M_n)$ for large n. This polynomial is more useful for dealing with graded modules of finite length, since it is constant with value equal to the length of the module, whereas the usual Hilbert polynomial is zero in that case.

Proposition 5.2.2 *Let M be a graded module over a $\mathbb{Z}$-graded ring A for which A_0 is Artinian. Let M have dimension d, and suppose the leading coefficient of its Hilbert polynomial is $e/d!$. Let $x_1, \ldots, x_d$ be homogeneous elements of A such that $M/(x_1, \ldots, x_d)M$ has finite length, and assume that x_i has degree $n_i > 0$ for each i. Then*

$$\chi(x_1, \ldots, x_d; M) = n_1 n_2 \cdots n_d e.$$

Proof We give the free modules in the Koszul complex on $x_1, \ldots, x_d$ the structure of graded modules by letting $e_{i_1} \wedge \cdots \wedge e_{i_j}$ have degree $n_{i_1} + \cdots + n_{i_j}$. Then, since multiplication by x_{i_k} raises the degree by n_{i_k}, the boundary maps become homogeneous of degree 0.

For integers i and j let $P_{i,j}$ be the Hilbert polynomial of the homology module $H_i(x_1, \ldots, x_j; M)$. We prove by induction that the alternating sum $\sum (-1)^i P_{i,j}$ is a polynomial of degree $d - j$ with leading coefficient $n_1 n_2 \cdots n_j e/(d - j)!$. For $j = 0$ we have $P_{i,j} = 0$ for $i > 0$, and $P_{0,j}$ is the Hilbert polynomial of M, so the result is true in this case.

Assume that this result is true for $j - 1$. The short exact sequence of complexes of Proposition 3.3.3 gives rise to a long exact sequence

$$\cdots \rightarrow H_{i+1}(x_1, \ldots, x_j; M) \rightarrow H_i(x_1, \ldots, x_{j-1}; M) \xrightarrow{x_j}$$

$$\rightarrow H_i(x_1, \ldots, x_{j-1}; M) \rightarrow H_i(x_1, \ldots, x_j; M) \rightarrow \cdots.$$

Since the first copy of $H_i(x_1, \ldots, x_{j-1}; M)$ in this sequence comes from the wedge product of the Koszul complex on $x_1, \ldots, x_{j-1}$ with e_j, the grading is shifted by n_j, and the Hilbert polynomial for this module in degree n is $P_{i,j-1}(n - n_j)$. Hence, using the additivity of Hilbert polynomials on exact sequences, we have

$$\sum (-1)^i P_{i,j}(n) = \sum (-1)^i P_{i,j-1}(n) - \sum (-1)^i P_{i,j-1}(n - n_j).$$

Since we are assuming that $\sum (-1)^i P_{i,j-1}(n)$ is a polynomial of degree $d - j + 1$ with leading coefficient $n_1 n_2 \cdots n_{j-1} e/(d - j + 1)!$, it now follows that $\sum (-1)^i P_{i,j}(n)$ is a polynomial of degree $d - j$ with leading coefficient $n_1 n_2 \cdots n_j e/(d-j)!$. For $j = d$, we conclude that the alternating sum of Hilbert polynomials is constant with value $n_1 n_2 \cdots n_d e$. Since the homology has finite length in this case, this means that the Euler characteristic $\chi(x_1, \ldots, x_d)$ is equal to $n_1 n_2 \cdots n_d e$. $\qquad\square$

We next prove another additivity result for Euler characteristics of Koszul complexes.

Proposition 5.2.3 *We have*

$$\chi\left(x_1 x_1', x_2, \ldots x_k; M\right) = \chi(x_1, x_2, \ldots, x_k; M) + \chi\left(x_1', x_2, \ldots x_k; M\right).$$

Proof We first prove this result in the case in which $k = 1$. Consider the complex $E_\bullet$ with $E_0 = E_1 = A \oplus A$, $E_i = 0$ for $i \neq 0$ or 1, and d_1^E defined by the matrix $\left(\begin{smallmatrix} x_1 & 1 \\ 0 & x_2 \end{smallmatrix}\right)$. There is a short exact sequence

$$0 \to K_\bullet(x_1) \to E_\bullet \to K_\bullet(x_2) \to 0,$$

in which $K(x_1)$ is embedded by the first summands of E_0 and E_1, (both E_0 and E_1 are isomorphic to $A \oplus A$), and $K(x_2)$ is the projection onto the second summands. Hence if M is a module such that $M/x_1 M$ and $M/x_2 M$ have finite length, we have $\chi(E_\bullet \otimes M) = \chi(x_1; M) + \chi(x_2; M)$.

On the other hand, there is a map from $K_\bullet(x_1 x_2)$ to $E_\bullet$, which is the first map in the following short exact sequence of complexes:

$$
\begin{array}{ccccccccc}
& & 0 & & 0 & & 0 & & \\
& & \downarrow & & \downarrow & & \downarrow & & \\
& & & \left(\begin{smallmatrix} -1 \\ x_1 \end{smallmatrix}\right) & & (x_1 \ \ 1) & & & \\
0 & \to & A & \to & A \oplus A & \to & A & \to & 0 \\
& & x_1 x_2 \downarrow & & \downarrow \left(\begin{smallmatrix} x_1 & 1 \\ 0 & x_2 \end{smallmatrix}\right) & & \downarrow 1 & & \\
& & & \left(\begin{smallmatrix} 0 \\ 1 \end{smallmatrix}\right) & & (1 \ \ 0) & & & \\
0 & \to & A & \to & A \oplus A & \to & A & \to & 0 \\
& & \downarrow & & \downarrow & & \downarrow & & \\
& & 0 & & 0 & & 0 & &
\end{array}.
$$

The vertical complex on the right is exact, so the map from $K_\bullet(x_1 x_2)$ into $E_\bullet$ is a quasi-isomorphism. Thus, we have $\chi(E_\bullet \otimes M) = \chi(x_1 x_2; M)$, and, putting this result together with the first equality, we have the required formula in the case $k = 1$.

If $k > 1$, we can take the tensor product of the above complexes and exact sequences with $K_\bullet(x_2, \ldots, x_k) \otimes M$. The resulting sequences are still exact, and we may thus conclude that the equality holds in the general case. $\qquad\square$

Corollary 5.2.4 *We have*

$$\chi\left(x_1^{m_1}, \ldots, x_k^{m_k}; M\right) = m_1 m_2 \cdots m_k \chi(x_1, \ldots, x_k; M)$$

Proof This result follows immediately from Proposition 5.2.1.   □

Our aim is to compare the Euler characteristic of the Koszul complex on a system of parameters with the Samuel multiplicity of the ideal generated by the system of parameters. Let $x_1, \ldots, x_d$ be a system of parameters for A, and let $\mathfrak{a}$ be the ideal generated by $x_1 \ldots, x_d$. We recall from Proposition 2.4.1 that the Samuel multiplicity can be defined as

$$e(\mathfrak{a}) = (d!) \lim_{n \to \infty} \frac{\text{length}(A/\mathfrak{a}^n)}{n^d}.$$

On the other hand, Corollary 5.2.4 implies that the Euler characteristic $\chi(x_1, \ldots, x_d)$ can trivially be expressed as a limit

$$\chi(x_1, \ldots, x_d) = \lim_{n \to \infty} \frac{\chi\left(x_1^n, \ldots, x_d^n\right)}{n^d}.$$

The next few propositions will show that this limit can be replaced by the limit of the length of $A/(x_1^n, \ldots, x_d^n)A$ over n^d, since the contributions of the other homology modules of the Koszul complex go to zero in the limit.

Proposition 5.2.5 *Let M be a finitely generated module of dimension d. Let $x_1, \ldots, x_k$ be a sequence of elements of the maximal ideal of A such that $M/(x_1, \ldots, x_k)M$ has finite length. Then for every integer i between 0 and k there is a constant c such that* $\text{length}(H_i(x_1^n, \ldots, x_k^n; M)) < cn^d$ *for all n.*

Proof The proof is by induction on d. If $d = 0$, then M has finite length, and the powers of x_i eventually annihilate M; thus the length of $(H_i(x_1^n, \ldots, x_k^n; M))$ is eventually constant (equal to $\binom{k}{i}(\text{length}(M))$, and the theorem is true in this case.

If the dimension is greater than one, we first take a filtration by submodules to reduce to the case in which $M = A/\mathfrak{p}$ for some prime ideal $\mathfrak{p}$. If $0 \to M' \to M \to M'' \to 0$ is a short exact sequence of modules, the short exact sequence of associated Koszul complexes gives rise to an exact sequence

$$H_i\left(K_\bullet\left(x_1^n, \ldots, x_k^n; M'\right)\right) \to H_i\left(K_\bullet\left(x_1^n, \ldots, x_k^n; M\right)\right)$$

$$\to H_i\left(K_\bullet\left(x_1^n, \ldots, x_k^n; M''\right)\right)$$

for each i. Hence if the lengths of the outer terms of this sequence are bounded by certain constants times n^d then the middle term is also, so if the theorem is true for M' and M'', it is true for M. Thus, we may assume that $M = A/\mathfrak{p}$ and that $A/\mathfrak{p}$ has dimension $d > 0$.

Since $M/(x_1, \ldots, x_k)M$ has finite length, at least one of the x_j is not in $\mathfrak{p}$; assume by renumbering if necessary that $x_1 \notin \mathfrak{p}$. We apply induction to $x_2, \ldots, x_k$ on the module $M/x_1M = (A/\mathfrak{p})/x_1(A/\mathfrak{p})$. The dimension of M/x_1M is $d - 1$, so the homology modules are bounded in length by cn^{d-1} for some constant c. Now $M/x_1^n M$ has a filtration of length n with quotients that are isomorphic to M/x_1M, so the long exact sequences associated to this filtration now imply that the length of $H_i(x_1^n, \ldots, x_k^n; M)$ is bounded by $nc(n^{d-1}) = cn^d$. $\qquad \square$

From Proposition 5.2.5 we can use a dualizing complex to show the stronger result that there is a polynomial of degree $k - i$ that bounds the length of $H_i(x_1^n, \ldots, x_k^n; M)$.

Proposition 5.2.6 *Let $x_1, \ldots x_k$ be a sequence of elements of A such that $M/(x_1, \ldots, x_k)M$ has finite length, and let i be an integer between 0 and k. Then there exists a constant c such that the length of the homology module $H_i(x_1^n, \ldots, x_k^n; M)$ is bounded by cn^{k-i} for all n.*

Proof As in the proof of Proposition 5.2.5, we may assume that M is of the form $A/\mathfrak{p}$ for some prime ideal $\mathfrak{p}$; thus, we may assume that $M = A$ and that A is an integral domain.

The proof is a variant of the proof of Theorem 4.6.6. Let d be the dimension of A, and let $D^\bullet$ be a dualizing complex for A. Let $K_\bullet^n$ denote the Koszul complex $K_\bullet(x_1^n, \ldots, x_k^n)$. We consider the double complex $\mathrm{Hom}(K_\bullet^n, D^\bullet)$:

$$
\begin{array}{ccc}
0 & & 0 \\
\downarrow & & \downarrow \\
0 \to \mathrm{Hom}(K_0^n, D^0) \to \cdots \to \mathrm{Hom}(K_0^n, D^d) \to 0 \\
\downarrow & & \downarrow \\
0 \to \mathrm{Hom}(K_1^n, D^0) \to \cdots \to \mathrm{Hom}(K_1^n, D^d) \to 0 \\
\downarrow & & \downarrow \\
\vdots & & \vdots \\
\downarrow & & \downarrow \\
0 \to \mathrm{Hom}(K_k^n, D^0) \to \cdots \to \mathrm{Hom}(K_k^n, D^d) \to 0 \\
\downarrow & & \downarrow \\
0 & & 0
\end{array}
$$

Since for $i < d$, D^i is a sum of $A_\mathfrak{p}$-modules with $\mathfrak{p} \neq \mathfrak{m}$, and since $K_\bullet^n$ has homology of finite length, all of the columns of this double complex are exact except the one on the right, which is $\mathrm{Hom}(K_\bullet^n, E(k))$. The homology of the dth column in degree $d + i$ is $\text{Hom}(H_i(K_\bullet^n), E(k))$, which by Proposition 4.6.5 has the same length as $H_i(K_\bullet^n)$. By Proposition 3.2.3, we thus have that the homology of the total complex of $\text{Hom}(K_\bullet^n, D^\bullet)$ in degree $i + d$ has length equal to the length of $H_i(K_\bullet^n)$.

By Theorem 3.2.1 there exists a filtration of the homology of the total complex with quotients that are isomorphic to subquotients of $H^{2,1}(\text{Hom}(K_\bullet^n, D^\bullet))$. For each m, since K_m^n is a free module, the homology of the mth row in the $(m + j)$th position is $\text{Hom}(K_m^n, H^j(D^\bullet))$. Thus, the complex induced on homology in the jth column is the complex $\text{Hom}(K_\bullet^n, H^j(D^\bullet))$, which by Proposition 3.3.5 is isomorphic to the Koszul complex $K_\bullet^n \otimes H^j(D^\bullet)$. As n varies, Proposition 5.2.5 implies that the length of the homology of this column is bounded by a polynomial of degree equal to the dimension of $H^j(D^\bullet)$.

Putting these two computations together, we have a filtration of the module $\text{Hom}(H_i(K_\bullet^n), E(k))$ with quotients that are isomorphic to subquotients of $H_m(K_\bullet^n \otimes H^j(D^\bullet))$ for j and m such that $j + m = d + i$. The highest value of m for which $K_m \neq 0$ is $m = k$, so the lowest value of j for which there exists a j with $j + m = d + i$ and $H_m(K_\bullet^n \otimes H^j(D^\bullet)) \neq 0$ is $j = d + i - k$. The dimension of $H^j(D^\bullet)$ for $j \geq d + i - k$ is at most $k - i$. Thus, as n varies, the length of each of these homology modules $H_m(K_\bullet^n \otimes H^j(D^\bullet))$ is bounded by a polynomial of degree at most $k - i$ by Proposition 5.2.5. Hence the length of the homology $H_i(K_\bullet^n)$ is also bounded by a polynomial of degree $k - i$. $\square$

It follows from Proposition 5.2.6 that if $x_1, \ldots, x_d$ forms a system of parameters for A, then the limit of $\chi(x_1^n, \ldots, x_d^n)/n^d$ as n goes to infinity is equal to the limit of the length of $A/(x_1^n, \ldots, x_d^n)$ divided by n^d.

Proposition 5.2.7 *Let $\mathfrak{a}$ be an ideal generated by a system of parameters of A. Then*

$$\text{length}(A/\mathfrak{a}) \geq e(\mathfrak{a}).$$

Proof Let $\mathfrak{a}$ be generated by the system of parameters $x_1, \ldots, x_d$, where d is the dimension of A. Let $X_1, \ldots, X_n$ be indeterminates, and let ϕ be the homomorphism from the polynomial ring $A/\mathfrak{a}[X_1, \ldots, X_d]$ to the associated graded ring of $\mathfrak{a}$ that takes X_i to the image of x_i in degree 1. The multiplicity of $\mathfrak{a}$ is $(d - 1)!$ times the leading coefficient of the Hilbert polynomial of the associated graded ring of $\mathfrak{a}$. Since ϕ is surjective, this implies that $e(\mathfrak{a})$ is less than or equal to $(d - 1)!$ times the leading coefficient of the Hilbert polynomial of $A/\mathfrak{a}[X_1, \ldots, X_d]$. The leading coefficient of this polynomial is $\text{length}(A/\mathfrak{a})/(d - 1)!$, so thus we have $\text{length}(A/\mathfrak{a}) \geq e(\mathfrak{a})$. $\square$

We next prove a result of Lech [41] that states that the limit of the length of $A/(x_1^n, \ldots, x_d^n)$ divided by n^d is equal to the multiplicity of the ideal $(x_1, \ldots, x_d)$.

Theorem 5.2.8 (Lech's Lemma) *Let* $\mathfrak{a}$ *be the ideal generated by the elements* $x_1, \ldots x_d$, *and let M be a finitely generated module such that* $M/(x_1, \ldots x_d)M$ *has finite length. Then*

$$\lim_{n \to \infty} \frac{\text{length}\big(M/\big(x_1^n, \ldots, x_d^n\big)M\big)}{n^d} = (d!) \lim_{n \to \infty} \frac{\text{length}(M/\mathfrak{a}^n M)}{n^d}.$$

Proof The condition that $M/(x_1, \ldots x_d)M$ has finite length implies that the dimension of M is at most d. Both sides are additive on short exact sequences of modules of dimension at most d, the left-hand side by Proposition 2.3.4, and the right-hand side by Proposition 5.2.6 together with the fact that $\chi(x_1, \ldots x_d; M)$ is additive on short exact sequences. It also follows that both sides are zero if the dimension of M is less than d. Hence we can assume that $M = A$, where A is a local ring of dimension d, so that $x_1, \ldots, x_d$ is a system of parameters for A.

We also know from the above results and those of Chapter 2 that both limits exist.

Fix n, and let $\mathfrak{b} = (x_1^n, \ldots, x_d^n)$. It follows from Proposition 5.2.7 that $e(\mathfrak{b}) \leq \text{length}(A/\mathfrak{b})$. Furthermore, since for all $x \in \mathfrak{a}^n$, we have $x^n \in \mathfrak{b}^n$, $\mathfrak{a}^n$ is integral over $\mathfrak{b}$, and thus, by Proposition 2.4.4 $e(\mathfrak{a}^n) = e(\mathfrak{b})$, so we have $e(\mathfrak{a}^n) \leq \text{length}(A/\mathfrak{b})$. Dividing by n^d and expressing $e(\mathfrak{a}^n)$ as a limit, we have

$$\frac{\text{length}\big(A/\big(x_1^n, \ldots, x_d^n\big)\big)}{n^d} \geq \frac{d!}{n^d} \lim_{m \to \infty} \frac{\text{length}(A/(\mathfrak{a}^n)^m)}{m^d}$$

$$= d! \lim_{m \to \infty} \frac{\text{length}(A/\mathfrak{a}^{mn})}{(mn)^d} = e(\mathfrak{a}).$$

Taking the limit as $n \to \infty$, we thus have

$$\lim_{n \to \infty} \frac{\text{length}\big(A/\big(x_1^n, \ldots, x_d^n\big)\big)}{n^d} \geq e(\mathfrak{a}).$$

To prove the opposite inequality, we let $G_\mathfrak{a}$ be the associated graded ring of $\mathfrak{a}$. Let $\overline{x}_i$ denote the image of x_i in $G_\mathfrak{a}$ in degree 1. Let $\overline{\mathfrak{b}}$ be the ideal of $G_\mathfrak{a}$ generated by $\overline{x}_1^n, \ldots, \overline{x}_d^n$, and let $\mathfrak{b} = (x_1^n, \ldots, x_d^n)$ as above. We claim that $\text{length}(A/\mathfrak{b}) \leq \text{length}(G_\mathfrak{a}/\overline{\mathfrak{b}})$. To see this, consider the natural map $\phi_k : (G_\mathfrak{a})_k = \mathfrak{a}^k/\mathfrak{a}^{k+1} \to (\mathfrak{a}^k + \mathfrak{b})/(\mathfrak{a}^{k+1} + \mathfrak{b})$ for each k. Denote the component of $\overline{\mathfrak{b}}$ of degree k by $\overline{\mathfrak{b}}_k$. If $\overline{x} \in \overline{\mathfrak{b}}_k$, then $\overline{x}$ can be written as $\sum \overline{a}_i \overline{x}_i^n$, where $\overline{a}_i \in \mathfrak{a}^{n-k}$. Then, lifting the elements back to A, we have that $x - \sum a_i x_i^n \in \mathfrak{a}^{k+1}$. Since $\sum a_i x_i^n \in \mathfrak{b}$, we thus have $\phi_k(x) = 0$. Hence we have a surjective map from $(G_\mathfrak{a})_k/\overline{\mathfrak{b}}_k$ to $(\mathfrak{a}^k + \mathfrak{b})/(\mathfrak{a}^{k+1} + \mathfrak{b})$ for each k, and since the latter quotients define a filtration of $A/\mathfrak{b}$, we can conclude that $\text{length}(G_\mathfrak{a})/\overline{\mathfrak{b}} = \sum_k \text{length}(G_\mathfrak{a})_k/\overline{\mathfrak{b}}_k \geq \text{length}(A/\mathfrak{b})$.

We now let n vary, and denote $\overline{\mathfrak{b}}$ by $\overline{\mathfrak{b}}_n$. To conclude the proof it suffices to show that we have

$$\lim_{n\to\infty} \frac{\text{length}(G_\mathfrak{a}/\overline{\mathfrak{b}}_n)}{n^d} \leq e(\mathfrak{a}).$$

We next approximate the length of $(G_\mathfrak{a}/\overline{\mathfrak{b}}_n)$ by the Euler characteristic of the Koszul complex on $\overline{x}_1^n, \ldots, \overline{x}_d^n$. By Proposition 5.2.6, the length of the homology in each degree greater than zero is bounded by a polynomial of degree at most $d-1$, and hence when divided by n^d it goes to zero in the limit (while Theorem 2.1.5 was proven for local rather than graded rings, it applies in this case by localizing at the unique graded maximal ideal of $G_\mathfrak{a}$). Hence it suffices to show that

$$\lim_{n\to\infty} \frac{\chi\left(\overline{x}_1^n, \ldots, \overline{x}_d^n\right)}{n^d} \leq e(\mathfrak{a}). \qquad (*)$$

By Proposition 5.2.2, since $\overline{x}_i^n$ has degree n for each i, we have

$$\chi\left(\overline{x}_1^n, \ldots, \overline{x}_d^n\right) = n^d e$$

where the leading coefficient of the Hilbert polynomial of $G_\mathfrak{a}$ is $e/(d-1)!$. Since $G_\mathfrak{a}$ is the associated graded ring of $\mathfrak{a}$, the leading coefficient of the Hilbert polynomial of $G_\mathfrak{a}$ is $e(\mathfrak{a})/(d-1)!$. Thus, the inequality $(*)$ is actually an equality, so this completes the proof. $\square$

Thus, we have shown that the Euler characteristic of the Koszul complex gives the Samuel multiplicity for ideals generated by systems of parameters. An immediate consequence of this theorem is the following.

Corollary 5.2.9 *Let $x_1, \ldots, x_d$ be a sequence of elements in the maximal ideal of A, and let M be a finitely generated A-module such that $M/(x_1, \ldots, x_d)M$ has finite length. Then*

(i) $\dim(M) \leq d$.
(ii) *If* $\dim(M) < d$, *then* $\chi(x_1, \ldots, x_d; M) = 0$.
(iii) *If* $\dim(M) = d$, *then* $\chi(x_1, \ldots, x_d; M) > 0$.

Proof The first statement is a consequence of the definition of dimension in terms of systems of parameters discussed in Chapter 2. The second and third follow from the representation of $\chi(x_1, \ldots, x_d; M)$ as the coefficient of the term of degree d of the Hilbert-Samuel polynomial in Theorem 5.2.8. $\square$

Another corollary is a characterization of the rank of a module over an integral domain; this result will be used in later chapters.

Corollary 5.2.10 *Let A be a local domain, and let $x_1, \ldots, x_d$ be a system of parameters for A. Then, if M is any finitely generated A-module, we have*

$$\text{rank}(M) = \frac{\chi(x_1, \ldots, x_d; M)}{\chi(x_1, \ldots, x_d; A)}.$$

Proof This corollary follows from Corollary 5.2.9 together with the fact that M has a filtration by submodules with rank(M) quotients isomorphic to A and all other quotients of dimension less than the dimension of A. □

We next show that the Euler characteristic of $K_\bullet(x_1, \ldots, x_d)$ is also equal to the result of intersection with the principal divisors defined by the elements $x_1, \ldots, x_d$. We begin by recalling a Proposition from Chapter 1 that relates the action of the divisor (x) to the result of dividing by x in the case of a cycle defined by a module. Let M be a module of dimension d, and let x be an element that is contained in no minimal prime ideal in the support of M of dimension d. Then Proposition 1.2.2 states that

$$(x) \cap [M]_d = [M/xM]_{d-1} - [_xM]_{d-1},$$

where $_xM$ is the set of elements of M annihilated by x.

Proposition 5.2.11 *Let A be a local or graded ring, let M be a module of dimension d, and let $x_1, \ldots, x_k$ be elements of the maximal ideal of A (or elements of positive degree in the graded case) such that $M/(x_1, \ldots x_k)$ has dimension $d - k$. Then*

$$(x_1) \cap \cdots \cap (x_k) \cap [M]_d = \sum_{i=0}^{k} (-1)^i [H_i(x_1, \ldots x_k; M)]_{d-k}.$$

Proof We use induction on k; the case $k = 1$ is Proposition 1.2.2.

Assume that $k > 1$ and that the result holds for $k - 1$; we apply the induction hypothesis to $x_2, \ldots, x_k$. First, $M/(x_2, \ldots, x_k)$ must have dimension $d-k+1$, since if its dimension were greater than $d - k + 1$, $M/(x_1, \ldots x_k)$ would have dimension greater than $d - k$. Hence the hypothesis of the theorem holds with k replaced by $k - 1$, so by induction we have

$$(x_2) \cap \cdots \cap (x_k) \cap [M]_d = \sum_{i=0}^{k-1} (-1)^i [H_i(x_2, \ldots x_k; M)]_{d-k+1}.$$

Now, denoting $H_i(x_2, \ldots x_k; M)$ and $H_i(x_1, \ldots x_k; M)$ by $H_i(k-1)$ and $H_i(k)$, respectively, we have a long exact sequence

$$\cdots H_i(k-1) \xrightarrow{x_1} H_i(k-1) \to H_i(k) \to H_{i-1}(k-1) \xrightarrow{x_1} H_{i-1}(k-1) \cdots$$

and thus, for each i we have a short exact sequence

$$0 \to H_i(k-1)/x_1 H_i(k-1) \to H_i(k) \to {}_{x_1}H_{i-1}(k-1) \to 0.$$

Since all of these modules have dimension at most $d-k$, localizing at prime ideals of dimension $d-k$ gives

$$[H_i(k)]_{d-k} = [H_i(k-1)/x_1 H_i(k-1)]_{d-k} + [{}_{x_1}H_{i-1}(k-1)]_{d-k}.$$

Thus, we have

$$\sum_{i=0}^{k}(-1)^i[H_i(k)]_{d-k}$$

$$= \sum_{i=0}^{k}(-1)^i([H_i(k-1)/x_1 H_i(k-1)]_{d-k} + [{}_{x_1}H_{i-1}(k-1)]_{d-k})$$

$$= \sum_{i=0}^{k}(-1)^i([H_i(k-1)/x_1 H_i(k-1)]_{d-k} - [{}_{x_1}H_i(k-1)]_{d-k})$$

$$= \sum_{i=0}^{k-1}(-1)^i((x_1) \cap [H_i(k-1)]_{d-k}.$$

By induction, we thus have

$$\sum_{i=0}^{k}(-1)^i[H_i(k)]_{d-k} = (x_1) \cap \left(\sum_{i=0}^{k-1}(-1)^i[H_i(k-1)]_{d-k}\right)$$

$$= (x_1) \cap (x_2) \cap \cdots \cap (x_k) \cap [M]_d.$$

□

We summarize these results in the following Theorem.

Theorem 5.2.12 *Let M be a finitely generated A-module of dimension d, and let $x_1, \ldots, x_d$ be a system of parameters for M. Let $\mathfrak{a}$ be the ideal generated by $x_1, \ldots, x_d$. Then the following are equal:*

  (i) *The Euler characteristic $\chi(x_1, \ldots, x_d; M)$.*
  (ii) *The Samuel multiplicity $e(\mathfrak{a}, M)$.*
  (iii) *The coefficient of $[A/\mathfrak{m}]$ in the intersection product $(x_1) \cap (x_2) \cap \cdots \cap (x_d) \cap [M]_d$ in $A_0(\mathrm{Spec}(A/\mathfrak{m}))$.*

Another consequence of Proposition 5.2.11 is that intersection with divisors commutes on the level of cycles if the codimension is correct.

Corollary 5.2.13 *If x and y are elements of A, and $\mathfrak{p}$ is a prime ideal of A such that the dimension of $A/\mathfrak{p}/(x, y)A/\mathfrak{p}$ is two less than the dimension of $A/\mathfrak{p}$, then $(x) \cap (y) \cap [A/\mathfrak{p}] = (y) \cap (x) \cap [A/\mathfrak{p}]$.*

Proof Both sides of the equation are the cycles defined by the Koszul complex on x and y as in Proposition 5.2.11. $\qquad\qquad\qquad\qquad\qquad\qquad\qquad$ □

5.3 Serre Intersection Multiplicity

In the first section we showed how to define an intersection operator for a divisor, and in the second section we described how this definition could be extended to an ideal generated by a system of parameters. The definition in the latter case was through Euler characteristics, and Theorem 5.2.12 shows that this gives a reasonable extension of the definition of intersection with divisors.

Serre [66] generalized the definition in terms of Euler characteristics to arbitrary ideals in a regular local ring. If $\mathfrak{a}$ and $\mathfrak{b}$ are ideals of a regular local ring A such that $\mathfrak{a} + \mathfrak{b}$ is primary to the maximal ideal, he defined

$$\chi(A/\mathfrak{a}, A/\mathfrak{b}) = \sum_{i=0}^{d} (-1)^i \, \text{length}\big(\text{Tor}_i^A(A/\mathfrak{a}, A/\mathfrak{b})\big).$$

This definition can be extended to modules; if M and N are finitely generated modules such that $M \otimes_A N$ is a module of finite length, then we let

$$\chi(M, N) = \sum_{i=0}^{d} (-1)^i \, \text{length}\big(\text{Tor}_i^A(M, N)\big).$$

If $\mathfrak{a}$ is an ideal generated by part of a system of parameters, so that $A/\mathfrak{a}$ is a complete intersection, and if $\mathfrak{b}$ is an ideal such that $\dim(A/\mathfrak{a}) + \dim(A/\mathfrak{b}) = \dim(A)$, then a free resolution for $A/\mathfrak{a}$ over A is a Koszul complex on a set of elements $x_1, \ldots, x_r$. Hence we have

$$\chi(A/\mathfrak{a}, A/\mathfrak{b}) = \chi(K_\bullet(x_1, \ldots, x_r; A/\mathfrak{b}))$$

and, as we discussed above, this definition is a natural generalization of intersection with divisors. We will also show in Chapter 8 that Bézout's theorem holds if this definition of intersection multiplicity is used. However, there are other properties that are not obvious, for example, that the intersection multiplicity is nonnegative. We discuss these questions further in the next chapter.

Exercises

5.1 Let $A = k[X, Y, Z]$, and let η be the cycle $[A]$. Compute the cycles $(XY) \cap (XZ) \cap \eta$ and $(XZ) \cap (XY) \cap \eta$ and show that they are not equal.

5.2 Let $A = k[X, Y]/(XY, X^2)$. Compute $H_1(K_\bullet(Y^n))$ for all n.

5.3 Let $A = k[[X, Y]]$, and let $\mathfrak{a}$ be the ideal (X^2, XY, Y^2). Show that the multiplicity of $\mathfrak{a}$ is greater than the length of $A/\mathfrak{a}$.

5.4 With A and $\mathfrak{a}$ as in the previous problem, show that the multiplicity of $\mathfrak{a}$ is not equal to the limit

$$\lim_{n \to \infty} \frac{\text{length}(A/((X^2)^n, (XY)^n, (Y^2)^n)}{n^2}.$$

5.5 Let $E_\bullet$ be a bounded complex of finitely generated A-modules such that the dimension of $H_i(E_\bullet)$ is less than d for all i, and let $x_1, \ldots, x_d$ be elements of A such that the quotient module $H_i(E_\bullet)/(x_1, \ldots, x_d)H_i(E_\bullet)$ has finite length for all i. Show that

(a) The complex $E_\bullet \otimes K_\bullet(x_1, \ldots, x_d)$ has homology of finite length.

(b) $\chi(E_\bullet \otimes K_\bullet(x_1, \ldots, x_d)) = 0$.

6

The Homological Conjectures

In this chapter we discuss several conjectures that arose from the theory of intersection multiplicities and that have come to be known as the homological conjectures. Some of these questions came directly from questions on Serre's intersection multiplicity, while others concern related problems on systems of parameters, modules of finite projective dimension, and properties of complexes.

Many of these conjectures can be verified directly in low dimension (usually dimension at most two). For proving them in higher dimension, three basic methods have been used. The first method, which goes back to the time before some of these questions were explicitly stated, is to use Cohen-Macaulay properties to show that certain homology groups must vanish. The second method is the application of the Frobenius map for rings of positive characteristic, which was introduced to this subject by Peskine and Szpiro in [50]. In addition to proving some of the conjectures in positive characteristic, they introduced a method of reduction from characteristic zero to positive characteristic; this method was completed by Hochster [23] to settle the characteristic zero case of several of these questions. The Frobenius methods will be discussed in the next chapter.

The third method is the most recent and is the use of local Chern characters, which can be used to prove a few of these conjectures in the most difficult case, that of mixed characteristic. We discuss the theory of local Chern characters at length in the later chapters of the book, and we will return to some of the questions discussed in this chapter in Chapter 13.

6.1 Conjectures on Intersection Multiplicities

In the last chapter we presented Serre's definition of intersection multiplicities. In this section we discuss several questions about multiplicities defined in this way.

Let A be a regular local ring of dimension d, and let M and N be finitely generated A-modules such that $M \otimes_A N$ has finite length. We defined the intersection multiplicity to be

$$\chi(M, N) = \sum_{i=0}^{d} (-1)^i \, \text{length}\big(\text{Tor}_i^A(M, N)\big).$$

The main questions about $\chi(M, N)$ concern whether it can be negative and the precise conditions for it to vanish. One reason to ask such questions, especially about vanishing, comes from an attempt to use this definition to define intersection multiplicities of cycles in the Chow group. Let M and N have dimension m and n, respectively. We have defined cycles associated to M and N by letting

$$[M]_m = \sum_{\dim(A/\mathfrak{p})=m} \text{length}(M_{\mathfrak{p}})[A/\mathfrak{p}]$$

and similarly for N. We wish to define an intersection product, which we denote $\cap$, on these cycles by letting

$$[M]_m \cap [N]_n = \chi(M, N)[A/\mathfrak{m}]$$

where $\mathfrak{m}$ is the maximal ideal of A. To make such a definition it is necessary that the intersection product defined by a cycle be independent of the module used to represent the cycle. That is, if M and M' are modules of dimension at most m such that $[M]_m = [M']_m$, we want to know that the intersection product defined using M coincides with that using M'. An obvious necessary condition for this to be true is that the above pairing vanishes whenever the dimension of M is less than m.

Serre [66] conjectured that the following should be true.

Let A be a regular local ring, and let M and N be finitely generated A-modules. Suppose that $M \otimes_A N$ is a module of finite length. Then

(i) $\dim(M) + \dim(N) \leq \dim(A)$.
(ii) $\chi(M, N) \geq 0$.
(iii) $\chi(M, N) \neq 0$ if and only if $\dim(M) + \dim(N) = \dim(A)$.

It is easy to see that the second and third conditions can be replaced by

(i) (Vanishing) If $\dim(M) + \dim(N) < \dim(A)$, then $\chi(M, N) = 0$.
(ii) (Positivity) If $\dim(M) + \dim(N) = \dim(A)$, then $\chi(M, N) > 0$.

While we have stated these as conjectures, in fact Serre proved the first one in general and the others in many cases. All three were proven in the case of equal characteristic using the technique of "reduction to the diagonal," which we describe briefly; the details can be found in Serre [66, Chapter V].

By completing and using the Cohen structure theorems (see [44, §29]), we may assume that the ring A is a power series ring $k[[X_1, \ldots X_d]]$ over a field k. Let M and N be A-modules of dimension m and n, respectively. We take another set of variables $Y_1, \ldots Y_d$; since $k[[X_1, \ldots X_d]]$ is isomorphic to $k[[Y_1, \ldots Y_d]]$, we may consider N as a $k[[Y_1, \ldots Y_d]]$-module. Serre then defines the complete tensor product over k, which is obtained by completing the usual tensor product, and which is denoted $M \hat{\otimes}_k N$. The module $M \hat{\otimes}_k N$ is a module over the ring

$$k[[X_1, \ldots X_d]] \hat{\otimes}_k k[[Y_1, \ldots Y_d]] \cong k[[X_1, \ldots X_d, Y_1, \ldots Y_d]],$$

and it has the property that

$$\dim(M \hat{\otimes}_k N) = \dim(M) + \dim(N).$$

The point of this construction is that now the tensor product of M and N over A can be obtained by setting $X_i = Y_i$, or, more precisely, by dividing by the ideal generated by $X_1 - Y_1, \ldots, X_d - Y_d$. That is, denoting $k[[X_1, \ldots X_d, Y_1, \ldots Y_d]]$ by S, we have

$$M \otimes_A N \cong (M \hat{\otimes}_k N) \otimes_S (S/(X_1 - Y_1, \ldots X_d - Y_d)).$$

In fact, the following theorem is true.

Theorem 6.1.1 *Let $A = k[[X_1, \ldots, X_d]]$, let M and N be A-modules, and let $S = k[[X_1, \ldots X_d, Y_1, \ldots Y_d]]$ as above. Then for all integers i we have*

$$\mathrm{Tor}_i^A(M, N) \cong \mathrm{Tor}_i^S(M \hat{\otimes}_k N, S/(X_1 - Y_1, \ldots, X_d - Y_d)).$$

For the proof we refer again to Serre [66, Chapter V]. Since a free resolution of $S/(X_1 - Y_1, \ldots, X_d - Y_d)$ over S is a Koszul complex, the above theorems now follow from the corresponding properties of the Koszul complex proven in Corollary 5.2.9.

In addition, Serre [66, Appendix II] showed that in the equal characteristic case, the partial Euler characteristics defined by two modules M and N are positive. Let M and N be two A-modules as above, and for each integer $i \geq 0$, let $\chi_i(M, N) = \sum_{j=0}^d (-1)^j \, \mathrm{length}(\mathrm{Tor}_{i+j}(M, N))$. Serre proved, again using the reduction to the Koszul complex, that $\chi_i(M, N) > 0$ for all $i < k$, where k is the smallest integer for which $\mathrm{Tor}_k(M, N) = 0$, and that $\mathrm{Tor}_m(M, N) = 0$ for all $m \geq k$. Although we do not prove this here, we point out one special case that follows from the results proven in Chapter 3. We showed in Theorem 3.3.4 that if $H_1(K_\bullet(x_1, \ldots, x_n; M)) = 0$, then $x_1, \ldots, x_n$ form a regular sequence on M, so that $H_i(K_\bullet(x_1, \ldots, x_n; M)) = 0$ for all $i > 0$. Thus, in the equicharacteristic regular case, if $\mathrm{Tor}_1(M, N) = 0$, then $\mathrm{Tor}_i(M, N) = 0$ for all $i > 0$. This property is known as the rigidity property of Tor. The rigidity property of Tor was conjectured to hold, first, for arbitrary regular local rings (not necessarily of equal characteristic), and, second, for all local rings if one of the modules was assumed to have finite projective dimension. Auslander [2] proved many of these statements for unramified regular local rings and pointed out that many of the consequences of regularity followed from this conjecture. The rigidity conjecture was proven by Lichtenbaum [42] for arbitrary regular local rings. The case in which A is not regular but M has finite projective dimension has recently been shown to be false by Heitmann [22].

As mentioned above, the first of the conjectures on intersection multiplicities was proven by Serre in the regular case and is easily seen to be false for general modules in the nonregular case (an example is given in the exercises). It is not known whether it holds if one of the modules is assumed to have finite projective dimension over an arbitrary local ring. The vanishing conjecture has been proven for arbitrary regular local rings using local Chern characters (Roberts [59, 60]) and independently using Adams operations (Gillet and Soulé [18, 19]). The proof using local Chern characters will be discussed in Chapter 13. The fact that $\chi(M, N) \geq 0$ has been proven recently by Gabber using de Jong's theorem on the existence of regular alterations; Gabber's result can be found in Berthelot [6], and de Jong's theorem can be found in de Jong [10]. The positivity conjecture is still open.

There have also been attempts to generalize these conjectures to the situation in which A is not assumed to be regular but M is assumed to have finite projective dimension. The vanishing conjecture has been shown to be false in this generality by an example of Dutta, Hochster, and MacLaughlin [12]. We discuss this example and its consequences for the general theory in Chapter 13.

6.2 The Peskine-Szpiro Intersection Theorem

The intersection theorem of Peskine and Szpiro [49, 50] arose from the problems on intersection multiplicities and related questions discussed in the previous section. In the second of the works cited above they list six conjectures. These include the first of Serre's multiplicity conjectures and the rigidity conjecture for the case in which one module has finite projective dimension as well as the intersection theorem that we discuss later in this section. Before proceeding with the discussion of the intersection theorem, we state the other three conjectures in their list. Let M be a module of finite projective dimension over a local ring A, and let N be a finitely generated module such that $M \otimes_A N$ has finite length. We define the *grade* of M to be the maximum length of a regular sequence on A contained in the annihilator of M.

(i) If $x_1, \ldots, x_k$ is a regular sequence on M, then $x_1, \ldots, x_k$ is a regular sequence on A.

(ii) We have $\dim(N) \le \text{grade}(M)$.

(iii) (Grade Conjecture) We have $\dim(M) + \text{grade}(M) = \dim(A)$.

Little is known about the last two of these conjectures except in the case of graded modules over a graded ring, which we discuss in Section 3. The first is a consequence of the intersection theorem, which we state next.

Theorem 6.2.1 (Peskine-Szpiro Intersection Theorem) *Let A be a local ring, and let M and N be modules such that $M \otimes_A N$ has finite length. Then*

$$\dim(N) \le \text{proj.dim.}(M).$$

The sense in which this theorem is a statement about intersections comes from the fact that the support of $M \otimes N$ is the intersection of the supports of M and N (Proposition 2.3.2). The first statement of Serre's conjectures (which is a theorem for regular rings) is the inequality

$$\dim(N) \le \dim(A) - \dim(M).$$

By Theorem 4.4.3, the projective dimension of M is equal to $\text{depth}(A) - \text{depth}(M)$, and thus an equivalent version of the intersection theorem states that, under the given hypotheses, we have

$$\dim(N) \le \text{depth}(A) - \text{depth}(M).$$

Thus, this theorem is analogous to Serre's theorem on the dimensions of M and N; however, while the depth is related to the dimension, there is no general inequality between these two differences, and the statements are not in any sense equivalent.

This theorem was proven for rings of positive characteristic or essentially of finite type over a field by Peskine and Szpiro [49, 50] and extended by Hochster [23] to the general equicharacteristic case. It was proven by Roberts [61, 62] for the mixed characteristic case. We will prove this theorem for equicharacteristic rings in the next chapter and in mixed characteristic in Chapter 13.

In addition to the relation with the previous conjectures shown above, the intersection theorem is important because it has several other consequences. Before discussing these, we state a more general version of the basic theorem.

Theorem 6.2.2 (New Intersection Theorem) *Let*

$$F_\bullet = 0 \to F_k \to \cdots \to F_0 \to 0$$

be a nonexact bounded free complex with homology of finite length. Then

$$k \geq \dim(A).$$

To see that this version of the intersection theorem implies the previous one, first take a filtration of N with quotients of the form $A/\mathfrak{p}$, and choose one such $\mathfrak{p}$ such that $\dim(N) = \dim(A/\mathfrak{p})$. The hypothesis that $M \otimes N$ has finite length implies that $M \otimes A/\mathfrak{p}$ has finite length. Now take a minimal free resolution $F_\bullet$ of M; the length of the complex $F_\bullet$, which we denote k, is the projective dimension of M. Our hypotheses imply that $F_\bullet \otimes A/\mathfrak{p}$ is a nonexact bounded free complex over $A/\mathfrak{p}$ of length k with homology of finite length. Thus, the new intersection theorem states that

$$\text{proj.dim.}(M) = k \geq \dim(A/\mathfrak{p}) = \dim(N)$$

as was to be shown.

We show that the intersection theorem implies Auslander's zero-divisor conjecture, which is (i) in the list of conjectures at the beginning of this section.

Theorem 6.2.3 *Suppose that the intersection theorem holds. Let M be a nonzero A-module of finite projective dimension. If x is not a zero-divisor on M, then x is not a zero-divisor on A.*

Proof Another way of stating the conclusion of the theorem is to say that every associated prime of A is contained in an associated prime of M. We prove this by induction on the dimension of M; if M has dimension zero, the maximal ideal is an associated prime of M, so the theorem is clear in that case.

Suppose M has dimension greater than zero, and let $\mathfrak{p}$ be an associated prime of A. If there exists a prime ideal $\mathfrak{q}$ other than the maximal ideal such that $\mathfrak{q} \supseteq \mathfrak{p}$ and $\mathfrak{q}$ is in the support of M, then we can localize at $\mathfrak{q}$ and, since $\dim(M_\mathfrak{q}) < \dim(M)$, conclude that there is an associated prime of $M_\mathfrak{q}$ that contains the localization of $\mathfrak{p}$ at $\mathfrak{q}$. Such an associated prime of $M_\mathfrak{q}$ must be the localization of an associated prime of M that contains $\mathfrak{p}$, so the theorem is true also in this case.

In the remaining case the only prime ideal in the support of M that contains $\mathfrak{p}$ is the maximal ideal $\mathfrak{m}$, so that $M \otimes A/\mathfrak{p}$ has finite length. The intersection theorem implies that the dimension of $A/\mathfrak{p}$ is less than or equal to the projective dimension of M. Since $\mathfrak{p}$ is an associated prime of A, it follows from Corollary 4.1.9 that we have $\text{depth}(A) \leq \dim(A/\mathfrak{p})$. Thus, the depth of A is less than or equal to the projective dimension of M. The Auslander-Buchsbaum theorem (Theorem 4.4.3) states that we have

$$\text{proj.dim.}(M) + \text{depth}(M) = \text{depth}(A).$$

Thus, we must have that the depth of A is equal to the projective dimension of M and the depth of M is zero. Hence $\mathfrak{m}$ is an associated prime of M, so there exists an associated prime of M contained in $\mathfrak{p}$ as was to be shown. $\square$

Another consequence of the intersection theorem is the following.

Proposition 6.2.4 *Again assume that the intersection theorem holds. If there exists an A-module M of finite length and finite projective dimension, then A is Cohen-Macaulay.*

Proof Applying the intersection theorem to the module M and the module $N = A$, we obtain the inequality $\dim(A) \leq \operatorname{proj.dim}(M)$. But we also know from Theorem 4.4.3 that $\operatorname{proj.dim}(M) \leq \operatorname{depth}(A) \leq \dim(A)$. Thus, these numbers must all be equal, and A is Cohen-Macaulay. $\square$

Peskine and Szpiro [50] also proved that the intersection theorem implies a conjecture of Bass, which states that if there exists a finitely generated A-module of finite injective dimension, then A is Cohen-Macaulay.

A longer list of homological conjectures, which includes the ones listed so far, was given by Hochster [24]. We state three of these conjectures here.

A stronger version of the new intersection theorem is as follows.

Conjecture 6.2.5 (Improved New Intersection Conjecture) *Let*

$$F_\bullet = 0 \to F_k \to \cdots \to F_0 \to 0$$

be a nonexact complex of free A-modules with homology of finite length except possibly for $H_0(F_\bullet)$. Assume there is a minimal generator of $H_0(F_\bullet)$ annihilated by a power of the maximal ideal of A. Then

$$k \geq \dim(A).$$

This conjecture was introduced by Evans and Griffith [15] as a lemma in the proof of their Syzygy theorem.

Conjecture 6.2.6 (Monomial Conjecture) *Let $x_1, \ldots, x_d$ be a system of parameters for a local ring A. Then for all integers $t > 0$, $x_1^t x_2^t \cdots x_d^t$ is not in the ideal generated by $x_1^{t+1}, \ldots, x_d^{t+1}$.*

Conjecture 6.2.7 (Canonical Element Conjecture) *Let $x_1, \ldots x_d$ be a system of parameters for a local ring A. Let $K_\bullet$ be the Koszul complex on $x_1, \ldots, x_d$, let $F_\bullet$ be a minimal free resolution of $A/(x_1, \ldots, x_d)$, and let $f_\bullet$ be a map of* *complexes from* $K_\bullet$ *to* $F_\bullet$ *that lifts the identity in degrees 0 and 1 (such a map exists by Proposition 3.1.4 since* $K_\bullet$ *is a free complex and* $F_\bullet$ *is exact). Then* $f_d(K_d) \not\subseteq \mathfrak{m}F_d$.

These three conjectures are known to be equivalent; see Hochster [25] and Dutta [11]. They can be proven for equicharacteristic rings (as we discuss in Chapter 7) and in certain special cases (see Koh [37]). An analytic proof can be found in Roberts [58].

6.3 The Multiplicity Conjectures for Graded Modules

In this section we prove that the strong results on multiplicities and the grade conjecture are true for graded modules over $\mathbb{Z}$-graded rings A such that the subring of elements of degree zero is Artinian. When we discuss projective schemes in Chapter 8, we will show that these results also imply Bézout's theorem for Serre's intersection multiplicity, which states that the sum of intersection multiplicities of two subschemes of projective space that meet at a finite number of points is equal to the product of degrees of the subschemes. The results of this section are taken from Peskine and Szpiro [51].

The reason one can prove strong results in the graded case is that free resolutions of graded modules can be constructed by using graded free modules. By a *graded free module* we mean a direct sum of modules of the form $A[k]$, where, as defined in Chapter 2, we denote by $A[k]$ the module A with degrees shifted by k, so that $A[k]_n = A_{n+k}$ for all n. Let A be a graded ring such that A_0 is Artinian. If we denote the Hilbert polynomial of a graded module M by P_M as in Chapter 2, we have $P_{A[k]}(n) = P_A(n+k)$. As in Proposition 5.2.2 from the last chapter, the Hilbert polynomial that we use in this section is the polynomial for which $P_M(n) = \text{length}(\oplus_{k \leq n} M_k)$ for large n. In particular, the degree of the Hilbert polynomial is the dimension of M. If M is a finitely generated graded module, then by choosing homogeneous generators for M, we can find a map from a direct sum of copies of $A[n_{0j}]$ onto M for various integers n_{0j}. Continuing to construct a free resolution in this way, we obtain a resolution $F_\bullet \to M$ in which F_i is a direct sum of copies of $A[n_{ij}]$. Many properties of M can be computed from the integers n_{ij}, as we show below.

First we expand $P_{A[k]}(n) = P_A(n+k)$ by Taylor's theorem to obtain

$$P_{A[k]}(n) = P_A(n) + kP'_A(n) + \cdots + (k^i/i!)P_A^{(i)}(n) + \cdots,$$

where $P_A^{(i)}$ denotes the ith derivative of P_A. Since P_A is a polynomial, this sum is finite.

We now let $F_\bullet$ be a bounded complex of graded free modules. Let $F_i = \oplus_j A[n_{ij}]$ for each i. For each $k = 0, 1, \ldots$ we let

$$c_k = \frac{1}{k!} \sum_{i,j} (-1)^i n_{ij}^k. \qquad (*)$$

Note that the homology modules of $F_\bullet$ are graded and thus have associated Hilbert polynomials.

Theorem 6.3.1 *Let $F_\bullet$ be a bounded complex of graded free modules $F_i = \oplus_j A[n_{ij}]$, and let c_k be defined as above. For each i, let $P_{H_i(F_\bullet)}$ be the Hilbert polynomial of the homology $H_i(F_\bullet)$. We then have*

$$\sum_i (-1)^i P_{H_i(F_\bullet)} = \sum_k c_k P_A^{(k)}.$$

Proof The additivity of Hilbert polynomials with respect to exact sequences implies that for all n we have the equality

$$\sum_i (-1)^i P_{H_i(F_\bullet)}(n) = \sum_i (-1)^i P_{F_i}(n).$$

Using the decomposition of F_i into a sum of modules $A[n_{ij}]$ and applying equation $(*)$ together with the expansion by Taylor's theorem we obtain

$$\sum_i (-1)^i P_{F_i}(n) = \sum_{i,j} (-1)^i P_A(n + n_{ij}) = \sum_{i,j,k} (-1)^i P_A^{(k)}(n) \left(\frac{n_{ij}^k}{k!} \right)$$

$$= \sum_k \frac{P_A^{(k)}(n)}{k!} \left(\sum_{i,j} (-1)^i n_{ij}^k \right) = \sum_k c_k P_A^{(k)}(n).$$

This proves the theorem. $\square$

There are several important consequences of this formula. We first consider the case in which $F_\bullet$ is a free resolution of a graded module M. In this case the alternating sum of Hilbert polynomials of the homology is simply the Hilbert polynomial of M. Recall that the degree of the Hilbert polynomial and its leading coefficient are two basic invariants of the module. The degree of the polynomial P_M is equal to the dimension of M, and the leading coefficient determines intersection multiplicities. If the Hilbert polynomial has degree d, we refer to $d!$ times the leading coefficient of P_M as the degree of the graded module M. Since the polynomials $P_A^{(k)}$ are polynomials of different degrees, both of these invariants are determined by the values of the c_k together with the leading coefficient of P_A. Let k_0 be the smallest integer k such that $c_k \neq 0$, and let d be the dimension of A, so that d is the degree of P_A. It then follows from the formula that the degree of P_M is $d - k_0$. Furthermore, let e be the degree of the graded module A, so that the leading coefficient of P_A is $e/d!$. Then the leading coefficient of $P_A^{(k_0)}$ is $e/(d - k_0)!$, so it now follows from the formula of Theorem 6.3.1 that the degree of the module M is $c_{k_0}e$.

The second important fact about the formula in Theorem 6.3.1 is that it remains valid after tensoring with another graded module N if we replace P_A by P_N. If we replace $F_\bullet$ by the complex $F_\bullet \otimes_A N$, the module in degree i is now a direct sum of the modules $N[n_{ij}]$ and, using the fact that the homology of $F_\bullet \otimes N$ in degree i is the graded module $\text{Tor}_i(M, N)$, the proof of Theorem 6.3.1 shows that we have the formula

$$\sum_i (-1)^i P_{\text{Tor}_i(M,N)} = \sum_k c_k P_N^{(k)}. \qquad (**)$$

From these considerations it is not difficult to show that the strong versions of the intersection conjectures hold for graded modules.

Theorem 6.3.2 *Suppose that M and N are graded modules over a graded ring A with A_0 of finite length. Suppose that $M \otimes_A N$ is a module of finite length and that M is a module of finite projective dimension. Then*

(i) $\dim(M) + \dim(N) \leq \dim(A)$.

(ii) *If* $\dim(M) + \dim(N) < \dim(A)$, *then* $\chi(M, N) = 0$.

(iii) *For a graded module Q, let $e(Q)$ denote the degree of Q. If* $\dim(M) + \dim(N) = \dim(A)$, *then*

$$\chi(M, N) = \frac{e(M)e(N)}{e(A)} \neq 0.$$

Proof Let $F_\bullet$ be a free resolution of M by graded free modules $F_i = \oplus A[n_{ij}]$, and define $c_k = \frac{1}{k!} \sum_{i,j} (-1)^i n_{ij}^k$ as above. Let k be the smallest integer such that $c_k \neq 0$. As we have seen, we then have $k = \dim(A) - \dim(M)$.

We now tensor $F_\bullet$ with N. Since $M \otimes N$ has finite length, so does $\text{Tor}_i(M, N)$ for every i, and all of the Hilbert polynomials occurring in the above formula are constant with values equal to the lengths of $\text{Tor}_i(M, N)$. Furthermore, their alternating sum is $\chi(M, N)$. Thus, formula $(**)$ implies that the smallest k such that $c_k \neq 0$ must be at least $\dim(N)$, and it is equal to $\dim(N)$ if and only if $\chi(M, N) \neq 0$. Let k_0 denote the smallest k such that $c_k \neq 0$. Since k_0 is also $\dim(A) - \dim(M)$, we obtain the first inequality.

It remains to prove the formula in statement (iii). The fact that $F_\bullet$ is a graded free resolution of M implies that $e(M) = c_{k_0}e(A)$. Furthermore, after tensoring with N, since we are assuming that k_0 is the dimension of N, we have that

$\chi(M, N) = c_{k_0} e(N)$. Hence

$$\chi(M, N) = \frac{e(M)e(N)}{e(A)}.$$

$\square$

It is a remarkable fact that the vanishing part of Theorem 6.3.2 also implies the grade conjecture, and thus that the grade conjecture is true for graded modules. This result is also due to Peskine and Szpiro [51]. We first prove that the grade conjecture follows from a result on the heights of prime ideals in the support of a module of finite projective dimension.

Proposition 6.3.3 *Suppose that the module of finite projective dimension M satisfies the condition that for all prime ideals $\mathfrak{p}$ in the support of M we have* height($\mathfrak{p}$) $\geq$ dim(A) − dim(M). *Then*

$$\text{grade}(M) = \dim(A) - \dim(M).$$

Proof We note first that if the annihilator of M contains a regular sequence of length t, then every prime ideal in the support of M has height at least t, so that $t + \dim(M) \leq \dim(A)$. Thus, grade(M) $\leq$ dim(A) − dim(M), and it remains to prove the opposite inequality.

If $\mathfrak{a}$ is the annihilator of M, this theorem states that there is a regular sequence of length dim(A) − dim(M) in $\mathfrak{a}$. Suppose first that M has finite length. Then dim(M) = 0, so the annihilator of M is an $\mathfrak{m}$-primary ideal. The theorem states that there is a regular sequence of length dim(A) in the annihilator of M, or, in other words, that A is a Cohen-Macaulay ring. Thus, in this case the theorem follows from Proposition 6.2.4, which is a consequence of the intersection theorem.

In fact, more can be deduced from the intersection theorem. Suppose that there is a maximal regular sequence for A contained in the annihilator of M. In other words, if r is the depth of A, the annihilator of M contains a regular sequence of length r. Let d be the dimension of M, and let $y_1, \ldots y_d$ be a sequence of elements of the maximal ideal of A such that $M/(y_1, \ldots y_d)$ has finite length. Let $B = A/(y_1, \ldots y_d)$. Letting $r = \text{depth}(A)$ as above, we know from the Auslander-Buchsbaum theorem that the projective dimension of M is less than or equal to r. Let $F_\bullet$ be a minimal free resolution of M, and consider the tensor product $F_\bullet \otimes_A B$. Since $M \otimes_A B = M/(y_1, \ldots y_d)$ has finite length, $F_\bullet \otimes_A B$ is a bounded complex of free B-modules with homology of finite length and thus satisfies the hypothesis of the new intersection theorem. The length of the complex $F_\bullet \otimes_A B$ is the projective dimension of M, which is at most r, so we have

$$r \geq \dim(B) = \dim(A) - d.$$

Since we are assuming that r is also the grade of M, and since d is the dimension of M, we thus have $\mathrm{grade}(M) \geq \dim(A) - \dim(M)$, so the theorem is true also in this case.

We now prove the general case. Let M be any module of finite projective dimension and let r denote the grade of M. We are assuming that for each minimal prime ideal $\mathfrak{p}$ in the support of M we have $\mathrm{height}(\mathfrak{p}) \geq t$. If $x_1, \ldots x_r$ is a maximal regular sequence in the annihilator of M, we can find an associated prime ideal $\mathfrak{p}$ of $A/(x_1, \ldots x_r)$ containing the annihilator of M; localizing at $\mathfrak{p}$, we are in the situation where $x_1, \ldots x_r$ is a maximal regular sequence for $A_\mathfrak{p}$, so, by the above discussion, we have $r = \dim(A_\mathfrak{p}) - \dim(M_\mathfrak{p}) \geq \mathrm{height}(\mathfrak{p}) \geq t$. Thus, again the grade conjecture holds. $\square$

It follows from Proposition 6.3.3 that the grade conjecture is really a question on the height of minimal prime ideals in the support of a module of finite projective dimension. In particular, if A is a catenary domain, the hypothesis of Proposition 6.3.3 holds automatically. The conjecture can be reduced to the catenary case by completion, so a counterexample would have to involve a ring with components of different dimension such that the heights of the minimal primes in the support of M in some components were less than $\dim(A) - \dim(M)$.

Theorem 6.3.4 *Let M be a module of finite projective dimension over a graded ring A for which A_0 is Artinian. Then the grade conjecture holds.*

Proof Let $t = \dim(A) - \dim(M)$. It follows from Proposition 6.3.3 that it suffices to show that every minimal prime in the support of M has height $\geq t$. Since A is a ring of finite type over an Artinian ring, A is catenary, so the only problem is that there could possibly be components of $\mathrm{Spec}(A)$ such that the intersection of the support of M with these components has codimension less than t. Assuming that such components exist, we define an ideal $\mathfrak{a}$ below such that the support of M intersected with $\mathrm{Spec}(A/\mathfrak{a})$ has codimension less than t in $\mathrm{Spec}(A/\mathfrak{a})$. The idea of the proof is to apply equation $(**)$ to the modules M and $A/\mathfrak{a}$ and show that this leads to a contradiction. We now work out the details.

Let $\mathfrak{p}_1, \ldots, \mathfrak{p}_j$ be the minimal primes in the support of M such that $\mathrm{height}(\mathfrak{p}_i) < t$; our assumption is that there is at least one such $\mathfrak{p}_i$. Let S be the complement of the union of the $\mathfrak{p}_i$, and let $\mathfrak{a}$ be the ideal of elements that are annihilated by an element of S. Then $\mathfrak{a}$ is the kernel of the map from A to the localization A_S, so we have $A_S = (A/\mathfrak{a})_S$, and the map from $A/\mathfrak{a}$ to $(A/\mathfrak{a})_S$ is injective. Thus, S cannot meet any associated prime of $A/\mathfrak{a}$ and hence has trivial intersection with every minimal prime of $A/\mathfrak{a}$. It follows that the minimal prime ideals of $A/\mathfrak{a}$ correspond to the minimal primes of A_S and are the minimal prime ideals of A that do not meet S; that is, the minimal primes of $A/\mathfrak{a}$ are the minimal primes of A that are contained in some $\mathfrak{p}$. Since the height of $A_{\mathfrak{p}}$ is less than t for all i, it then follows that we have

$$\dim(A/\mathfrak{a}) - \dim(M \otimes A/\mathfrak{a}) < t.$$

Let r be the maximum of the dimensions of $A/\mathfrak{p}_i$. We next wish to show that any prime ideal $\mathfrak{q}$ in the support of $M \otimes A/\mathfrak{a}$ with $\dim(A/\mathfrak{q}) \geq r$ is one of the $\mathfrak{p}_i$. Let $\mathfrak{q}$ be such a prime ideal; we may assume that $\mathfrak{q}$ is minimal in the support of $M \otimes A/\mathfrak{a}$. Then $M_{\mathfrak{q}} \otimes (A/\mathfrak{a})_{\mathfrak{q}}$ has finite length, so it follows from the intersection theorem that

$$\dim((A/\mathfrak{a})_{\mathfrak{q}}) \leq \text{proj.dim.}(M_{\mathfrak{q}}).$$

On the other hand, since the depth of a ring is less than or equal to the dimension of the quotient of the ring modulo any minimal prime by Corollary 4.1.9, and, since there is at least one minimal prime ideal of A in the support of $A/\mathfrak{a}$, we have

$$\text{depth}(A_{\mathfrak{q}}) \leq \dim(A/\mathfrak{a})_{\mathfrak{q}}.$$

The Auslander-Buchsbaum theorem implies that we have the inequality $\text{proj.dim.}(M_{\mathfrak{q}}) \leq \text{depth}(A_{\mathfrak{q}})$. Combining these inequalities, we have

$$\dim((A/\mathfrak{a})_{\mathfrak{q}}) = \text{proj.dim.}(M_{\mathfrak{q}}) = \text{depth}(A_{\mathfrak{q}}).$$

In particular, we have $\dim(A/\mathfrak{a})_{\mathfrak{q}} = \text{depth}(A_{\mathfrak{q}})$. If $\mathfrak{q}$ did not contain any $\mathfrak{p}_i$, every minimal prime in the support of $M_{\mathfrak{q}}$ would have height at least t, so it would follow from Proposition 6.3.3 that $\text{depth}(A_{\mathfrak{q}}) \geq t$. Thus,

$$\dim(A/\mathfrak{a}) \geq \dim(A/\mathfrak{q}) + \dim(A/\mathfrak{a})_{\mathfrak{q}} \geq \dim(M \otimes A/\mathfrak{a}) + t > \dim(A/\mathfrak{a}).$$

This contradiction shows that every prime ideal in the support of $M \otimes A/\mathfrak{a}$ with dimension $\geq r$ is one of the $\mathfrak{p}_i$. Let $\mathfrak{p}_i$ be one of these. Then, since $\mathfrak{p}$ does not meet S, we have

$$\text{Tor}_i^{A_{\mathfrak{p}_i}}\left(M_{\mathfrak{p}_i}, (A/\mathfrak{a})_{\mathfrak{p}_i}\right) = \text{Tor}_i^{A_{\mathfrak{p}_i}}\left(M_{\mathfrak{p}_i}, A_{\mathfrak{p}_i}\right) = 0$$

for $i > 0$. Thus, $M \otimes A/\mathfrak{a}$ has dimension r and $\text{Tor}_i(M, A/\mathfrak{a})$ has dimension less than r for $i > 0$, and the polynomial $\sum_i (-1)^i P_{\text{Tor}_i(M,N)}$ has degree r. Hence the lowest k for which $c_k \neq 0$ is $k = \dim(A/\mathfrak{a}) - \dim(M \otimes A/\mathfrak{a})$. But this integer is also equal to $\dim(A) - \dim(M) > \dim(A/\mathfrak{a}) - \dim(M \otimes A/\mathfrak{a})$.

Hence we must have had height($\mathfrak{p}$) $\geq t$ for all $\mathfrak{p}$ in the support of M, so M satisfies the grade conjecture.   □

While the hypothesis that $\chi(M, N) = 0$ when $\dim(M) + \dim(N) < \dim(A)$ does not hold in general, it does hold when the dimension of N is zero or one, and it can be shown that the grade conjecture holds in these cases. Putting these results together, one can deduce that the lowest unknown case is when A has dimension 4 and depth 3, and M has grade 2, dimension 1, and depth 0. The conjecture is unknown even for cyclic modules in this case.

For another interpretation of the grade conjecture, see Avramov and Foxby [3].

Exercises

6.1 Show that if the dimension of A is d and there exists a finitely generated Cohen-Macaulay module of dimension d, then the monomial conjecture holds.

6.2 Show that the monomial conjecture holds if and only if it holds for all integral domains.

6.3 Let A be an integral domain of dimension 2. Let B be the integral closure of A in its quotient field. Show that if B is a finitely generated A-module, then it is Cohen-Macaulay, so that the monomial conjecture holds for A.

6.4 Show that the canonical element conjecture holds for a local ring A if and only if it holds for the completion of A.

6.5 Let $A = k[[X, Y, Z, W]]/(XY - ZW)$, $M = A/(X, Z)$, and $N = A/(Y, W)$. Show that $M \otimes_A N$ has finite length, but that $\dim(M) + \dim(N) > \dim(A)$. Show that M and N do not have finite projective dimension.

7

The Frobenius Map

Let A be a ring of positive characteristic p, so that p is a prime number and $pa = 0$ for every $a \in A$. We define the *Frobenius map* F from A to A by letting $F(a) = a^p$. Since A has characteristic p, we have that $(a+b)^p = a^p + b^p$ for all a and b in A, so the Frobenius map is a ring homomorphism.

The systematic use of the Frobenius map has become one of the basic methods for studying rings of characteristic p, and it is particularly useful in studying homological questions. In this chapter we describe three ways the Frobenius map can be applied. One of the simplest is to prove theorems by the following method: If there were a counterexample to a given theorem, one could construct new counterexamples by taking the tensor product with the Frobenius map. In many cases one can show that the new counterexamples could not exist, and therefore the original one could not have existed either. We give two specific examples of this method in the second section.

In the third section we discuss limits taken over powers of the Frobenius map. For example, if $\mathfrak{a}$ is an $\mathfrak{m}$-primary ideal of A, the ideal generated by pth powers of elements of $\mathfrak{a}$ is an ideal contained in $\mathfrak{a}^p$. In Chapter 2 we discussed multiplicities, which are defined in terms of the lengths of A-modulo powers of $\mathfrak{a}$; one can define *Hilbert-Kunz* multiplicities, which are defined by using similar limits over Frobenius powers. Similarly, if $E_\bullet$ is a bounded free complex with homology of finite length, one can define *Dutta* multiplicities, which are limits of Euler characteristics of $E_\bullet$ tensored with powers of the Frobenius map. In some respects these asymptotic Euler characteristics have better properties than the Euler characteristic itself. We will see in Chapter 12 that they are also closely related to local Chern characters.

A third approach is to define a closure operation using the Frobenius map. This technique was introduced more recently by Hochster and Huneke [26] and is called *tight closure*. The relations between multiplicities and integral closure of ideal discussed in Chapter 2 have analogies in relations between Hilbert-Kunz multiplicities and tight closure. This theory has numerous ramifications for rings of positive characteristic; we limit ourselves to outlining a couple of the basic ideas in Section 4.

Many of the results proven in this chapter can be proven also in the case of equal characteristic zero by reduction to positive characteristic. In the fifth section of this chapter we give a brief outline of how this is done.

7.1 General Properties of the Frobenius Map

As above, let A be a ring of positive characteristic p, and let $F : A \to A$ be the Frobenius map. If M is an A-module, we may tensor M over F to obtain a new A-module. This gives a functor from the category of A-modules to itself, often called the *Peskine-Szpiro* functor. We denote this functor $\mathbf{F}$. We have $\mathbf{F}(A) = A$, and, to compute the result of applying $\mathbf{F}$ to a finitely generated module M, the simplest method is to take a presentation $A^n \to A^m \to M \to 0$. Since tensor product is right exact, applying $\mathbf{F}$ to this sequence gives a presentation of $\mathbf{F}(M)$. The free modules that occur in this presentation are the same as those for M, but, if (a_{ij}) is the matrix defining the map, the matrix defining a presentation of $\mathbf{F}(M)$ is (a_{ij}^p).

Similarly, if $F_\bullet$ is a complex of free modules, tensoring with F gives a new complex defined by matrices whose entries are the pth powers of those of the original complex. Repeating this process, applying $\mathbf{F}^e$ replaces the original entries by their (p^e)th powers.

If $\mathfrak{a}$ is an ideal of A, its image under the Frobenius map generates another ideal of A, which we denote $\mathfrak{a}^{[p]}$. The ideal $\mathfrak{a}^{[p]}$ is generated by all pth powers of elements of $\mathfrak{a}$, and it can easily be checked that when a set $\{x_1, \ldots x_k\}$ generates $\mathfrak{a}$, then $\{x_1^p, \ldots x_k^p\}$ generates $\mathfrak{a}^{[p]}$. We note that $\mathfrak{a}^{[p]}$ is contained in the ordinary power $\mathfrak{a}^p$, and usually these two ideals are not equal. Again, this process can be iterated, giving a decreasing sequence of ideals $\mathfrak{a}^{[p^e]}$.

7.2 Applications to the Homological Conjectures

In this section we give the proofs in positive characteristic of two of the homological conjectures discussed in the previous chapter. In both cases the method is to assume that there exists a counterexample and then to show that a high enough tensor power of the counterexample with the Frobenius map would violate a bound that depends only on the ring. Both of these theorems use Theorem 4.6.6 of Chapter 4, where it was proven that, if A is a local ring of dimension d, there exists a nonzero ideal that annihilates the homology of every bounded free complex $F_\bullet$ with homology of finite length in degrees greater than $k - d$, where k is the largest integer with $F_k \neq 0$.

We first prove the new intersection theorem in positive characteristic.

Theorem 7.2.1 (New Intersection Theorem) *Let A be a ring of characteristic p, and let*

$$F_\bullet = 0 \to F_k \to \cdots \to F_0 \to 0$$

be a nonexact bounded complex of free A-modules with homology of finite length. Then

$$k \geq \dim(A).$$

Proof Suppose that we had a complex $F_\bullet$ as in the statement of the theorem with k less than the dimension d of the ring. Since $F_\bullet$ is not exact, we may assume that $F_0 \neq 0$ and that $F_\bullet$ is minimal. By Theorem 4.6.6, there exists a nonzero ideal $\mathfrak{a}$ of A, independent of $F_\bullet$, which annihilates $H_{k-i}(F_\bullet)$ for all $i < d$. Since $k < d$, this implies that $\mathfrak{a}$ annihilates $H_0(F_\bullet)$. Now consider the result $\mathbf{F}^e(F_\bullet)$ of tensoring e times with the Frobenius map. We claim that $\mathbf{F}^e(F_\bullet)$ also satisfies the conditions of the theorem; that is, that its homology still has finite length. Let $\mathfrak{p}$ be a nonmaximal prime ideal. Then F^e maps the complement of $\mathfrak{p}$ to the complement of $\mathfrak{p}$, so it induces a map on $A_\mathfrak{p}$ that is simply the Frobenius map on $A_\mathfrak{p}$. Hence

$$\mathbf{F}^e(F_\bullet) \otimes A_\mathfrak{p} \cong \mathbf{F}^e(F_\bullet \otimes_A A_\mathfrak{p}).$$

Since $F_\bullet$ has homology of finite length, $F_\bullet \otimes_A A_\mathfrak{p}$ is split exact; hence $\mathbf{F}^e(F_\bullet) \otimes A_\mathfrak{p}$ is split exact and $\mathbf{F}^e(F_\bullet)$ has homology of finite length.

Since the map from $\mathbf{F}^e(F_1)$ to $\mathbf{F}^e(F_0)$ has entries in $\mathfrak{m}^{[p^e]}$, the annihilator of $H_0(\mathbf{F}^e(F_\bullet))$ is contained in $\mathfrak{m}^{[p^e]}$, which in turn is contained in $\mathfrak{m}^{p^e}$. But $\mathfrak{a}$ annihilates $H_0(\mathbf{F}^e(F_\bullet))$. Thus, $\mathfrak{a} \subset \mathfrak{m}^{p^e}$ for all e, so $\mathfrak{a} = 0$. This contradiction proves the theorem. $\qquad\square$

We next prove the canonical element conjecture in positive characteristic. For simplicity we prove it in the case of an integral domain; the general case can easily be reduced to this case.

Theorem 7.2.2 *Let A be a local integral domain of dimension d, and let $x_1, \ldots, x_d$ be a system of parameters for A. Let $K_\bullet$ be the Koszul complex on $x_1, \ldots, x_d$, and let $e_1, \ldots, e_d$ be a basis of a free module G such that $K_i = \Lambda^i(G)$ for each i. Let $F_\bullet$ be a minimal free resolution of $A/(x_1, \ldots, x_d)$,* *and let $f_\bullet : K_\bullet \to F_\bullet$ be a map of complexes that lifts the identity map in degrees 0 and 1. Then*

$$f_d(e_1 \wedge \cdots \wedge e_d) \notin \mathfrak{m} F_d.$$

Proof We note first that the map $f_\bullet$ exists by Proposition 3.1.4 and that K_d is free rank 1 with generator $e_1 \wedge \cdots \wedge e_d$. It follows from Theorem 4.6.6 that there exists a nonzero element $x \in A$ that annihilates the homology of the Koszul complex on any system of parameters of A in all degrees greater than 0.

Suppose first that $K_\bullet$ is exact in degrees greater than one (that is, suppose that A is Cohen-Macaulay). Of course, we know that in this case $K_\bullet$ is itself a free resolution of $A/(x_1, \ldots, x_d)$, and we know from Proposition 4.4.1 that $f_\bullet$ is an isomorphism, but we wish to give another proof of this result in this case that can be generalized to the case of a non-Cohen-Macaulay ring of positive characteristic. Since now $F_\bullet$ is a free complex and $K_\bullet$ is exact, it follows from Proposition 3.1.4 that there exists a map $g_\bullet$ from $F_\bullet \to K_\bullet$ lifting the identity in degrees 0 and 1, and that $g_\bullet$ is a homotopy inverse for $f_\bullet$. Let s denote a homotopy between $f_\bullet g_\bullet$ and the identity map on $F_\bullet$. We then have in particular

$$f_d g_d - 1_{F_\bullet} = d^F_{d+1} s_d + s_{d-1} d^K_d.$$

Now if $f_d(e_1 \wedge \cdots \wedge e_d) \in \mathfrak{m} F_d$, then $f_d g_d(F_d) \subseteq \mathfrak{m} f_d$, so $f_d g_d - 1_{F_\bullet}$ is the identity modulo $\mathfrak{m}$, and the right-hand side, since $F_\bullet$ and $K_\bullet$ are minimal complexes, maps F_d to $\mathfrak{m} F_d$. This contradiction proves the theorem if $K_\bullet$ is exact.

Suppose now that $K_\bullet$ is not exact and that x is a nonzero element that annihilates the homology of $K_\bullet$ in degrees greater than zero. Although we cannot construct a homotopy inverse $g_\bullet$ as before, it now follows from Proposition 3.1.5 that, since x annihilates the homology of $K_\bullet$, we can find a map $g_\bullet$ that extends the map given by multiplication by x^{d-1} in degrees 0 and 1 to a map of complexes from $F_\bullet$ to $K_\bullet$. Since $f_\bullet g_\bullet : F_\bullet \to F_\bullet$ is then a map of complexes that lifts multiplication by x^{d-1} in degrees 0 and 1, and since $F_\bullet$ is a complex of free modules that is exact in degrees greater than zero, we then conclude from Proposition 3.1.4 that $f_\bullet g_\bullet$ is homotopic to the map on $F_\bullet$ defined by multiplication by x^{d-1}.

At this point we use the assumption that A has characteristic p. It follows from Theorem 4.6.6 that there is a nonzero element x in A that annihilates the homology of the Koszul complex of any system of parameters for A. Suppose that the map $f_\bullet$ has the property that $f_d(e_1 \wedge \cdots \wedge e_d) \in \mathfrak{m} F_d$. We then tensor the whole diagram $f_\bullet : K_\bullet \to F_\bullet$ with the eth power of the Frobenius map. Note that $\mathbf{F}^e(K_\bullet)$ is the Koszul complex on $x_1^{p^e}, \ldots, x_d^{p^e}$. The construction of the previous paragraph can be carried out, and we conclude that there is a map $g_\bullet : \mathbf{F}^e(F_\bullet) \to \mathbf{F}^e(K_\bullet)$ such that $\mathbf{F}^e(f_\bullet) g_\bullet$ is homotopic to multiplication by x^{d-1}. But the boundary maps of the complexes $\mathbf{F}^e(F_\bullet)$ and $\mathbf{F}^e(K_\bullet)$ as well as the map $\mathbf{F}^e(f_d)$ are zero modulo $\mathfrak{m}^{p^e}$. Thus, arguing as in the case when $K_\bullet$ is exact, we can conclude that x^{d-1} is in $\mathfrak{m}^{p^e}$ for all e. Since we are assuming that A is an integral domain, $x^{d-1} \neq 0$, and this contradiction proves the theorem. $\square$

7.3 Hilbert-Kunz Multiplicity and Dutta Multiplicity

Both of the multiplicities considered in this section are defined as limits of lengths of modules tensored with powers of the Frobenius map. We first prove a general theorem on the existence of limits of this type, following the treatment of Seibert [65]. In what follows we assume that A is a local ring of positive characteristic p, and we make the further assumptions that A is finite over A^p, where A^p is the image of the Frobenius map (or the subring of pth powers) and that the residue field of A is perfect. If A has perfect residue field, the condition that A is finite over A^p will be satisfied for complete local rings and for localizations of rings of finite type over a field.

In what follows, rather than looking at the result of tensoring over the Frobenius map, we consider the functor defined by restriction of scalars. That is, for each positive integer e we let ^eM be M as a module over A through restriction by the eth power of the Frobenius map. Thus, if m_e denotes an element m of M considered as an element of ^eM, we have $r \cdot m_e = (r^{p^e}m)_e$. The functor that sends M to ^eM is clearly exact.

We first work out the case in which $M = A$ and A is a regular local ring of dimension d. Since A^p is regular and A is Cohen-Macaulay, A is free over A^p. Since the residue field is assumed to be perfect, the rank can be computed by finding the length of A modulo the image of $\mathfrak{m}$ under the Frobenius map; since this image is generated by the pth powers of a set of generators for $\mathfrak{m}$, this length is p^d. Thus, A is free of rank p^d over A^p. Then ^1A in this notation becomes a free module of rank p^d over A.

Assume next that A is an integral domain of dimension d. Then A is a torsion-free module over A^p, but not necessarily free. To compute its rank, we let $x_1, \ldots, x_d$ be a system of parameters for A; then by Corollary 5.2.10 the rank of a module M is equal to the quotient of Euler characteristics $\chi(x_1, \ldots x_d; M)/\chi(x_1, \ldots x_d; A)$. In this case we have

$$\chi(x_1, \ldots x_d; {}^1A) = \chi\left(x_1^p, \ldots x_d^p; A\right) = p^d \chi(x_1, \ldots x_d; A)$$

by Corollary 5.2.4, so the rank is again p^d.

We prove in Theorem 7.3.2 that there is an analogue of Hilbert polynomials for Frobenius powers. That is, given an additive function g from the category of modules to the integers, the function $\phi(e)$ defined by $\phi(e) = g(^e M)$ can be expressed as a polynomial in p^e. As in the case of Hilbert polynomials, the existence of such a polynomial is proven by induction on dimension. However, the inductive step is more complicated, and we prove it separately.

Lemma 7.3.1 *Let ϕ be a real-valued function defined on the set of nonnegative integers. Let d and p be positive integers. Suppose that for all integers $e > 0$ we have*

$$\phi(e) - p^d \phi(e-1) = b_{d-1} p^{e(d-1)} + b_{d-2} p^{e(d-2)} + \cdots + b_0.$$

Then

$$\phi(e) = \phi(0) p^{de} + \psi_{d-1}(e) + \psi_{d-2}(e) + \cdots + \psi_0(e) \qquad (*)$$

where for $i = 0, \ldots, d-1$ we have

$$\psi_i(e) = \left(\frac{b_i}{p^{d-i} - 1} \right) (p^{ed} - p^{ei}).$$

Proof Let $\mu(e)$ denote the sum of functions on the right-hand side of equation $(*)$. To prove this result, it suffices to prove that $\mu(0) = \phi(0)$ and that $\mu(e)$ satisfies the correct recursion relation. If $e = 0$, then $\psi_i(e) = 0$ for all i, and we are left with $\mu(0) = \phi(0) p^0 = \phi(0)$, so the functions agree for $e = 0$.

It remains to check the recursion relation $\mu(e)$. We compute the corresponding relations separately for $\phi(0) p^{de}$ and for each of the functions ψ_i and then add the results.

For $\phi(0) p^{de}$, we have

$$\phi(0) p^{de} - p^d \phi(0) p^{d(e-1)} = \phi(0) (p^{de} - p^{de}) = 0.$$

For the functions ψ_i, we have

$$\psi_i(e) - p^d \psi_i(e-1) = \left(\frac{b_i}{p^{d-i} - 1} \right) \left[(p^{ed} - p^{ei}) - p^d \left(p^{(e-1)d} - p^{(e-1)i} \right) \right]$$

$$= \left(\frac{b_i}{p^{d-i} - 1} \right) (p^{ed} - p^{ei} - p^{ed} + p^{d+ei-i})$$

$$= \left(\frac{b_i}{p^{d-i} - 1} \right) (-p^{ei} + p^{ei} p^{d-i}) = b_i p^{ei}.$$

Hence we have $\psi_i(e) - p^d \psi_i(e-1) = b_i p^{ei}$ for each i, so, adding the contributions for all i, we conclude that the sum of the functions $\phi(0) p^{de}$ and the ψ_i satisfy the same recursion relation as ϕ. Since we have verified that they agree for $e = 0$, we thus have $\mu(e) = \phi(e)$ for all $e \geq 0$. $\qquad \square$

We note that each of the functions ψ_i in the above Lemma is a polynomial of degree d in p^e and that, if the coefficients b_i in the recursion relation are rational numbers, then the coefficients of ψ_i are also rational numbers.

Theorem 7.3.2 *Let C be a category of finitely generated A-modules of dimension at most d that has the property that, if $0 \to M' \to M \to M'' \to 0$ is a short exact sequence, then M is in C if and only if M' and M'' are in C. Let g be a function from C to the integers such that for every short exact sequence as above, we have $g(M) = g(M') + g(M'')$. Then*

(i) *If $M \in C$, then $^e M \in C$ for all e.*
(ii) *For all $M \in C$ there is a polynomial in p^e such that*

$$g(^e M) = a_d p^{ed} + a_{d-1} p^{e(d-1)} + \cdots + a_0$$

for all $e \geq 0$. The coefficients a_i are in $\mathbb{Q}$.

Proof The conditions on the category C imply that a module M is in C if and only if $A/\mathfrak{p}$ is in C for every prime ideal $\mathfrak{p}$ in the support of M. The first assertion thus follows from the fact that the support of $^e M$ is the same as the support of M.

We next prove the second assertion by induction on the dimension d. If $d = 0$, any module of dimension d has finite length, and since the residue field of A is assumed to be perfect, the length of $^e M$ is equal to the length of M for all e. Thus, the statement is true with $a_0 = \text{length}(M) g(A/\mathfrak{m})$.

Let $d > 0$, and let C^* be the category of modules in C of dimension less than d; by induction statement (ii) is true for $M \in C^*$. Using the additivity of g, we may reduce to the case in which $M = A/\mathfrak{p}$, where $A/\mathfrak{p}$ has dimension d. Thus, since $A/\mathfrak{p}$ is a torsion-free module of rank p^d over $(A/\mathfrak{p})^p$, we have a short exact sequence

$$0 \to M^{p^d} \to {}^1 M \to Q \to 0$$

for some module $Q \in C^*$. This then implies that we have a short exact sequence

$$0 \to {}^{e-1} M^{p^d} \to {}^e M \to {}^{e-1} Q \to 0$$

for all e.

From the above exact sequences, for all $e > 0$ we have

$$g(^e M) = p^d g(^{e-1} M) + g(^{e-1} Q).$$

By induction, there are rational numbers $c_{d-1}, \ldots c_0$ such that

$$g(^{e-1} Q) = c_{d-1} p^{(e-1)(d-1)} + \cdots + c_0$$

for all e. If we let $b_i = c_i/p^i$, we then have $g(^{e-1}Q) = b_{d-1}p^{e(d-1)} + \cdots + b_0$, so that for all e we have

$$g(^e M) - p^d g(^{e-1}M) = b_{d-1}p^{e(d-1)} + \cdots + b_0.$$

This is exactly the form of the recursion relation in Lemma 7.3.1, so it follows from that Lemma that the function $\phi(e) = g(^e M)$ can be written as a polynomial in p^e with rational coefficients. Hence there exist rational numbers $a_0, \ldots, a_d$ such that

$$g(^e M) = a_d p^{ed} + a_{d-1}p^{e(d-1)} + \cdots + a_0$$

for all $e \geq 0$.   □

We next show that even if the function g is only subadditive on short exact sequences, the limit of $g(^e M)/p^{ed}$ as e goes to infinity exists and is additive on short exact sequences.

Theorem 7.3.3 *Let C be a category of finitely generated A-modules of dimension at most d satisfying the conditions of Theorem 7.3.2, and let g be a function from C to the integers that is additive on direct sums and such that for every short exact sequence $0 \to M' \to M \to M'' \to 0$ we have $g(M) \leq g(M') + g(M'')$. Then*

(i) *If $M \in C$, the limit*

$$c(M) = \lim_{e \to \infty} \frac{g(^e M)}{p^{ed}}$$

exists. More precisely, there exists a real number $c(M)$ and a constant u (which may also depend on M) such that

$$|p^{ed}c(M) - g(^e M)| \leq u p^{e(d-1)}$$

for all $e \geq 0$.

(ii) *If $0 \to M' \to M \to M'' \to 0$ is a short exact sequence, we have*

$$c(M) = c(M') + c(M'').$$

Proof We again use induction on the dimension, let C^* be the category of modules in C with dimension less than d, and assume that the result is true for $M \in C^*$. If $d = 0$, we let C^* be the empty category. We first prove the existence of $c(M)$ when M is of the form $A/\mathfrak{p}$ for a prime ideal $\mathfrak{p}$. In this case $A/\mathfrak{p}$ is an integral domain, and ^1M is a torsion-free $A/\mathfrak{p}$ module of rank p^d.

Thus, as in the case of Theorem 7.3.2, have short exact sequences

$$0 \to {}^e M \to {}^{e-1}M^{p^d} \to {}^{e-1}Q \to 0$$

and

$$0 \to {}^{e-1}M^{p^d} \to {}^e M \to {}^{e-1}Q' \to 0$$

for all $e > 0$, where Q and Q' have dimension at most $d - 1$ and so are in C^*. These exact sequences together with the assumed properties of g give inequalities

$$p^d g({}^{e-1}M) \le g({}^e M) + g({}^{e-1}Q) \quad \text{and} \quad g({}^e M) \le p^d g({}^{e-1}M) + g({}^{e-1}Q').$$

By the induction assumption applied to Q and Q' we thus have that there exists a constant C such that

$$|g({}^e M) - p^d g({}^{e-1}M)| \le C p^{(d-1)e}$$

for all e. We now consider the sequence $n_e = \frac{g({}^e M)}{p^{de}}$. The above inequality implies that $|n_e - n_{e-1}| \le C p^{-e}$. Thus, n_e is a Cauchy sequence, so it has a limit $c(M)$.

We next show that this limit satisfies the inequality in the statement of the theorem with $u = 2C$. Suppose that $|p^{ed}c(M) - g({}^e M)| \ge 2C p^{e(d-1)}$, or, equivalently, that $|n_e - c(M)| \ge 2C/p^e$. Then for all $m \ge 1$ we have

$$|n_{e+m} - n_e| \le |n_{e+m} - n_{e+m-1}| + \cdots + |n_{e+1} - n_e|$$

$$\le C \left(\frac{1}{p^{e+m}} + \cdots + \frac{1}{p^{e+1}} \right) = C \frac{p^m - 1}{(p^{e+m})(p - 1)}$$

$$\le \frac{C}{(p^e)(p - 1)} \le \frac{C}{p^e}.$$

Thus, if $|n_e - c(M)| \ge 2C/p^e$, we would have

$$|n_{e+m} - c(M)| \ge |n_e - c(M)| - |n_{e+m} - n_e| \ge \frac{2C}{p^e} - \frac{C}{p^e} = \frac{C}{p^e}$$

for all $m \ge 1$, contradicting the fact that the limit of n_{e+m} is $c(M)$.

We have so far shown that the limit $c(M)$ exists for modules of the form $A/\mathfrak{p}$ of dimension d; if M is in C^* the induction hypothesis implies that the required formula is correct if we let $c(M) = 0$. We next prove a Lemma that will allow us to extend the definition to all modules in C of dimension d and prove additivity on short exact sequences.

Lemma 7.3.4 *Let M and N be modules in C of dimension d. Let S be a multiplicatively closed subset of A that does not meet any prime ideals $\mathfrak{p}$ that are in the support of M or N, and such that $\dim(A/\mathfrak{p}) = d$. If the localizations M_S and N_S are isomorphic, then the limit of $g({}^e M)/p^{ed}$ exists if and only if the limit of $g({}^e N)/p^{ed}$ exists, and in this case the two limits are equal.*

Proof Let Q be the submodule of M consisting of elements annihilated by elements of S. The hypothesis implies the the dimension of Q is at most $d - 1$ so that $Q \in C^*$. Let $K = M/Q$. The short exact sequence $0 \to Q \to M \to K \to 0$ implies that

$$g(^e M) \leq g(^e Q) + g(^e K)$$

for all e. Let $E(Q)$ denote the injective hull of Q. The imbedding of Q into $E(Q)$ can be extended to a map ϕ from M to $E(Q)$ since $E(Q)$ is an injective module. Let P denote the image of ϕ. Since P is a submodule of $E(Q)$, the support of P is contained in the support of Q by Proposition 4.5.3, and P is in C^*. Furthermore, the maps ϕ and the map onto K define a map from M to $P \oplus K$ such that we have an exact sequence

$$0 \to M \to P \oplus K \to Q' \to 0$$

where Q' is also in C^*. Thus, using the additivity of g on direct sums together with the subadditivity on short exact sequences, we have

$$g(^e P) + g(^e K) \leq g(^e M) + g(^e Q')$$

for all e. Dividing these two equations by p^{ed} and using the inductive hypothesis on Q, P, and Q', we conclude that there exists a constant C such that

$$\frac{g(^e M)}{p^{ed}} \leq \frac{g(^e K)}{p^{ed}} + \frac{C}{p^e} \quad \text{and} \quad \frac{g(^e K)}{p^{ed}} \leq \frac{g(^e M)}{p^{ed}} + \frac{C}{p^e}$$

for all e. It then follows that one of the limits exists if and only if the other one does and that they are then equal.

We now return to the modules M and N. The above argument shows that we can divide by the submodules of elements annihilated by elements of S and assume that the maps from M to M_S and from N to N_S are injective. Since M_S and N_S are assumed to be isomorphic, and M is finitely presented, there is a map ψ from M to N that induces an isomorphism after localizing at S. Since the map from M to M_S is injective, ψ is also injective, and since S does not meet any prime ideals in the support of M or N of dimension d, the cokernel of ψ lies in C^*. Hence we have a short exact sequence

$$0 \to M \to N \to Q \to 0$$

with $Q \in C^*$. Similarly, we have a short exact sequence

$$0 \to N \to M \to Q' \to 0$$

with $Q' \in C^*$. These short exact sequences imply inequalities for the limits of $g(^e M)/p^{ed}$ and of $g(^e N)/p^{ed}$, and, following the argument of the previous paragraph, we can conclude that one of the limits exists if and only if the other one does, and that if the limits exist they are equal. $\qquad\square$

We can now complete the proof of Theorem 7.3.3. Let M be a module in C of dimension d, and let S be the complement of the union of the set of prime ideals of dimension d in the support of M. The localization M_S is then a module of finite length over the localization A_S. Let $\mathfrak{p}_1, \ldots, \mathfrak{p}_k$ be the prime ideals of A whose localizations at S are in the support of M_S, and let $\mathfrak{a}_S = (p_1)_S \cap \cdots \cap (\mathfrak{p}_k)_S$. Then some power of $\mathfrak{a}$ annihilates M_S. So, since the Frobenius map sends elements of $\mathfrak{a}$ to elements of the pth power of $\mathfrak{a}$, there exists an f such that $\mathfrak{a}$ annihilates $^f M_S$. Hence the module $^f M_S$ is isomorphic to a direct sum of copies of the modules $A_S/(\mathfrak{p}_i)_S$. Applying Lemma 7.3.4 to $^f M$ and a direct sum of copies of $A/\mathfrak{p}_i$, we deduce that the limit

$$\lim_{e \to \infty} \frac{g^{(e+f}M)}{p^{ed}}$$

exists and is equal to

$$\sum_{i=1}^{k} \text{length}((^f M)_{\mathfrak{p}_i}) \left(\lim_{e \to \infty} \frac{g^{(e}(A/\mathfrak{p}_i))}{p^{ed}} \right).$$

It follows immediately that the limit defining $c(M)$ exists. Furthermore, since the rank of $^f(A/\mathfrak{p}_i)$ over $A/\mathfrak{p}_i$ is equal to p^{fd} and length is additive on short exact sequences, we have that $\text{length}(^f (M)_{\mathfrak{p}_i})$ is equal to $p^{fd} \text{length}(M_{\mathfrak{p}_i})$ for each i. Thus, we have

$$c(M) = \lim_{e \to \infty} \frac{g^{(e}M)}{p^{ed}} = \lim_{e \to \infty} \frac{g^{(e+f}M)}{p^{(e+f)d}} = \frac{1}{p^{fd}} \lim_{e \to \infty} \frac{g^{(e+f}M)}{p^{ed}}$$

$$= \frac{1}{p^{fd}} \sum_{i=1}^{k} \text{length}((^f M)_{\mathfrak{p}_i}) \left(\lim_{e \to \infty} \frac{g^{(e}(A/\mathfrak{p}_i))}{p^{ed}} \right)$$

$$= \sum_{i=1}^{k} \text{length}(M_{\mathfrak{p}_i}) \left(\lim_{e \to \infty} \frac{g^{(e}(A/\mathfrak{p}_i))}{p^{ed}} \right) = \sum_{i=1}^{k} \text{length}(M_{\mathfrak{p}_i}) c(A/\mathfrak{p}_i).$$

It follows from the last equality that $c(M)$ is additive on short exact sequences. $\square$

The two applications of this theorem of most interest to us are the definitions of Hilbert-Kunz multiplicities and of the asymptotic Euler characteristic χ_∞ defined by Dutta. We first consider a more general situation. Let $F_\bullet = F_{i+1} \to F_i \to F_{i-1}$ be a complex of three free modules, and suppose that for all modules M (or for all modules in a suitable category C) we have that $H_i(F_\bullet \otimes M)$ has finite length. Then the function $g(M) = \text{length}(H_i(F_\bullet \otimes M))$ satisfies the conditions of Theorem 7.3.3. Thus, we have a function $c(M)$ that is additive on short exact sequences given by the limit. In this example $g(^eM)$ is the length of the homology of $F_\bullet \otimes {}^eM$, which can be computed by representing the maps from F_{i+1} to F_i and from F_i to F_{i-1} by matrices, and then computing the homology of the complex with the same modules as $F_\bullet \otimes M$ but with the original entries in the matrices replaced by their pth powers. Thus, this complex is isomorphic to $\mathbf{F}^e(F_\bullet) \otimes M$.

Suppose now that the complex $F_\bullet$ of the previous paragraph is of the form $A^n \to A \to 0$, where the image of the map from A^n to A is an $\mathfrak{m}$-primary ideal $\mathfrak{a}$ of A and A has dimension d. Then the homology of $F_\bullet \otimes M$ has finite length for all M, and the limit is the limit of the length of $M/\mathfrak{a}^{[p^e]}M$ divided by p^{de} as e goes to infinity. In the case where $M = A$, this limit is called the *Hilbert-Kunz multiplicity* of $\mathfrak{a}$. Since $\mathfrak{a}^{[p^e]} \subseteq \mathfrak{a}^{p^e}$ for all e, comparison with the ordinary multiplicity of $\mathfrak{a}$ shows that the Hilbert-Kunz multiplicity is positive. These numbers, in particular where $\mathfrak{a}$ is the maximal ideal of A, have been studied by Monsky [46] and others, but they remain quite mysterious. In particular, it is not known whether the Hilbert-Kunz multiplicity is always rational, even for the maximal ideal of a local ring.

In the next application, we let $F_\bullet$ be a bounded free complex of arbitrary length with homology of finite length over a local ring of dimension d. For every finitely generated module M, we let

$$\chi_\infty(F_\bullet; M) = \lim_{e \to \infty} \frac{\chi(\mathbf{F}^e(F_\bullet) \otimes M)}{p^{ed}}.$$

Since the hypothesis on $F_\bullet$ implies that $F_\bullet \otimes M$ has homology of finite length for all M, Theorem 7.3.3 implies that this limit exists. In fact, since the Euler characteristic is additive on short exact sequences, Theorem 7.3.2 implies that there exist rational numbers $a_0, \ldots, a_k$, where k is the dimension of M such that

$$\frac{\chi(\mathbf{F}^e(F_\bullet) \otimes M)}{p^{ed}} = a_k p^{ek} + \cdots + a_0$$

for all $e \geq 0$. In the case in which $M = A$, we write $\chi_\infty(F_\bullet; A) = \chi_\infty(F_\bullet)$. We will show later that this number can also be interpreted by using Chern characters and that, in particular, its denominator is at most $d!$. Dutta [13] has developed a more extensive theory, applying this idea not only to the Euler characteristic of a complex, but also to intersection multiplicities of modules of the type described in Chapter 5.

We prove here a positivity property of χ_∞ that is closely related to Proposition 5.2.6 and that will be used in Chapter 13.

Theorem 7.3.5 *Let A be a ring of positive characteristic of dimension d, and let $F_\bullet$ be a complex of free modules of length d with homology of finite length and such that $H_0(F_\bullet) \neq 0$. Then $\chi_\infty(F_\bullet) > 0$.*

Proof We may assume that $F_\bullet$ is minimal and the assumption that $H_0(F_\bullet) \neq 0$ implies that $F_0 \neq 0$.

This theorem is proven by considering the limit

$$\lim_{e \to \infty} \frac{\text{length}(H_i(\mathbf{F}^e(F_\bullet)))}{p^{de}} \qquad (**)$$

for each i. If $i = 0$, then the homology is the cokernel of the map from $\mathbf{F}^e(F_1)$ to $\mathbf{F}^e(F_0)$. Since this map has entries in $\mathfrak{m}^{p^e}$ and the limit

$$\lim_{e \to \infty} \frac{\text{length}(A/\mathfrak{m}^{p^e})}{p^{de}}$$

is positive, we deduce that the limit $(**)$ is positive for $i = 0$.

To conclude the proof we show that the limit $(**)$ is zero for $i > 0$. The proof is similar to that of Proposition 5.2.6. We consider the double complex obtained by taking $\text{Hom}(\mathbf{F}^e(F_\bullet), D^\bullet)$, where $D^\bullet$ is a dualizing complex for A. We thus have a diagram

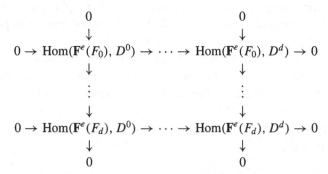

The remainder of the proof follows the same argument as Proposition 5.2.6, so we sketch it briefly. As in that proof, we show that all columns of the double complex are exact except for the one on the right, so that the homology of the associated total complex is isomorphic to the homology of the column in degree d, which has the same length as the homology of $\mathbf{F}^e(F_\bullet)$. Also following the proof of Proposition 5.2.6, we show that there exists a filtration of the homology of the total complex with quotients that are subquotients of the homology modules $H_i(\mathbf{F}^e(F_\bullet) \otimes H^j(D^\bullet))$. Thus, since $H^j(D^\bullet)$ has dimension at most $d - j$, it remains to show that, if M is a finitely generated module of dimension less than d, we have

$$\lim_{e\to\infty} \frac{\text{length}(\mathbf{F}^e(F_\bullet) \otimes M)}{p^{ed}} = 0.$$

However, if M is a module of dimension less that d, there is a polynomial in p^e of degree less than d whose value is equal to the length of $\mathbf{F}^e(F_\bullet) \otimes M$ for large e, so this limit is zero. Hence the only i for which the limit $(**)$ is not zero is $i = 0$, so we have $\chi_\infty(F_\bullet) > 0$ □

Generalizations of this theorem can be found in Dutta [14].

7.4 Tight Closure

Recently, Hochster and Huneke have developed a new method for using the properties of the Frobenius map, called *tight closure*. This concept was introduced in Hochster and Huneke [26], and we refer to the book by Huneke [28] for more on the theory as well as a complete bibliography. Tight closure is a closure operation on ideals and on submodules of a module that has roughly the same relation to integral closure as Hilbert-Kunz mutiplicities have to Samuel multiplicities. We present in this section a small part of this theory, which is related to other properties discussed in this book. To simplify the discussion, we assume throughout this section that A is an integral domain of positive characteristic p, and it is usually a local domain.

Let $\mathfrak{a}$ be an ideal in an integral domain of positive characteristic. We say that $x \in A$ is in the *tight closure* of $\mathfrak{a}$ if there exists an element $c \neq 0$ in A such that $cx^{p^e} \in \mathfrak{a}^{[p^e]}$ for all integers e. The set of such elements is denoted $\mathfrak{a}^*$.

Proposition 7.4.1 *The tight closure of $\mathfrak{a}$ is an ideal of A and is tightly closed (i.e., $(\mathfrak{a}^*)^* = \mathfrak{a}^*$).*

Proof The set $\mathfrak{a}^*$ is clearly closed under multiplication by elements of A. If a and b are in $\mathfrak{a}^*$, let c and d be nonzero elements such that ca^{p^e} and db^{p^e} are in $\mathfrak{a}^{[p^e]}$ for all e. Then $cd(a+b)^{p^e} = cd(a^{p^e} + b^{p^e}) = d(ca^{p^e}) + c(db^{p^e})$ is in $\mathfrak{a}^{[p^e]}$ for all e, so $a + b$ is in the tight closure of $\mathfrak{a}$.

To show that $\mathfrak{a}^*$ is tightly closed, let $a \in (\mathfrak{a}^*)^*$. Let c be a nonzero element such that $ca^{p^e} \in (\mathfrak{a}^*)^{p^e}$ for all e. Since A is Noetherian, the ideal $\mathfrak{a}^*$ is finitely generated, so there is a single element c' that satisfies the condition that $c'b^{p^e} \in (\mathfrak{a}^*)^{p^e}$ for all elements b of $\mathfrak{a}^*$. We then have $(c'c)a^{p^e} = c'ca^{p^e} \in c'(\mathfrak{a}^*)^{p^e} \subseteq \mathfrak{a}^{p^e}$ for all e, so a is in the tight closure of $\mathfrak{a}$.   □

The next proposition implies that the tight closure of $\mathfrak{a}$ is contained in the integral closure of $\mathfrak{a}$.

Proposition 7.4.2 *Let x be an element of A and let $\mathfrak{a}$ be an ideal of A. Then the following are equivalent:*

(i) *x is integral over $\mathfrak{a}$.*

(ii) *There exists a nonzero element c in A such that $cx^n \in \mathfrak{a}^n$ for all $n \geq 1$.*

(iii) *There exists a nonzero element c in A such that $cx^n \in \mathfrak{a}^n$ for an infinite number of n.*

Proof Assume first that x is integral over $\mathfrak{a}$. We can then write

$$x^k = a_1 x^{k-1} + \cdots + a_k$$

with $a_i \in \mathfrak{a}^i$. Thus, in particular, we have $x^k \in \mathfrak{a}$. We next show by induction that $x^r \in \mathfrak{a}^{r-k+1}$ for all $r \geq k$. Assuming that this statement is true for r, we multiply the above equation of integral dependence by x^{r-k+1} to obtain

$$x^{r+1} = a_1 x^r + \cdots + a_j x^{r-j+1} + \cdots + a_k x^{r-k+1}.$$

We claim that each term of the expression on the right-hand side is in $\mathfrak{a}^{r-k+2}$. If $r - j + 1 \geq k$, then the induction hypothesis implies that $a_j x^{r-j+1} \in \mathfrak{a}^j \mathfrak{a}^{r-j+1-k+1} = \mathfrak{a}^{r-k+2}$. If $r - j + 1 < k$, then $j \geq r - k + 2$, so a_j itself is in $\mathfrak{a}^{r-k+2}$. Thus, each term is in $\mathfrak{a}^{r-k+2}$, so $x^r \in \mathfrak{a}^{r-k+2}$ for all $r \geq k$. Hence if we let $c = x^{k-1}$, we have $cx^n = x^{n+k-1} \in \mathfrak{a}^n$ for all $n \geq 1$.

It is clear that (ii) implies (iii). Suppose that there exists a $c \neq 0$ such that $cx^n \in \mathfrak{a}^n$ for an infinite number of n. We use the criterion that x is integral over $\mathfrak{a}$ if and only if xT is integral over the Rees ring $R(\mathfrak{a})$ of $\mathfrak{a}$ considered as a subring of $A[T]$ (see Proposition 2.4.3). Let $R(\mathfrak{a}) = B$ and let $y = xT$. If y is not integral over B, then y is not in the subring $B[y^{-1}]$ of the quotient field generated by y^{-1}, so y^{-1} is not a unit in $B[y^{-1}]$. The hypothesis implies that $cy^n \in B \subseteq B[y^{-1}]$ for infinitely many n. Hence $c \in (y^{-1})^n B[y^{-1}]$ for infinitely many n. But since y^{-1} generates a proper ideal of the Noetherian domain $B[y^{-1}]$, we must have $c = 0$. Thus, we must have had that y is integral over B, and thus x is integral over $\mathfrak{a}$. $\square$

In Chapter 2 we showed that if $\mathfrak{a} \subseteq \mathfrak{b}$ are $\mathfrak{m}$-primary ideals such that $\mathfrak{b}$ is integral over $\mathfrak{a}$, then the multiplicities of $\mathfrak{a}$ and $\mathfrak{b}$ are equal. The next proposition is an analogous result for Hilbert-Kunz multiplicities.

Proposition 7.4.3 *Suppose that $\mathfrak{a}$ and $\mathfrak{b}$ are $\mathfrak{m}$-primary ideals such that $\mathfrak{a} \subseteq \mathfrak{b} \subseteq \mathfrak{a}^*$. Then the Hilbert-Kunz multiplicities of $\mathfrak{a}$ and $\mathfrak{b}$ are equal.*

Proof Let d be the dimension of A. The hypothesis implies that there exists a nonzero element $c \in A$ such that $\mathfrak{b}^{p^e}/\mathfrak{a}^{p^e}$ is annihilated by c for all e. Let $b_1, \ldots, b_k$ generate $\mathfrak{b}$. Then for each e there is a surjective map

$$\left(\oplus_{i=1}^k (A/cA) \right) / \mathfrak{a}^{p^e} \left(\oplus_{i=1}^k (A/cA) \right) \to \mathfrak{b}^{p^e}/\mathfrak{a}^{p^e}$$

sending the ith generator to $b_i^{p^e}$. Thus, we have

$$\text{length}(A/\mathfrak{a}^{p^e}) - \text{length}(A/\mathfrak{b}^{p^e}) = \text{length}(\mathfrak{b}^{p^e}/\mathfrak{a}^{p^e})$$
$$\leq \text{length}\left(\oplus_{i=1}^k (A/cA) \right) / \mathfrak{a}^{p^e} \left(\oplus_{i=1}^k (A/cA) \right)$$

for all e. Since the dimension of $\oplus_{i=1}^k (A/cA)$ is $d - 1$, by Theorem 7.3.3 this difference is bounded by a constant times $p^{e(d-1)}$, and hence the Hilbert-Kunz multiplicities of $\mathfrak{a}$ and $\mathfrak{b}$ are equal. $\qquad \square$

We conclude this section with a brief description of some related results and conjectures.

Hochster and Huneke [27] have shown using ideas similar to those described here that the integral closure of an integral domain of positive characteristic in the algebraic closure of its quotient field is Cohen-Macaulay, giving simple proofs and stronger versions of many of the homological conjectures in this case.

The concept of tight closure can be generalized to submodules of a free module F by letting $M^{[p^e]}$ denote the submodule of F generated by elements obtained from elements of M by taking the (p^e)th powers of the coordinates with respect to a basis of F. Then m is in the tight closure of M if there is a $c \neq 0$ in A such that $cm^{p^e} \in M^{[p^e]}$ for all e, where m^{p^e} is the element whose coordinates are the (p^e)th powers of the coordinates of m.

Consider a bounded free complex $F_\bullet = 0 \to F_d \to \cdots \to F_0 \to 0$ with homology of finite length, where d is the dimension of the ring. Then it follows from Theorem 4.6.6 that there is a non-zero element c which annihilates the homology of $\mathbf{F}^e(F_\bullet)$ for all e, and therefore in positive characteristic the kernel of d_i^F is contained in the tight closure of the image of d_{i+1}^F for all $i > 0$. It is conjectured that the kernel of d_i^F is integral over the image of d_{i+1}^F for all $i > 0$ in any characteristic, where an element of a free module F is said to be integral over a submodule M if the corresponding element of the polynomial ring $A[T_1, \ldots, T_k]$ in degree 1 is integral over the Rees ring of M. This conjecture is known for rings of positive characteristic and in low dimension but not in general, and it would imply several of the conjectures in Chapter 6 (see the exercises for one example). A study of this conjecture has led to new examples of non-Noetherian symbolic Rees algebras (Roberts [56]).

7.5 Reduction from Characteristic Zero
to Positive Characteristic

While the method of the Frobenius map works only for rings of positive characteristic, some of the results follow in the case of equal characteristic zero by reducing to the case of positive characteristic p for a suitable prime number p. This method was developed by Peskine and Szpiro [50] for the case of finitely generated algebras over a field and certain complete local rings, and it was completed by Hochster [23, 24], who showed how to use the Artin approximation theorem to reduce from the complete case to the finitely generated over a field case. We give a short outline of one example of this method, which illustrates its main points, and we refer to the works cited above for the details.

Assume that we had a counterexample to the new intersection theorem over a ring A, which is the localization of a ring that is finitely generated over a field K of characteristic zero. Then K contains $\mathbb{Q}$, and the first step is to adjoin enough elements of A to $\mathbb{Q}$ so that the example can be lifted to the algebra they generate over $\mathbb{Q}$. This requires several finiteness conditions, including the condition that the example involves only a finite number of elements of A (which it does). One then adjoins a finite number of inverses of elements of $\mathbb{Z}$ so that the example is contained from an example over $\mathbb{Z}$ by localization. Finally, one shows that for all but finitely many primes $p \in \mathbb{Z}$, the reduction modulo p remains a counterexample over $\mathbb{Z}/p\mathbb{Z}$. Thus, the original counterexample could not have existed.

The reduction from the complete case to the case of a ring finitely generated over a field is done by using the Artin approximation theorem, which states that a solution to a set of equations over a complete ring can be lifted to a solution over an algebraic extension. The idea is to include the equations defining the ring over a power series ring into the equation and reduce to a localization of a polynomial ring. For details we refer to Hochster [24] and Szpiro [68].

Exercises

7.1 Give a direct proof of Theorem 7.3.3 for the case of dimension 0.

7.2 Show that the Hilbert-Kunz multiplicity of an $\mathfrak{m}$-primary ideal $\mathfrak{a}$ in a regular local ring A is equal to the length of $A/\mathfrak{a}$.

7.3 Show that if $\mathfrak{a}$ is an ideal generated by a system of parameters for A, then the Hilbert-Kunz multiplicity of $\mathfrak{a}$ is equal to the Samuel multiplicity of $\mathfrak{a}$.

7.4 Give an example to show that the conclusion of the previous problem is not true for all $\mathfrak{m}$-primary ideals.

7.5   Let A be a local ring that is finitely generated as a module over a regular
      local subring R of dimension 3. Let x_1, x_2, x_3 be a regular system of
      parameters for R, so also a system of parameters for A.
      (a) Compute a minimal free resolution $F_\bullet$ of $R/\mathfrak{a}$ over R, where $\mathfrak{a}$ is the
          ideal generated by $x_1^t, x_2^t, x_3^t, x_1^{t-1}x_2^{t-1}x_3^{t-1}$.
      (b) Show that, if in the complex $F_\bullet \otimes_R A$ the kernel of d_1 is integral over
          the image of d_2, then the monomial conjecture holds for this system
          of parameters. (See Katz [31] for a discussion of this conjecture.)
7.6   Show that Z is in the tight closure of the ideal generated by X and Y in
      the ring $k[X, Y, Z]/(X^7 + Y^3 + Z^2)$.
7.7   Show that the definition of $M^{[p^e]}$ for a submodule M of a free module F
      does not depend on the choice of basis of F.

Part Two

Chern Classes

8

Projective Schemes

In Chapter 1 we defined Spec(A), for a Noetherian ring A, to be the set of all prime ideals of A. From the set of prime ideals of A we formed a group of cycles $Z_*(A)$ and the Chow group $A_*(A)$. In addition, we defined a topology on Spec(A) in which the closed sets are given by

$$V(\mathfrak{a}) = \{\mathfrak{p} \in \mathrm{Spec}(A) \mid \mathfrak{a} \subseteq \mathfrak{p}\}$$

for every ideal $\mathfrak{a}$. In this chapter we carry out similar constructions for a graded ring. These constructions are essential for the definition of Chern classes, and they are necessary even for the applications to modules over local rather than graded rings. We do not use the general theory of schemes, for which we refer to Hartshorne [21]. In fact, we develop some of the theory for multi-graded rings differently than usual; we define the associated projective scheme of such a ring in terms of multi-graded prime ideals satisfying certain properties. While this involves some extra complications, it often makes it simpler to treat the multi-graded rings and modules that arise in our algebraic applications. In addition, for specific computations it is often easier to deal directly with multi-graded rings.

In the first section we define the projective scheme corresponding to a $\mathbb{Z}$-graded ring. We then carry out the analogous construction for a $\mathbb{Z}^n$-graded ring and introduce a more general theory of Hilbert polynomials for $\mathbb{Z}^n$-graded rings that will be used in later applications. The remainder of the chapter is devoted to the Chow group, intersection operations defined by divisor and hyperplane sections, and functorial properties of the objects and operations we define. The last section gives a proof of a classical version of Bézout's theorem, completing a discussion of Euler characteristics for graded modules initiated in Section 6.3.

141

8.1 Projective Schemes Defined by $\mathbb{Z}$-graded Rings

Let A be a $\mathbb{Z}$-graded ring. We assume, as in Chapter 2, that A is Noetherian and is finitely generated as an algebra over A_0 by elements of degree 1. We recall that the ideal $\oplus_{i>0} A_i$ of A is called the irrelevant ideal of A.

Definition 8.1.1 *Let A be a $\mathbb{Z}$-graded ring such that A is generated over A_0 by elements of degree 1. We define* $\operatorname{Proj}(A)$ *to be the set of all graded prime ideals of A that do not contain the irrelevant ideal of A.*

We call $\operatorname{Proj}(A)$ a *projective scheme*. In this context we will often refer to Spec of a ring as an *affine scheme*.

The topology on $\operatorname{Proj}(A)$ is defined in the same way as the Zariski topology for $\operatorname{Spec}(A)$; we define a set to be closed if it is of the form $V(\mathfrak{a}) = \{\mathfrak{p} \in \operatorname{Proj}(A) \mid \mathfrak{a} \subset \mathfrak{p}\}$, where $\mathfrak{a}$ is a graded ideal. In particular, if x is a homogeneous element of degree 1, the set U_x consisting of graded prime ideals that do not contain x is the complement of $V(\mathfrak{a})$ where $\mathfrak{a}$ is the ideal generated by x and is thus open. The condition that graded prime ideals in $\operatorname{Proj}(A)$ are not allowed to contain the irrelevant ideal, together with the assumption that A is generated over A_0 by elements of degree 1, imply that $\operatorname{Proj}(A)$ is covered by open sets of the form U_x.

Let S be a multiplicatively closed subset of A consisting of homogeneous elements. If M is a graded A-module, we denote by $M_{(S)}$ the set of m/s, where m and s are homogeneous of the same degree. Thus, $M_{(S)}$ consists of the elements of degree 0 in the localization M_S. If $\mathfrak{p}$ is a graded prime ideal, we denote by $A_{(\mathfrak{p})}$ the localization $A_{(S)}$, where S is the set of homogeneous elements that are not in $\mathfrak{p}$, and similarly for $M_{(\mathfrak{p})}$. If x is a homogeneous element of A, we denote $A_{(x)}$ and $M_{(x)}$ the elements of degree 0 in the localizations A_x and M_x. We note that for any such multiplicatively closed set S, $A_{(S)}$ is a ring and $M_{(S)}$ is an $A_{(S)}$-module.

Proposition 8.1.1 *Let A be a $\mathbb{Z}$-graded ring, and let x be an element of degree 1. Then there is a natural bijection between U_x and $\operatorname{Spec}(A_{(x)})$.*

Proof Define the map $\phi : U_x \to \operatorname{Spec}(A_{(x)})$ by letting $\phi(\mathfrak{p})$ be the intersection of the localization $\mathfrak{p}_x$ with the subring $A_{(x)}$ of A_x. First of all, from Proposition 1.1.2, localization defines a one-one correspondence between prime ideals of A_x and prime ideals of A that do not contain x. Since x is homogeneous, the localization of a graded prime ideal is graded. Conversely, assume that $\mathfrak{p}_x$ is graded, and let $a = \sum a_i$ be an element of $\mathfrak{p}$, where a_i is the homogeneous component of degree i. Since $\mathfrak{p}_x$ is graded, $a_i/1$ is in $\mathfrak{p}_x$ for each i, so $x^k a_i$

is in $\mathfrak{p}$ for some k; since $x \notin \mathfrak{p}$, this implies that $a_i \in \mathfrak{p}$. Hence $\mathfrak{p}$ is graded. Thus, there is a one-one correspondence between graded prime ideals of A_x and graded prime ideals of A that do not contain x, the set we are denoting U_x.

We next show that intersection with $A_{(x)}$ defines a one-one correspondence between the graded prime ideals of A_x and prime ideals of $A_{(x)}$. In fact, an inverse to this map is constructed by sending a prime ideal $\mathfrak{q}$ of $A_{(x)}$ to $\mathfrak{q}A_x$. Since $\mathfrak{q}$ consists of elements of degree 0, $\mathfrak{q}A_x$ is the direct sum of the submodules $\mathfrak{q}((A_x)_n)$ over all indices n, and it is clear that $(\mathfrak{q}A_x) \cap A_{(x)} = \mathfrak{q}A_{(x)} = \mathfrak{q}$.

To complete the proof it suffices to show that, for every graded prime ideal $\mathfrak{p}$ of A_x, we have $\mathfrak{p} = (\mathfrak{p} \cap A_{(x)})A_x$. Since $\mathfrak{p}$ is graded, it is enough to show that every homogeneous element a of $\mathfrak{p}$ belongs to $(\mathfrak{p} \cap A_{(x)})A_x$. If the degree of a is i, then $a = (a/x^i)x^i$ and a/x^i is in $A_{(x)}$, so we have the required inclusion. Thus, we have shown that ϕ is a bijection. $\qquad\square$

Let $X = \text{Proj}(A)$. If $\mathfrak{p}$ is a graded prime ideal of A that does not contain the irrelevant ideal, we let the dimension of $\mathfrak{p}$, or the dimension of $A/\mathfrak{p}$, be the dimension of the corresponding prime ideal of $A_{(x)}$ for any x not in $\mathfrak{p}$. Since dimension is not affected by localization (see Section 4.2), the dimension of $\mathfrak{p}$ does not depend on which x is chosen. We let the group of cycles of X be the graded group whose component in degree k is the free Abelian group on all elements $\mathfrak{p}$ of X such that the dimension of $A/\mathfrak{p}$ is k. Note that this definition is similar to that for a nongraded ring, except that we restrict to graded ideals that do not contain the irrelevant ideal. The definition of rational equivalence is also similar, although here the difference from the nongraded case is more substantial. Let $\mathfrak{q}$ be a graded prime ideal with $\dim(A/\mathfrak{q}) = k + 1$. Let x and y be nonzero homogeneous elements of $A/\mathfrak{q}$ of the same degree. We then define $\text{div}(\mathfrak{q}, x/y)$ to be the cycle $[A/\mathfrak{q}/xA/\mathfrak{q}]_k - [A/\mathfrak{q}/yA/\mathfrak{q}]_k$, and we define rational equivalence to be the equivalence relation generated by $\text{div}(\mathfrak{q}, x/y)$ for all such $\mathfrak{q}$, x, and y.

Definition 8.1.2 *Let $X = \text{Proj}(A)$, where A is a $\mathbb{Z}$-graded ring. The* Chow group *of X is the quotient of the group of cycles of X modulo rational equivalence. The Chow group is denoted $A_*(X)$.*

For convenience, we will also use the notation $\text{div}(\mathfrak{q}, x)$ for any homogeneous element x not in $\mathfrak{q}$. Unless x has degree zero, however, $\text{div}(\mathfrak{q}, x)$ is not rationally equivalent to zero.

As before, we may define the Chow group of any open or closed subset of $\text{Proj}(A)$, and the theorems on restriction to an open set and imbedding of a closed set from Chapter 1 go through as before.

If M is a finitely generated graded A-module, we define the dimension of M to be the supremum of the dimensions of $A/\mathfrak{p}$ for prime ideals $\mathfrak{p}$ in the support of M. If M has dimension at most k we define a cycle $[M]_k$ as in Chapter 1 by letting

$$[M]_k = \sum \text{length}\big(M_{(\mathfrak{p})}\big)[A/\mathfrak{p}],$$

where the sum is taken over all $\mathfrak{p}$ in Proj(A) in the support of M with $\dim(A/\mathfrak{p}) = k$. We note that the length of $M_{(\mathfrak{p})}$ over $A_{(\mathfrak{p})}$ is the same as the length of $M_{\mathfrak{p}}$ over $A_{\mathfrak{p}}$; this equality is certainly true when $M = A/\mathfrak{p}$, and since both lengths are additive on short exact sequences it is true for all M.

As in the nongraded case, we have the following proposition.

Proposition 8.1.2 *Let M be a finitely generated graded A-module of dimension at most $i + 1$. Let x be a homogeneous element of A that is contained in no minimal prime ideal of dimension $i + 1$ in the support of M. Then*

$$[M/xM]_i - [_xM]_i = \sum_{\{\mathfrak{p}|\dim(\mathfrak{p})=i+1\}} \text{length}(M_{\mathfrak{p}})\text{div}(\mathfrak{p}, x)$$

where $_xM$ is the set of elements of M annihilated by x.

Proof The proof is the same as that of Proposition 1.2.2. □

Let A be a graded ring, and let A_0 be the subring of elements of degree 0. We then have a map from Proj(A) to Spec(A_0) obtained by intersecting a graded prime ideal with A_0. A map obtained in this way will be called a *projective* map. We say that this map is *flat of relative dimension k* if, for any element $x \in A_1$, the map from A_0 to $A_{(x)}$ is flat of relative dimension k. As in the affine case treated in Chapter 1, a flat map of relative dimension k induces a map on Chow groups.

Theorem 8.1.3 *Let $\phi : X = \text{Proj}(A) \to \text{Spec}(A_0) = Y$ be a flat projective map of dimension k. Then the map ϕ^* defined by letting*

$$\phi^*([A_0/\mathfrak{p}]) = \big[(A_0/\mathfrak{p}) \otimes_{A_0} A\big]_{i+k}$$

induces a map on Chow groups from $A_i(Y)$ to $A_{i+k}(X)$ for each i.

Proof The hypothesis implies that the map ϕ^* defines a map from $Z_i(Y)$ to $Z_{i+k}(X)$ for each i, and we have to show that it preserves rational equivalence. If $\mathfrak{q}$ is a prime ideal of A_0 of dimension $i + 1$ and x is an element of A_0 not in q, then we have to show that $\phi^*(\mathrm{div}(q, x))$ is rationally equivalent to zero. Let B denote $(A_0/q) \otimes_{A_0} A$. From Proposition 8.1.2 we have

$$[B/xB]_i - [_xB]_i = \sum_{\{\mathfrak{p}|\dim(\mathfrak{p})=i+1\}} \mathrm{length}(M_\mathfrak{p})\mathrm{div}(\mathfrak{p}, x).$$

Since x has degree zero, this equation implies that $[B/xB]_i - [_xB]_i$ is rationally equivalent to zero. The module $_xB$ may not be zero, but since we have that $A_{(y)}$ is flat over A_0 for all y of degree 1, $_xB_{(y)}$ is zero for all such y, so that $_xB$ has no prime ideals of $\mathrm{Proj}(A)$ in its support. Thus, the class $[_xB]_i$ is zero and $[B/xB]$ is rationally equivalent to zero.

It remains to show that $[B/xB] = \phi^*(\mathrm{div}(q, x))$. This argument is the same as that of Theorem 1.2.3, and it proceeds by taking a filtration for $(A_0/q)/x(A_0/q)$ and showing that it induces a filtration on B/xB up to factors whose support has empty intersection with $\mathrm{Proj}(A)$ and can thus be omitted. $\qquad \square$

As in the affine case, we call the map ϕ^* the *flat pull-back* induced by ϕ. We show in a later section that a projective map also induces a map from $A_i(X)$ to $A_i(Y)$ for each i.

8.2 Projective Schemes Defined by $\mathbb{Z}^n$-graded Rings

In Chapter 2 we defined graded rings over $\mathbb{Z}^n$ for the purpose of defining Hilbert polynomials and mixed multiplicities. It was always assumed that graded rings were generated by elements of total degree 1 and had no elements of negative degree. Since this condition is satisfied by Rees rings and associated graded rings of ideals, it sufficed for the purposes of that chapter. However, for later applications we will need to use graded rings that do not satisfy this condition but that are still special enough so that the theory of Hilbert polynomials works in pretty much the same way.

The main situation for which we need $\mathbb{Z}^n$-graded rings for $n > 1$ is for the construction of Rees rings and symmetric algebras of ideals and modules over a graded ring. As an example, let $\mathfrak{a}$ be a graded ideal of a $\mathbb{Z}$-graded ring. There is no reason why $\mathfrak{a}$ should have any elements of degree 0, and, if it does not, the bigraded Rees algebra of $\mathfrak{a}$ will not be generated by elements of degrees $(1, 0)$ and $(0, 1)$. For example, if $A = k[X, Y]$ with the usual grading and $\mathfrak{a}$ is the ideal generated by X, $\mathfrak{a}$ will have no elements of degree 0. If we give the Rees ring $R(\mathfrak{a})$ the grading in which the first component is the degree as an A-module and the second is the power of $\mathfrak{a}$, then $R(\mathfrak{a})$ will contain no nonzero elements of degree $(0, 1)$. Thus, we cannot impose the condition that A be generated over A_0 by elements of total degree 1 if we wish to allow this example. The condition we use instead is constructed so that it will be satisfied by Rees and symmetric algebras and so that a modified version of the theory of Hilbert polynomials still works. If A is a $\mathbb{Z}^n$-graded ring, for each $j = 0, \ldots, n$ we let $A_{[j]}$ denote the subring of A consisting of elements of degrees of the form $(i_1, \ldots, i_j, 0, \ldots, 0)$. The condition we impose is that for each $j = 0, \ldots,$ $n - 1$, $A_{[j+1]}$ must be finitely generated over $A_{[j]}$ by elements of degrees of the form $(i_1, \ldots, i_j, 1, 0, \ldots, 0)$ for various j-tuples $i_1, \ldots, i_j$. It is clear that if a $\mathbb{Z}^n$-graded ring has this property, then Rees rings and symmetric algebras considered as $\mathbb{Z}^{n+1}$-graded rings in the usual way also have this property. In Chapter 2 we considered rings that were graded over A_0 by elements of degree e_i, where $e_i = (0, \ldots, 0, 1, \ldots 0)$ with 1 in the ith position. It is clear that if A is generated over A_0 by elements of degree e_i for various i, then A satisfies the new condition. For $\mathbb{Z}$-graded rings, these two conditions are equivalent.

In the remainder of the chapter, "graded ring" will mean "$\mathbb{Z}^n$-graded ring such that $A_{[j+1]}$ is finitely generated over $A_{[j]}$ by elements of degree $(i_1, \ldots, i_j, 1, 0, \ldots, 0)$ for each j." We note that this condition implies that for each fixed i_{j+1}, there are finitely many monomials in the generators of $A_{[j+1]}$ such that the $j + 1$ component of their degree is i_{j+1}, so that the $A_{[j]}$-submodule of such elements is finitely generated.

One exception to the above convention is that we will also have to consider localizations of graded rings at sets of homogeneous elements. These localizations are usually not generated by elements of the type described above. All the graded rings we consider will either satisfy the condition of the previous paragraph or be localizations of rings satisfying this condition at multiplicative sets of homogeneous elements.

As mentioned above, one of the properties we wish to establish for graded rings is the existence of Hilbert polynomials. However, Hilbert polynomials as defined previously do not exist; to see this, we consider the simple example of the bigraded ring $A = k[X, Y]$, where X has degree $(1, 0)$ and Y has degree $(1, 1)$. Let $H(i, j)$ be the Hilbert function of A, so that $H(i, j)$ is the length of A_{ij}, which is equal to 1 if $i \geq j \geq 0$ and zero otherwise. There do not exist integers i_0 and j_0 such that the function $H(i, j)$ is a polynomial for $i \geq i_0$ and $j \geq j_0$. However, there is a cone of pairs of integers on which this function is a polynomial (a constant polynomial with constant value equal to 1). We next formalize the condition for a set of n-tuples of integers to form a cone on which the Hilbert function is a polynomial; this definition will also be used in our definition of projective schemes in the multi-graded case.

We say that a subset C of $\mathbb{Z}^n$ is a *positive cone* if there are nonnegative integers $k_1, \ldots, k_{n-1}, k_n$ such that C is the set of n-tuples $(i_1, \ldots, i_n)$ such that $i_j \geq k_j i_{j+1}$ for $j = 1, \ldots, n - 1$ and such that $i_n \geq k_n$. If all k_j are equal to zero, we obtain as a special case the set of all n-tuples with nonnegative components. We define a *positive cone with vertex at zero* to be a positive cone for which $k_n = 0$.

In Chapter 2 we showed that the Hilbert function of a graded module is given by a polynomial on a set of the form $u + C$, where C is the positive cone with vertex at zero consisting of elements of $\mathbb{Z}^n$ with nonnegative components. We will show here that Hilbert functions of our more general graded rings are polynomials on a positive cone of the type we have just defined.

Proposition 8.2.1

(i) *If C and D are positive cones, then $C \cap D$ is a positive cone.*

(ii) *A positive cone generates $\mathbb{Z}^n$ as an Abelian group.*

Proof Let C and D be defined by the integers $k_1, \ldots, k_n$ and $k'_1, \ldots, k'_n$, respectively. If we then let k''_i be the greater of k_i and k'_i for each i, $C \cap D$ is equal to the positive cone defined by $k''_1, \ldots, k''_n$. This proves (i).

Let C be the positive cone defined by integers $k_1, \ldots, k_n$. To prove that C generates $\mathbb{Z}^n$ as an Abelian group, we first note that if we have integers $i_m, i_{m+1}, \ldots i_n$ such that $i_j \geq k_j i_{j+1}$ for $j = m, \ldots, n-1$ and $i_n \geq k_n$, then we can choose $i_{j-1}, \ldots i_1$ successively so that they are large enough so that $(i_1, \ldots, i_n)$ is in C. In particular, C is not empty. Let $(i_1, \ldots, i_n)$ be in C, and let j be an integer between 1 and n. From the above discussion there exist $m_1, \ldots, m_{j-1}$ such that $(m_1, \ldots, m_{j-1}, i_j + 1, \ldots, i_n)$ is in C. We then have that

$$(m_1, \ldots, m_{j-1}, i_j + 1, \ldots, i_n) - (i_1, \ldots, i_{j-1}, i_j, \ldots, i_n)$$
$$= (r_1, \ldots, r_{j-1}, 1, 0, \ldots, 0)$$

is in the Abelian group generated by C for some integers $r_1, \ldots, r_{j-1}$. Since there exists an element of this form for each j, C generates $\mathbb{Z}^n$ as an Abelian group. $\qquad\square$

We remark that if C is a positive cone with vertex at zero, then the construction in the proof of the second part of Proposition 8.2.1 shows that there exists an element in C of the form $(i_1, \ldots, i_{j-1}, 1, 0, \ldots, 0)$ for each j between 1 and n.

We next define projective schemes corresponding to $\mathbb{Z}^n$-graded rings.

Definition 8.2.1 *Let A be a $\mathbb{Z}^n$-graded ring. $\mathrm{Proj}(A)$ is the set of graded prime ideals $\mathfrak{p}$ of A such that the set of indices of homogeneous elements not in $\mathfrak{p}$ contains a positive cone with vertex at zero in $\mathbb{Z}^n$.*

As before, we call $\mathrm{Proj}(A)$ a *projective scheme*. To avoid special cases, we include the case where $n = 0$; in this case what we denote $\mathrm{Proj}(A)$ is just $\mathrm{Spec}(A)$. If A is a $\mathbb{Z}$-graded ring, then the only positive cone with vertex at zero is the set of all nonnegative integers, so that a graded prime ideal is in $\mathrm{Proj}(A)$ if and only if there is an element of degree 1 not in $\mathfrak{p}$, and this definition agrees with the previous one.

More generally, let A be generated over A_0 by elements of degree e_i for various i as in the situation considered in Chapter 2. We recall that for each i we define the irrelevant ideal $\mathfrak{a}_i$ to be the ideal generated by all elements of degree e_i. If $\mathfrak{p}$ contains no irrelevant ideal, the condition above is satisfied, with C the cone of elements with nonnegative components. On the other hand, if $\mathfrak{p}$ does contain an irrelevant ideal, the set of indices i for which there is an element of degree i not in $\mathfrak{p}$ is contained in a subgroup of the form $\mathbb{Z}^{n-1}$ so that this set of indices cannot generate $\mathbb{Z}^n$ as an Abelian group and $\mathfrak{p}$ is not in $\mathrm{Proj}(A)$. Thus, if A is generated over A_0 by elements of total degree 1, a prime ideal $\mathfrak{p}$ is in $\mathrm{Proj}(A)$ if and only if it contains no irrelevant ideal.

For each homogeneous element x, let U_x denote the set of $\mathfrak{p} \in \mathrm{Proj}(A)$ such that $x \notin \mathfrak{p}$. The sets U_x, where x is any homogeneous element, form a basis for the topology on $\mathrm{Proj}(A)$, and we refer to them as basic open sets. If $x = x_1 x_2 \cdots x_n$, where the indices of x_j are of the form $(i_1, \ldots, i_j, 0, \ldots, 0)$ with $i_j > 0$, we call x a special homogeneous element and we call U_x a *special* basic open set. If $x = x_1 x_2 \cdots x_n$ where the indices of x_j are of the form $(i_1, \ldots, 1, 0, \ldots, 0)$ with 1 in the jth position, we call x a unit special homogeneous element.

Proposition 8.2.2 *For a graded prime ideal $\mathfrak{p}$ of A, the following are equivalent:*

(i) $\mathfrak{p}$ *is in* $\mathrm{Proj}(A)$.
(ii) *The set of indices of homogeneous elements not in $\mathfrak{p}$ contains a positive cone.*
(iii) *For all $j = 0, \ldots, n - 1$, there is a generator of $A_{[j+1]}$ over $A_{[j]}$ of degree $(i_1, \ldots, i_j, 1, 0 \ldots, 0)$ that is not in $\mathfrak{p}$.*
(iv) *There exists a special homogeneous element not in $\mathfrak{p}$.*
(v) *There exists a unit special homogeneous element not in $\mathfrak{p}$.*

Proof Since we have shown that a positive cone with vertex at zero contains an element of the form $(i_1, \ldots, i_{j-1}, 1, 0, \ldots, 0)$ for each j, we have that (i) implies (v), and it is clear that (v) implies (iv). It is also clear that (i) implies (ii) and that (iii) implies (v). Thus, it remains to see that (ii) implies (iii), that (iv) implies (iii), and that (v) implies (i).

Suppose that for some j, $\mathfrak{p}$ contains all the generators of $A_{[j+1]}$ over $A_{[j]}$. Then every element not in $\mathfrak{p}$ that is in $A_{[j+1]}$ must have degree with $j + 1$ component equal to zero. On the other hand, an element of A that is not in $\mathfrak{p}$ and not in $A_{[j+1]}$ must have degree $(i_1, \ldots, i_n)$ with $i_k \neq 0$ for some $k > j+1$. Hence there is no element not in $\mathfrak{p}$ with degree $(i_1, \ldots, i_{j+1}, 0, \ldots, 0)$ with $i_{j+1} > 0$. Thus, (iv) implies (iii). In addition, since every positive cone contains an element $(i_1, \ldots, i_{j+1}, 0, \ldots, 0)$ with $i_{j+1} > 0$, this also shows that (ii) implies (iii).

To prove that (v) implies (i) we first note that the subset of $\mathbb{Z}^n$ consisting of indices of homogeneous elements that are not in $\mathfrak{p}$ is closed under addition. Thus, we must show that if a subset of $\mathbb{Z}^n$ that is closed under addition contains an element of the form $(i_1, \ldots, i_j, 0, \ldots, 0)$ with $i_j = 1$ for all j, then it contains a positive cone with vertex at zero. Let S be such a subset, and suppose that for each j it contains the element $a_j = (i_{j1}, \ldots, i_{jj}, 0 \ldots, 0)$ with $i_{jj} = 1$. Let $b = (b_1, \ldots, b_n)$ be in $\mathbb{Z}^n$. We can write b uniquely as a linear combination $b = r_1 a_1 + \cdots r_n a_n$ with integer coefficients r_i. Furthermore, we can solve for $r_1, \ldots, r_n$ and obtain solutions in the form $r_n = b_n$, $r_{n-1} = b_{n-1} - s_{n-1,n} b_n, \ldots, r_1 = b_1 - s_{12} b_2 - \cdots - s_{1n} b_n$ for some (positive or negative) integers s_{ij}. We claim that there is a positive cone C with vertex at zero such that if $b \in C$, then each r_i is nonnegative. To define C, for each $i = 1, \ldots, n - 1$ choose m_i with $m_i \geq 1$ and such that $m_i \geq s_{ji}$ for all j. Let $k_i = m_{i+1} + \cdots + m_n$ for $i = 1, \ldots, n - 1$, let $k_n = 0$, and let C be the positive cone defined by $k_1, \ldots, k_{n-1}, k_n$. Let $b \in C$, and let $r_1, \ldots, r_n$ be defined by the above equations; we claim that $r_i \geq 0$ for all i. We have $b_n \geq 0$, so $r_n \geq 0$. Since $k_1 \geq 1$ for each i, we have $b_1 \geq b_2 \geq \cdots \geq b_n$; in particular, we have $b_i \geq 0$ for each i. Since $m_i \geq s_{ji}$ for all i and j and the b_i are nonnegative, we have

$$r_i = b_i - s_{i,i+1} b_{i+1} - \cdots - s_{in} b_n \geq b_i - m_{i+1} b_{i+1} - \cdots - m_n b_n.$$

Since $b \in C$, we have

$$b_i \geq k_i b_{i+1} \geq (m_{i+1} + \cdots + m_n) b_{i+1} \geq m_{i+1} b_{i+1} + \cdots + m_n b_n.$$

Hence, combining this inequality with the previous one, we have $r_i \geq 0$. Thus, every element of C can be written as a sum of a_i with nonnegative coefficients, as was to be shown. $\qquad\square$

The next proposition is a variant of Proposition 1.1.5.

Proposition 8.2.3 *Let $\mathfrak{p}_1, \ldots, \mathfrak{p}_k$ be prime ideals in* Proj(A). *Then there is a special homogeneous element that is contained in none of the $\mathfrak{p}_i$.*

Proof Since a special homogeneous element is by definition a product of elements of degree $(i_1, \ldots i_j, 0, \ldots, 0)$ with $i_j > 0$, it suffices to show that we may find such an element not in any of the $\mathfrak{p}_i$ for each j. Assume that such elements exist for $1, \ldots, j-1$. For each i, we may assume by induction on i that there exists an element x_i of the correct type that does not lie in $\mathfrak{p}_j$ for all $j \neq i$. If $x_i \notin \mathfrak{p}_i$ we are done. Otherwise, for each i we let y_i be the product of x_j for $j \neq i$. Then y_i is still homogeneous of the correct form and y_i lies in all of the $\mathfrak{p}_j$ except $\mathfrak{p}_i$. The sum of the y_i will lie in none of the $\mathfrak{p}_i$ but will not necessarily be homogeneous. The remainder of the proof consists of modifications of the y_i to make them homogeneous of the same degree.

We first replace the y_i by suitable powers so that they are of degree $(i_1, \ldots, i_j, 0, \ldots, 0)$ with $i_j > 0$ and such that i_j is the same for all i. Next, let $u_1, \ldots u_{j-1}$ be elements that lie in none of the $\mathfrak{p}_i$ and such that u_k is of degree $(i_1, \ldots, i_k, 0, \ldots, 0)$ with $i_k > 0$ for each k. We wish to multiply the y_i by suitable powers of $u_1, \ldots, u_{j-1}$ so that the resulting products will be homogeneous of the same degree. We first modify the $j-1$ component of the degree. Let u_{j-1} have degree $(i_1, \ldots, i_{j-1}, 0, \ldots, 0)$. We replace the y_i by a power of y_i, the same power for each i, so that the $j-1$ components of their degrees are divisible by i_{j-1}. Then we can multiply each y_i by a power of u_{j-1} so that the $j-1$ components of their degrees are equal. This process does not change the fact that the j components of the degrees of the y_i are equal, and now the $j-1$ components are equal as well. We continue this process for $j-2, \ldots, 1$ and thus produce a new set of y_i such that y_i is in all the $\mathfrak{p}_j$ except $\mathfrak{p}_i$ and such that all the elements of the set are homogeneous of the same degree. The sum $y_1 + \cdots + y_k$ now lies in none of the $\mathfrak{p}_j$ and has degree $(i_1, \ldots, i_j, 0 \ldots, 0)$ for some $i_1, \ldots, i_j$ with $i_j > 0$. $\qquad\square$

As in the case of $\mathbb{Z}$-graded rings, if x is a homogeneous element of A, we let $A_{(x)}$ be the subring of the localization A_x consisting of elements of degree 0. Similarly, if M is a graded A-module, we let $M_{(x)}$ denote the set of elements of M_x of degree 0; $M_{(x)}$ is then an $A_{(x)}$-module. If S is a multiplicative set of homogeneous elements, $M_{(S)}$ denotes the set of elements of degree zero in M_S. If $\mathfrak{p}$ is a graded prime ideal, then we denote $M_{(S)}$ by $M_{(\mathfrak{p})}$, where S is the set of homogeneous elements not in $\mathfrak{p}$. We let $k(\mathfrak{p})$ denote $(A/\mathfrak{p})_{(\mathfrak{p})}$.

Proposition 8.2.4 *Let A be a $\mathbb{Z}^n$-graded ring, and let $T_1, \ldots, T_k$ be indeterminates. Then the inclusion of A into $A[T_1, T_1^{-1}, \ldots T_k, T_k^{-1}]$ induces a homeomorphism between the projective schemes $\mathrm{Proj}(A)$ and $\mathrm{Proj}(A[T_1, T_1^{-1}, \ldots T_k, T_k^{-1}])$, where $A[T_1, T_1^{-1}, \ldots T_k, T_k^{-1}]$ is given the natural grading over $\mathbb{Z}^{n+k}$ by letting T_j have degree e_{n+j}.*

Proof Let B denote $A[T_1, T_1^{-1}, \ldots T_k, T_k^{-1}]$. If $\mathfrak{a}$ is a graded ideal of A, then its extension to B is a graded ideal whose intersection with A is $\mathfrak{a}$. Conversely, if $\mathfrak{b}$ is a graded ideal of B, and if x is a homogeneous element of $\mathfrak{b}$ of degree $(i_1, \ldots, i_n, i_{n+1}, \ldots, i_{n+j})$, then we have

$$x = \left(T_1^{i_{n+1}} \cdots T_j^{i_{n+j}}\right)\left(T_1^{-i_{n+1}} \cdots T_j^{-i_{n+j}} x\right)$$

and, since $T_1^{-i_{n+1}} \cdots T_j^{-i_{n+j}} x$ is in $A \cap \mathfrak{b}$, we see that x is in the ideal generated by $A \cap \mathfrak{b}$ in B. Thus, the correspondences that map a graded ideal of A to its extension to B, and that map a graded ideal of B to its intersection with A, are inverse bijections that preserve prime ideals and inclusions. In addition, since no prime ideal of B can contain any monomials in T_i and T_i^{-1}, the set of indices of elements not in a given prime ideal $\mathfrak{p}$ necessarily contains the summand $\mathbb{Z}^k$ of $\mathbb{Z}^{n+k} = \mathbb{Z}^n \oplus \mathbb{Z}^k$, so a prime ideal is in $\mathrm{Proj}(A[T_1, T_1^{-1}, \ldots T_k, T_k^{-1}])$ if and only if its intersection with A is in $\mathrm{Proj}(A)$. Hence this correspondence defines a homeomorphism between the associated projective schemes. $\qquad\square$

Proposition 8.2.5 *Let A be a $\mathbb{Z}^n$-graded ring. Then the unit special basic open sets U_x cover $\mathrm{Proj}(A)$. For every such element x, there is a homeomorphism between U_x and $\mathrm{Spec}(A_{(x)})$ in the Zariski topology.*

Proof The first statement follows from Proposition 8.2.2.

For the second statement, let $x = x_1 x_2 \cdots x_n$ be a unit special homogeneous element, and let $y_1, \ldots, y_n$ be monomials in the x_i (possibly with some negative exponents) such that y_j has degree e_j for each j; the existence of such y_j follows from the fact that the indices of the x_i generate $\mathbb{Z}^n$. The map that sends T_j to y_j induces an isomorphism of graded rings from $A_{(x)}[T_1, T_1^{-1}, \ldots, T_n, T_n^{-1}]$ to A_x, so it follows that there is a bijection between U_x and $\mathrm{Proj}(A_{(x)}[T_1, T_1^{-1}, \ldots, T_n, T_n^{-1}])$, and hence it now follows from Proposition 8.2.4 that $\mathrm{Spec}(A_{(x)})$ is homeomorphic to U_x. $\qquad\square$

We remark that the second part of the above proposition holds for any multiplicative set of homogeneous elements that contains a unit special homogeneous element. That is, if S is such a multiplicative set, there is an isomorphism $A_S \cong A_{(S)}[T_1, T_1^{-1} \ldots, T_n, T_n^{-1}]$ and a resulting correspondence of prime ideals.

8.3 Hilbert Polynomials and Dimension

While the graded rings we are now considering are more general than those discussed in the second chapter, the condition we impose on generators of A

is enough to imply that Hilbert polynomials of graded modules can be defined as long as the homogeneous components of the module have finite length. Let M be a finitely generated graded module over a graded ring such that the component M_i of degree i is of finite length for each i. We let the Hilbert function $H_M(i)$ be the function whose value is the length of M_i for all $i \in \mathbb{Z}^n$. We will show that there exists a polynomial P such that $P(i) = H_M(i)$ for all i in some positive cone in $\mathbb{Z}^n$.

Theorem 8.3.1 *Let A be a graded ring, and let M be a finitely generated A-module such that M_i has finite length as an A_0-module for all i. Then there exists a positive cone C and a polynomial P such that we have $H_M(i) = P(i)$ for all $i \in C$.*

Proof As before, we use double induction on n and m, where n is the integer for which A is a $\mathbb{Z}^n$-graded ring, and where m is an integer for which A is generated over $A_{[n-1]}$ by m elements. If $n = 1$, the ring A is an ordinary $\mathbb{Z}$-graded ring, so the theorem follows from Theorem 2.1.5 of Chapter 2.

We assume that Hilbert polynomials exist for lower values of n and m. Let $\mathbb{Z}^{n-1}$ denote the subgroup of $\mathbb{Z}^n$ consisting of elements whose last component is zero, and let $A_{[n-1]}$ denote the subring of elements whose indices lie in $\mathbb{Z}^{n-1}$ as defined above. Our assumptions imply that A is generated over $A_{[n-1]}$ by elements of degree $j = (j_0, \ldots, j_{n-1}, 1)$ for various $(n-1)$-tuples $j_1, \ldots, j_{n-1}$. Let B be a subring of A containing $A_{[n-1]}$ and generated over $A_{[n-1]}$ by $m-1$ elements, and assume that A is generated over B by an element a of degree $j = (j_0, \ldots, j_{n-1}, 1)$. We have an exact sequence of graded modules

$$0 \to {}_a M[-j] \to M[-j] \xrightarrow{a} M \to M/aM \to 0,$$

where ${}_a M[-j]$ denotes the submodule of elements of $M[-j]$ annihilated by a. Since ${}_a M[-j]$ and M/aM are annihilated by a, they are finitely generated as B-modules, so by induction on m there exists a positive cone C on which the Hilbert functions of ${}_a M$ and M/aM are given by polynomials. Thus, using the above exact sequence, we can conclude that the difference $H_M(i) - H_M(i - j)$ is given by a polynomial on the set of i such that i and $i - j$ lie in C. Let C be defined by the integers $k_1, \ldots, k_n$. Let N denote the sub-$A_{[n-1]}$-module of M consisting of elements of degree $(i_1, \ldots, i_n)$ with $i_n = k_n$. Then N is a finitely generated $A_{[n-1]}$-module, so there is a positive cone D in $\mathbb{Z}^{n-1}$ on which $H_N(i_1, \ldots, i_{n-1}, k_n)$ is a polynomial in $i_1, \ldots, i_{n-1}$. By shrinking C and D if necessary, we can assume that D is the set of elements in C whose nth component is k_n; denote this set C_{k_n}.

To summarize, we now have a positive cone C and polynomials P_0 and Q in n and $n-1$ variables, respectively, such that $P_0(i) = H_M(i) - H_M(i-j)$ for all i such that i and $i-j$ are in C and such that $Q(i_1, \ldots, i_{n-1}, k_n) = H_M(i_1, \ldots, i_{n-1}, k_n)$ whenever $(i_1, \ldots, i_{n-1}, k_n) \in C$. Consider the set of all n-tuples of the form $m + rj$, where $m \in C_{k_n}$ and r is a nonnegative integer such that $m + ij \in C$ for all i between 0 and r. We can now deduce from Lemma 2.1.6 that H_M is a polynomial $P(m_1, \ldots, m_{n-1}, r)$ for $m = (m_1, \ldots, m_{n-1}, k_n)$ and r for all such elements. We claim that the set of such elements $m + rj$ contains a positive cone C' and that, for $i \in C'$, $H_M(i)$ is a polynomial in i.

Thus, we must find a positive cone C' such that if $i = (i_1, \ldots, i_n)$ is in C', then $i_n \geq k_n$, and for all $s = 0, 1, \ldots, i_n - k_n$ we have $i - sj \in C$. As above, let C be defined by the integers $k_1, \ldots, k_n$. Choose $l_1, \ldots, l_n$ such that $l_n \geq k_n$, such that $l_m \geq 1$ for all m, and such that $l_m \geq k_m + (j_m - k_m j_{m+1})$ for $m = 1, \ldots, n-1$. Let C' be the positive cone defined by $l_1, \ldots, l_n$. We claim that if $i \in C'$, then $i_n \geq k_n$, and for $s = 0, \ldots, i_n - k_n$ we have $i - sj \in C$.

Let $i = (i_1, \ldots, i_n) \in C'$. The condition that $i_n \geq k_n$ follows since $l_n \geq k_n$. We note that the condition that $l_m \geq 1$ implies that $i_m \geq i_n$ for $m = 1, \ldots, n-1$. Let s be an integer between 0 and $i_n - k_n$. To show that $i - sj$ is in C we must show that

$$(i_m - sj_m) \geq k_m(i_{m+1} - sj_{m+1})$$

for $m = 1, \ldots, n-1$. We rewrite this inequality:

$$i_m \geq sj_m + k_m(i_{m+1} - sj_{m+1}) = k_m i_{m+1} + s(j_m - k_m j_{m+1}).$$

Since $k_n \geq 0$ and $s \leq i_n - k_n$, we have $i_n \geq s$, and we have shown that $i_{m+1} \geq i_n$; hence $i_{m+1} \geq s$. Our assumption that $i \in C'$ and our assumptions on the integers l_i defining C' imply that

$$i_m \geq l_m i_{m+1} \geq (k_m + (j_m - k_m j_{m+1}))i_{m+1}$$
$$\geq k_m i_{m+1} + (j_m - k_m j_{m+1})i_{m+1} \geq k_m i_{m+1} + (j_m - k_m j_{m+1})s,$$

which is the required inequality.

We now know that there is a positive cone C' consisting of elements of the form $m + rj$ for $r \geq 0$ and that in C', $H_M(i)$ is a polynomial in m and r. That it is also a polynomial in i, where $i = m + rj$, follows from the fact that the components of m and r can be expressed as linear functions of the components of i by letting $r = i_n - k_n$ and $m_k = i_k - j_k i_n$ for $k = 1, \ldots, n-1$. $\qquad\square$

If $P(i_1, \ldots, i_n)$ is a Hilbert polynomial of degree d, we call the homogeneous component of degree d the *leading polynomial* of P.

Lemma 8.3.2 *Let* $P(i_1, \ldots, i_n)$ *be a polynomial of degree* d, *and let* $(j_1, \ldots, j_n)$ *be an* n-*tuple of integers. Let* P^d *be the component of* P *of degree* d. *Let* $Q(i_1, \ldots, i_n) = P(i_1, \ldots, i_n) - P(i_1 - j_1, \ldots, i_n - j_n)$. *For each* k, *let* P_k^d *denote the partial derivative of* P^d *with respect to the* kth *variable. Then* Q *is a polynomial of degree at most* $d - 1$ *with component of degree* $d - 1$ *equal to* $\sum_{k=1}^{n} j_k P_k^d$.

Proof That Q is a polynomial is clear. It suffices to prove that Q has degree at most $d - 1$ with degree $d - 1$ component given by the above formula for a monomial of degree d, for which it is a straightforward computation. $\square$

If M is a graded module, we define the *dimension* of M to be the maximum of the dimensions of the $A_{(x)}$-modules $M_{(x)}$, where x runs over all special homogeneous elements of A. Since a graded prime ideal $\mathfrak{p}$ is in Proj(A) if and only if there exists a special homogeneous element not in $\mathfrak{p}$, and since the dimension of the module $M_{(x)}$ is the supremum of the dimensions of the localizations $(M_{(x)})_\mathfrak{p}$ for prime ideals in Spec$(A_{(x)})$, it follows that the dimension is also the supremum of the dimensions of $M_{(\mathfrak{p})}$ for $\mathfrak{p} \in$ Proj(A). In particular, if the support of M contains no prime ideals in Proj(A), its dimension is the same as that of the zero module. In the next chapter we introduce an equivalence relation on modules in which all such modules are equivalent to zero.

Proposition 8.3.3 *The Hilbert polynomial of a module* M *is constant if and only if* M *has dimension* 0. *In this case, there exists a positive cone* C *such that if* x *is a special homogeneous element of* A *such that the degree of* x *is in* C *and such that* x *lies in no prime ideal in* Proj(A) *in the support of* M, *the constant is equal to the length of the module* $M_{(x)}$ *over* A_0.

Proof Suppose first that the Hilbert polynomial of M is defined on the positive cone C and is constant. By taking a filtration of M and using the additivity of the Hilbert function together with the fact that the Hilbert function of a module cannot be negative, we can assume that the module M is of the form $A/\mathfrak{p}$ for some graded prime ideal $\mathfrak{p}$ and that $A = A/\mathfrak{p} = M$. In particular, A_0 is an integral domain of finite length and so is a field.

Let n be the constant value of $H_M(i)$. Let x be an element of degree j not in $\mathfrak{p}$, where $j \in C$. Then, since a positive cone is closed under addition, $rj \in C$ for all integers $r > 0$. The map from M_{rj} to $M_{(r+1)j}$ defined by multiplication by x must be an isomorphism for all $r > 0$, since it is injective and the length of M_{rj} is equal to n for all $r > 0$. If $m_1, \ldots, m_k$ are elements of M_{rj}, then $m_1, \ldots, m_k$ are linearly independent over the field A_0 if and only if $m_1/x^{rj}, \ldots, m_k/x^{rj}$ are linearly independent over A_0 in $M_{(x)}$. Hence $M_{(x)}$ has finite length equal to the length of M_{rj} for large r, and this length is equal to n, the constant value of the Hilbert polynomial.

Conversely, if $M_{(x)}$ has finite length for all special homogeneous elements x, we can reverse the above argument and deduce that M_{rj} has constant length for large r. Thus, the Hilbert polynomial is constant. $\square$

Theorem 8.3.4 *The degree of the Hilbert polynomial of a module M is equal to the dimension of M.*

Proof The case of dimension zero is Proposition 8.3.3. We prove the theorem in general by induction on the dimension. Taking a filtration, we reduce as usual to the case where M is of the form $A/\mathfrak{p}$.

We claim that we can find a homogeneous element x such that the dimension of M/xM is equal to $\dim(M) - 1$. Let y be an element not in $\mathfrak{p}$ such that the dimension of $M_{(y)} = (A/\mathfrak{p})_{(y)}$ is equal to the dimension of M. We can find an element $x/y^m \in (A/\mathfrak{p})_{(y)}$ such that the dimension of $(A/\mathfrak{p})_{(y)}/(x/y^m)(A/\mathfrak{p})_{(y)}$ is equal to $\dim((A/\mathfrak{p})_{(y)}) - 1$. Then x satisfies the condition that $\dim(M/xM) = \dim(M) - 1$.

Let $j = (j_1, \ldots, j_n)$ be the degree of x. We wish to show that the degree of the Hilbert polynomial of M/xM is equal to one less than the degree of the Hilbert polynomial of M. Let P_M be the Hilbert polynomial of M, and let d be its degree. Using the exact sequence

$$0 \to M[-j] \overset{x}{\to} M \to M/xM \to 0$$

we see that the Hilbert polynomial $P_{M/xM}$ of M/xM is equal to $P_M(i) - P_M(i - j)$. Lemma 8.3.2 implies that the degree of $P_{M/xM}$ is at most $d - 1$, and that the component of degree $d - 1$ is equal to $\sum j_k (P_M^d)_k$, where $(P_M^d)_k$ denotes the partial derivative of the component of P_M of degree d with respect to the kth variable. It does not follow immediately that this polynomial does not vanish; however, we can replace x by xz for any z not in $\mathfrak{p}$, and, since the degrees of homogeneous elements not in $\mathfrak{p}$ form a positive cone, we may find a choice of xz of degree j such that the linear combination of partial derivatives of Lemma 8.3.2 is not zero. Thus, the degree of the Hilbert poynomial and the dimension both decrease by exactly one, so we can conclude by induction that the degree of P_M is equal to the dimension of M. $\square$

Using this definition of dimension, we define the Chow group $A_*(X)$ of $X = \mathrm{Proj}(A)$ in the same way as before, so that $A_i(X)$ is the free Abelian group on generators $[A/\mathfrak{p}]$, where $A/\mathfrak{p}$ has dimension i modulo the subgroup generated by $\mathrm{div}(\mathfrak{q}, x/y)$, where $\mathfrak{q}$ has dimension $i+1$ and x and y are nonzero homogeneous elements of $A/\mathfrak{q}$ of the same degree.

8.4 Proper Maps

If A and B are graded rings, we define a proper map from $\mathrm{Proj}(B)$ to $\mathrm{Proj}(A)$ to be the map induced by a map of graded rings from A to B with certain properties. More precisely, suppose that B is graded over $\mathbb{Z}^n$ and A is graded over $\mathbb{Z}^m$. By a map of graded rings we mean a ring homomorphism $f : A \rightarrow B$ such that there exists a group homomorphism $\eta : \mathbb{Z}^m \rightarrow \mathbb{Z}^n$ such that whenever a is homogeneous of degree i, then $f(a)$ is homogeneous in B of degree $\eta(i)$. A ring homomorphism f defines a map from the set of graded prime ideals of B to the set of graded prime ideals of A by taking inverse image, but in general there will be prime ideals $\mathfrak{p} \in \mathrm{Proj}(B)$ for which $f^{-1}(\mathfrak{p})$ is not in $\mathrm{Proj}(A)$. The set U consisting of those of $\mathfrak{p} \in \mathrm{Proj}(B)$ such that $f^{-1}(\mathfrak{p}) \in \mathrm{Proj}(A)$ is open, although it may be empty. We call the map induced from U to $\mathrm{Proj}(A)$ a *quasi-projective* map. We will sometimes refer to the induced map from U to $\mathrm{Proj}(A)$ as a quasi-projective map from $\mathrm{Proj}(B)$ to $\mathrm{Proj}(A)$.

The most common situation is where $n = m + k$ and η is the embedding along the first n coordinates. A map of this type will be called *projective*; the projective maps defined in Section 1 are the special case in which $m = 0$ and $k = n = 1$.

We define a quasi-projective map to be *proper* if it is a composition of maps ϕ satisfying one of the two following conditions:

(i) ϕ is projective.
(ii) $\phi : \mathrm{Proj}(B) \rightarrow \mathrm{Proj}(A)$ is a closed embedding defined by a map f of graded rings from A to B with the property that the induced map from $A_{(x)}$ to $B_{(f(x))}$ is surjective for all special homogeneous elements x in A.

The above definition is a very special case of the general class of proper maps in algebraic geometry, for which we refer to [21, Chapter II]. In general, a map of projective schemes is defined by giving local definitions on an open cover in such a way that they agree on intersections. We will usually consider only maps induced by homomorphisms of graded rings as defined here; however, we point out that two maps are considered to be equal when they agree on an open cover. We will only need the properties of proper maps for certain constructions, so we prove the necessary properties only in the cases we need. In addition, the general case of many of these theorems is proven by reducing to the projective case.

Proposition 8.4.1 *Let $f : A \to B$ be a projective map defined by the inclusion of $A = B_{[j]}$ into B. Then if $\mathfrak{p} \in \mathrm{Proj}(B)$, $f^{-1}(\mathfrak{p}) \in \mathrm{Proj}(A)$.*

Proof Assume that $B = B_{[j+1]}$; the general case follows by composing the successive inclusions. Let $\mathfrak{p}$ be in $\mathrm{Proj}(B)$. Since the intersection of a positive cone in $\mathbb{Z}^{j+1}$ is a positive cone in $\mathbb{Z}^j$, there is a positive cone of elements i in $\mathbb{Z}^j$ such that there is an element of degree i not in $\mathfrak{p}$. Hence by Proposition 8.2.2, $f^{-1}(\mathfrak{p}) \in \mathrm{Proj}(A)$. □

It follows from Proposition 8.4.1 that a proper map does induce a map from $\mathrm{Proj}(B)$ to $\mathrm{Proj}(A)$, since it is true by definition for a closed embedding.

As a particular example, let A be a $\mathbb{Z}^n$-graded ring and let E be a finitely generated graded module. We consider two examples that will be used throughout the next chapter: first, the symmetric algebra $S(E)$; and second, if E is a submodule of a free module, the Rees algebra $R(E)$, both with the usual $\mathbb{Z}^{n+1}$ grading. We recall that if E is a submodule of a graded free module F, then the Rees algebra of E is defined to be the subalgebra of the symmetric algebra on F generated by E. In both cases the inclusion of A defines a projective and hence proper map, where η is the imbedding of $\mathbb{Z}^n$ into $\mathbb{Z}^{n+1}$.

8.5 Proper Push-Forward of Cycles

Let $\phi : X \to Y$ be a proper map. In this section we define a map on Chow groups induced by ϕ. We define a map $\phi_* : Z_*(X) \to Z_*(Y)$ as follows: let $X = \mathrm{Proj}(B)$ and $Y = \mathrm{Proj}(A)$, and let $f : A \to B$ be the map of graded rings that induces ϕ. Let $\mathfrak{q} \in X$, and let $\mathfrak{p} = \phi(\mathfrak{q}) = f^{-1}(\mathfrak{q}) \in Y$. The inclusion of $A/\mathfrak{p}$ into $B/\mathfrak{q}$ defines an inclusion of fields from $k(\mathfrak{p})$ into $k(\mathfrak{q})$. We define $\phi_*([B/\mathfrak{q}])$ to be $[k(\mathfrak{q}) : k(\mathfrak{p})][A/\mathfrak{p}]$ if the degree of this field extension is finite and zero otherwise. The map ϕ_* is called *proper push-forward*. One of the fundamental results in this chapter is that this map of cycles preserves rational equivalence and thus induces a map on Chow groups.

If ϕ is a closed embedding, this result is immediate. Hence for the remainder of this section we assume that ϕ is projective; more precisely, we assume that A is a $\mathbb{Z}^{n-1}$-graded ring and that B is a $\mathbb{Z}^n$-graded ring with $B_{[n-1]} = A$.

Before proving that ϕ_* preserves rational equivalence, we establish some basic results on the relations between ideals locally defined by principal ideals in X and Y. Let $\mathfrak{q} \in X$ and let $\mathfrak{p} = \phi(\mathfrak{q}) = f^{-1}(\mathfrak{q})$ as above. The dimensions of $\mathfrak{p}$ and $\mathfrak{q}$ are defined to the dimensions of the rings $(A/\mathfrak{p})_{(x)}$ and $(B/\mathfrak{q})_{(y)}$ for suitable homogeneous elements x and y, and these dimensions are in turn defined in terms of the transcendence degree over a quotient of a regular ring

R (see Section 4.2). The prime ideals $\mathfrak{p}$ and $\mathfrak{q}$ contract to the same prime ideal of R, so the dimensions are equal if and only if the transcendence degrees are equal, which is true if and only if the extension of quotient fields $k(\mathfrak{q})/k(\mathfrak{p})$ is finite. Thus, in the definition of proper push-forward we could have stipulated that the dimension of $\mathfrak{p}$ is equal to the dimension of $\mathfrak{q}$ rather than defining the condition in terms of the finiteness of the associated field extension.

Assume now that the extension $k(\mathfrak{q})/k(\mathfrak{p})$ is finite, so that the dimensions of $\mathfrak{p}$ and $\mathfrak{q}$ are equal. We wish to compute the push-forward of a cycle on X of the form $\mathrm{div}(\mathfrak{q}, x)$ to Y and the pull-back of a cycle on Y of the form $\mathrm{div}(\mathfrak{p}, y)$ to X. By dividing by $\mathfrak{p}$ and $\mathfrak{q}$, we may reduce to the case where $\mathfrak{p}$ and $\mathfrak{q}$ are zero. Note that the map induced from $\mathrm{Proj}(B/\mathfrak{q})$ to $\mathrm{Proj}(A/\mathfrak{p})$ is projective.

Proposition 8.5.1 *Let* $f : A \to B$ *be an injective map of graded integral domains that defines a projective map* ϕ *from* $X = \mathrm{Proj}(B)$ *to* $Y = \mathrm{Proj}(A)$ *as specified above. Assume that the dimensions of A and B are both equal to d. Let $K = A_{(0)}$ and $L = B_{(0)}$, and let n be the degree of the field extension L/K. Let $\mathfrak{p}$ be an element of Y such that $\dim(A/\mathfrak{p}) = d - 1$.*

(i) *There are a finite number of prime ideals $\mathfrak{q}$ in $\mathrm{Proj}(B)$ such that $\phi(\mathfrak{q}) = \mathfrak{p}$.*
(ii) *Let x be a special homogeneous element of B such that $x \notin \mathfrak{q}$ for all $\mathfrak{q}$ with $\phi(\mathfrak{q}) = \mathfrak{p}$. Let S be the multiplicative set in B generated by x and the homogeneous elements in $A - \mathfrak{p}$. Then*
    (a) *$B_{(S)}$ is a finitely generated $A_{(\mathfrak{p})}$-module with quotient field L.*
    (b) *The graded prime ideals $\mathfrak{q}$ of $\mathrm{Proj}(B)$ with $\phi(\mathfrak{q}) = \mathfrak{p}$ are in one-one correspondence with the prime ideals of $B_{(S)}$ lying over $\mathfrak{p}$.*
    (c) *If the prime ideal $\mathfrak{q}$ in $\mathrm{Proj}(B)$ corresponds to the prime ideal $\tilde{\mathfrak{q}}$ of $B_{(S)}$ under this correspondence, then the degree $[k(\mathfrak{q}) : k(\mathfrak{p})]$ is equal to the dimension of $B_{(S)}/\tilde{\mathfrak{q}}$ as a vector space over $A_{(\mathfrak{p})}/\mathfrak{p}_{(\mathfrak{p})}$.*

Proof Let $\mathfrak{q}$ be a graded prime ideal of B such that $f^{-1}(\mathfrak{q}) = \mathfrak{p}$. Then we have an extension $k(\mathfrak{q}) \supseteq k(\mathfrak{p})$, so the dimension of $B/\mathfrak{q}$ is at least as large as the dimension of $A/\mathfrak{p}$, which is $d - 1$. On the other hand, the dimension of B is equal to d, so $\mathfrak{q}$ must be a minimal nonzero prime ideal of B. Hence if y is any nonzero element of $\mathfrak{p}$, $\mathfrak{q}$ is a minimal prime ideal over the ideal generated by y. Since there are a finite number of such minimal primes, this proves (i).

We now prove the statements of part (ii). We first note that the existence of an element x that is in no prime ideal lying over $\mathfrak{p}$ follows from Proposition 8.2.3. We localize B at the multiplicative set $A - \mathfrak{p}$ and take the subring of the localization consisting of elements of degree $(0, \ldots, 0, i_n)$ for arbitrary i_n. Denote this ring $\tilde{B}$; then $\tilde{B}$ is a $\mathbb{Z}$-graded ring. There exist homogeneous elements s and t of A that are not in $\mathfrak{p}$ such that $y = (s/t)x$ has degree of the form $(0, \ldots, 0, i_n)$ and is thus in $\tilde{B}$. Then $B_{(S)}$ is equal to the localization $\tilde{B}_{(y)}$. For the remainder of the argument we consider $\tilde{B}_{(y)}$ in place of $B_{(S)}$. The nonzero prime ideals of $\tilde{B}$ correspond to the prime ideals of B that contract to $\mathfrak{p}$. Since x is in no prime ideal of $\mathrm{Proj}(B)$ lying over $\mathfrak{p}$, there are no graded prime ideals of $\tilde{B}/y\tilde{B}$ except the irrelevant ideal. Thus, the reduction of $\tilde{B}/y\tilde{B}$ modulo $\mathfrak{p}_{(\mathfrak{p})}$ has only one prime ideal, so it is Artinian and thus a finite dimensional module over $A_{(\mathfrak{p})}/\mathfrak{p}_{(\mathfrak{p})}$. Hence $\tilde{B}/y\tilde{B}$ is a finitely generated $A_{(\mathfrak{p})}$-module.

Let n be the degree of y; n is some positive integer. Let $b_1, \ldots, b_k$ be a set of homogeneous elements of B that generate the sub-$A_{(\mathfrak{p})}$-module of $\tilde{B}/y\tilde{B}$ generated by all homogeneous elements with degrees that are multiples of n, and let the degree of b_i be nm_i. We claim that $b_1/y^{m_1}, \ldots, b_k/y^{m_k}$ generate $\tilde{B}_{(y)}$ as an $A_{(\mathfrak{p})}$-module. To show this we prove by induction on i that the homogeneous component of $\tilde{B}$ of degree ni is generated by all elements $b_j y^r$ that are of degree ni. In fact, since $b_1, \ldots, b_k$ generate the sum of homogeneous components of $\tilde{B}/y\tilde{B}$ with degrees that are multiples of n, $\tilde{B}_{ni}$ is generated by all the b_j of degree ni and $y\tilde{B}_{n(i-1)}$. By induction, $\tilde{B}_{n(i-1)}$ is generated by all elements $b_j x^r$ that are of degree $n(i - 1)$; multiplying these elements by y, we see that $\tilde{B}_{ni}$ is generated by all elements $b_j y^r$ that are of degree ni as claimed. Since any element of $\tilde{B}_{(y)}$ can be written in the form b/y^m, where b is a homogeneous element of degree nm, we can write b as sum $\sum a_i b_i y^{n_i}$ with $a_i \in A_{(\mathfrak{p})}$, and we then have $b/y^m = \sum a_i b_i/y^{m-n_i}$. By comparing degrees we see that this is equal to $\sum a_i b_i/y^{m_i}$. Thus, $b_1/y^{m_1}, \ldots, b_k/y^{m_k}$ generate $\tilde{B}_{(y)}$ as an $A_{(\mathfrak{p})}$-module.

To see that the quotient field of $B_{(S)}$ is L, let b/c be in L, and let the degrees of b and c be $i \in \mathbb{Z}^n$. Multiplying by an element d of the appropriate degree we have that $b/c = bd/cd = (bd/y^m)/(cd/y^m)$ and is thus in the quotient field of $B_{(S)}$.

Since none of the prime ideals lying over $\mathfrak{p}$ contains x, they correspond exactly to the prime ideals of $B_{(S)}$ lying over the localization of $\mathfrak{p}$. If $\mathfrak{q}$ is a prime ideal of B lying over $\mathfrak{p}$, then the corresponding prime ideal $\tilde{\mathfrak{q}}$ of $B_{(S)}$ consists of all elements q/s, where $s \in S$ and $q \in \mathfrak{q}$. We then have an isomorphism of localizations

$$B_{(\mathfrak{q})} \cong \left(B_{(S)}\right)_{(\tilde{\mathfrak{q}})}$$

and in particular the residue fields can be identified so their degrees over $k(\mathfrak{p})$ are equal. $\square$

Thus, we can study prime ideals of codimension 1 over a given prime by looking at the corresponding prime ideals in a finite extension of rings of dimension 1. We recall that it follows from Proposition 5.2.10 that if A is a one-dimensional local domain, if x is a nonzero element of A, and if M is a finitely generated A-module, then

$$\text{length}(M/xM) - \text{length}(_xM) = (\text{rank}(M))(\text{length}(A/xA)).$$

In particular, if M is torsion free, then the length of M/xM is equal to the length of A/xA times the rank of M.

Proposition 8.5.2 *Let $f : A \to B$ be an injective map of integral domains of the same dimension inducing a projective map ϕ from $\text{Proj}(B)$ to $\text{Proj}(A)$ as in Proposition 8.5.1. Let x be a nonzero homogeneous element of A, and denote the zero prime ideals of A and B by $(0)_A$ and $(0)_B$, respectively. Then we have an equality of cycles*

$$\phi_*(\text{div}((0)_B, f(x))) = [L : K]\text{div}((0)_A, x)).$$

Proof To show that the cycle $\phi_*(\text{div}((0)_B, f(x)))$ is the same as the cycle $[L : K]\text{div}((0)_A, x)$, it must be shown that for each prime ideal $\mathfrak{p}$ minimal over (x), we have

$$[L : K]\big(\text{length}((A/xA)_{(\mathfrak{p})})\big) = \sum_{\mathfrak{q}}[k(\mathfrak{q}) : k(\mathfrak{p})]\big(\text{length}\big((B/f(x)B)_{(\mathfrak{q})}\big)\big)$$

where the sum is taken over all $\mathfrak{q}$ minimal over $f(x)$ such that $f^{-1}(\mathfrak{q}) = \mathfrak{p}$. By Proposition 8.5.1, we may localize at $\mathfrak{p}$ and find a ring C contained in L with quotient field equal to L that is a finite module over $A_{(\mathfrak{p})}$ and such that the prime ideals $\mathfrak{q}$ minimal over $f(x)$ contracting to $\mathfrak{p}$ correspond to prime ideals of C contracting to $\mathfrak{p}$. We replace A by $A_{(\mathfrak{p})}$ and B by C and find homogeneous elements s and t not in $\mathfrak{p}$ such that $y = xs/t$ has degree 0 and so lies in $A_{(\mathfrak{p})}$. We must now show that if A is an integral domain of dimension 1 and B is a finite module over A with quotient field L, then

$$[L : K](\text{length}_A(A/yA)) = \sum_{\mathfrak{q}}[k(\mathfrak{q}) : k(\mathfrak{p})]\big(\text{length}_{B_{(\mathfrak{q})}}(B_{\mathfrak{q}}/xB_{\mathfrak{q}})\big)$$

where the sum is over all prime ideals $\mathfrak{q}$ contracting to $\mathfrak{p}$.

It follows from the remark preceding this proposition that, since the rank of B is $[L : K]$, the left-hand side of this equation is equal to the length of B/xB as an A-module. By the Chinese remainder theorem, we have

$$B/xB \cong \oplus_{\mathfrak{q}} B_{\mathfrak{q}}/xB_{\mathfrak{q}}.$$

The length of $B_{\mathfrak{q}}/xB_{\mathfrak{q}}$ as an A-module is equal to $[k(\mathfrak{q}) : k(\mathfrak{p})]$ times its length as a $B_{\mathfrak{q}}$-module. Combining this equality with the above expression for the length of B/xB then proves the result. $\square$

Proposition 8.5.3 *Let the notation be as in Proposition 8.5.2, and let x and y be nonzero homogeneous elements of B of the same degree. Then*

$$\phi_*(\text{div}((0)_B, x/y)) = \text{div}((0)_A, N(x/y))$$

where N is the norm from L to K.

Proof Following the same argument as in Proposition 8.5.2, we may assume that A is an integral domain of dimension 1 with quotient field K and that B is an extension that is a finite module over A with quotient field L. Furthermore, we can represent x/y as a quotient of elements in B, and, since the norm is multiplicative, we may thus reduce to the case where $x/y = b$ is in B.

Let $\mathfrak{p}$ be the maximal ideal of A. Using the argument at the end of Proposition 8.5.2, we have that the coefficient of $[A/\mathfrak{p}]$ in $\phi_*(\text{div}((0)_B, b))$ is the length of B/bB over A. Thus, we have to show that the length of B/bB over A is equal to the length of $A/N(b)$. We again use the remark preceding Proposition 8.5.2 and replace A by its integral closure, which is a discrete valuation ring. In this case B is free over A, and the norm of B is the determinant of any matrix that defines multiplication by b with respect to a basis for B over A. By choosing two bases for B with respect to which multiplication by b is represented by a diagonal matrix, which changes the determinant up to a unit in A, we see that both sides of the equation are given by the ideal generated by the product of diagonal entries. Hence we have

$$\text{length}_A(B/bB) = \text{length}_A(A/N(b)),$$

which proves the result. $\qquad\square$

We next consider the inverse image of an element of Y in the case in which the difference of dimensions is equal to 1. Let A, B, $\mathfrak{p}$, and $\mathfrak{q}$ be as above, except that we now assume that $\dim(B/\mathfrak{q}) = \dim(A/\mathfrak{p}) + 1$. Dividing by $\mathfrak{p}$ and $\mathfrak{q}$, we assume that we have an extension of domains. Let S be the set of nonzero homogeneous elements of A. We then have an injection $A_{(S)} \to \tilde{B}_{(S)}$, where we use $\tilde{B}_{(S)}$ to denote the subring of the localization B_S consisting of elements of degrees of the form $(0, \ldots, 0, i_n)$. Then $\text{Proj}(\tilde{B}_{(S)})$ is homeomorphic to the set of elements of $\text{Proj}(B)$ that contract to the zero ideal in A. We note that since $\dim(B) = \dim(A) + 1$, we have $\dim(\tilde{B}_{(S)}) = 1$.

Since B is assumed to satisfy the condition on generators, the ring $\tilde{B}_{(S)}$ is generated over $A_{\mathfrak{p}}$ by elements of degree 1, so it has a Hilbert polynomial $P(i_n)$. Since the dimension of $\text{Proj}(\tilde{B}_{(S)})$ is 1, the Hilbert polynomial is linear; let $P(i_n) = ai_n + b$.

Let x be a nonzero homogeneous element of B. We wish to consider the image of $\text{div}(\mathfrak{q}, x)$ under ϕ_*, where ϕ is the map from $\text{Proj}(B)$ to $\text{Proj}(A)$. Since div(q, x) is a cycle of dimension equal to the dimension of A, $\phi_*(\mathrm{div}(q, x))$ is a multiple of $[A]$, we can localize at S. We can adjust x by multiplying by a unit in $A_{(S)}$ and assume that x is in $\tilde{B}_{(S)}$. Let the degree of x in $\tilde{B}_{(S)}$ be j_n. It then follows from Lemma 8.3.2 (or from the results of Chapter 2) that the dimension of $(B/xB)_{(S)}$ as a vector space over $A_{(\mathfrak{p})}$ is equal to $a j_n$. In particular, it depends only on the degree of x.

We now prove the main result of this section.

Theorem 8.5.4 *Let* $X = \mathrm{Proj}(B)$ *and* $Y = \mathrm{Proj}(A)$, *and let* ϕ *be a projective map from* X *to* Y. *Then the map* ϕ_* *from* $Z_*(X)$ *to* $Z_*(Y)$ *preserves rational equivalence.*

Proof Let q be a prime ideal in B and let x and y be homogeneous elements in of B of the same degree that are not in q; we must show that $\phi_*(\mathrm{div}(q, x/y))$ is rationally equivalent to zero. Let $\mathfrak{p} = f^{-1}(q)$. By dividing by q and $\mathfrak{p}$ we may assume that A and B are domains, the map from A to B is injective, and $q = \mathfrak{p} = 0$. If, after these reductions, the dimension of B is greater than or equal to $\dim(A) + 2$, then $f_*(\mathrm{div}(\mathfrak{p}, x/y))$ is the zero cycle for dimension reasons, so we may assume that the difference in dimension is 0 or 1. There are thus two basic cases, where the difference in dimension is 1 and where it is 0.

Suppose that the difference in dimension is 1. Then the only prime ideal with possible nonzero coefficient in the image of div(q, x/y) is the zero prime ideal, so we can localize at $\mathfrak{p}$. We have shown that in this case $\phi_*\,\mathrm{div}(q, x)$ depends only on the degree of x. Hence $\phi_*\,\mathrm{div}(q, x) = \phi_*\,\mathrm{div}(q, y)$, and $\phi_*\,\mathrm{div}(q, x/y)$ $= \phi_*\,\mathrm{div}(q, x) - \phi_*\,\mathrm{div}(q, y) = 0$.

Assume now that A and B have the same dimension. Now Proposition 8.5.3 implies that

$$\phi_*(\mathrm{div}(q, x/y)) = \mathrm{div}(z/w)$$

where z and w are elements of the same degree in A such that the norm of x/y in the field extension $k(q)$ over $k(\mathfrak{p})$ is equal to z/w. Thus, $\phi_*(\mathrm{div}(q, x/y))$ is rationally equivalent to zero. $\square$

8.6 Products of Projective Schemes

Let $X = \mathrm{Proj}(A)$ and $Y = \mathrm{Proj}(B)$ be projective schemes, and let ϕ and ψ be projective maps to a projective scheme $Z = \mathrm{Proj}(C)$. We then define the product $X \times_Z Y$ to be the projective scheme associated to the graded ring $A \otimes_C B$. More precisely, since ϕ and ψ are projective maps, there exist k, m, and n such that C is $\mathbb{Z}^k$-graded and A and B are $\mathbb{Z}^{k+m}$- and $\mathbb{Z}^{k+n}$-graded, respectively. We then have a $\mathbb{Z}^{k+m+n}$-grading on $A \otimes_C B$ defined by letting the degree of $a \otimes b$ be equal to $\deg(a) + \deg(b)$ for homogeneous elements $a \in A$ and $b \in B$. We use this grading to define the product. Since $A \otimes_C B$ is generated as an algebra over C by the images of A and B, the condition on generators for $A \otimes_C B$ follows from the corresponding conditions for A and B.

If x, y, and z are special homogeneous elements of A, B, and C, respectively, then $x \otimes y$ is a special homogeneous element of $A \otimes_C B$. To describe the local situation, we restrict to a special case. Assume that $z = z_1 \cdots z_k$, $x = z_1 \cdots z_k x_1 \cdots x_m$, and $y = z_1 \cdots z_k y_1 \cdots y_n$ are the representations as special homogeneous elements. Then we have maps from $C_{(z)}$ to $A_{(x)}$ and from $C_{(z)}$ to $B_{(y)}$ that induce a map

$$\left(A_{(x)}\right) \otimes_{C_{(z)}} \left(B_{(y)}\right) \to (A \otimes_C B)_{(x \otimes y)}.$$

If we assume in addition that z is a unit special homogeneous element, then this map is an isomorphism. To see this, we note that if we take any element of the form $a \otimes b/(x \otimes y)^n$, where $a \otimes b$ and $(x \otimes y)^n$ are homogeneous of the same degree, we can find (possibly negative) integers $i_1, \ldots, i_k$ such that $a z_1^{i_1} \cdots z_k^{i_k}$ and x^n have the same degree and such that $b z_1^{-i_1} \cdots z_k^{-i_k}$ and y^n have the same degree. We can then find elements $a' \in A$ and $b' \in B$ and an integer m such that

$$\frac{a z_1^{i_1} \cdots z_k^{i_k}}{x^n} = \frac{a'}{x^m} \quad \text{and} \quad \frac{b z_1^{-i_1} \cdots z_k^{-i_k}}{y^n} = \frac{b'}{y^m}.$$

Then

$$\frac{a \otimes b}{(x \otimes y)^n} = \frac{a z_1^{i_1} \cdots z_k^{i_k}}{x^n} \otimes \frac{b z_1^{-i_1} \cdots z_k^{-i_k}}{y^n} = \frac{a'}{x^m} \otimes \frac{b'}{y^m}.$$

Hence the induced map on localizations is surjective and so is an isomorphism.

Theorem 8.6.1 *The product we have defined is the product in the category of projective schemes with quasi-projective maps. In addition,*

(i) *The projection maps from $X \times_Z Y$ to X and Y are proper.*

(ii) *If W is a projective scheme and $\alpha : W \to X$ and $\beta : W \to Y$ are proper maps, then the induced map from W to $X \times_Z Y$ is proper.*

Proof The first statement follows formally from the definitions and from the corresponding universal property for tensor products of rings. In fact, let $W = \text{Proj}(D)$ be another projective scheme, where D is a $\mathbb{Z}^r$-graded ring, and let $\alpha : W \to X$ and $\beta : W \to Y$ be quasi-projective maps such that $\phi \alpha = \psi \beta$. Then α and β are defined by maps of graded rings that agree on C; we denote these maps h and k, respectively. Let η and τ be the associated maps on indices from $\mathbb{Z}^{k+n}$ and $\mathbb{Z}^{k+m}$ to $\mathbb{Z}^r$. The maps h and k define a unique map on indices from $\mathbb{Z}^{k+m+n}$ to $\mathbb{Z}^r$, and, by using this map the unique map of rings from $A \otimes_C B$ to D whose restrictions to A and B are h and k, respectively, becomes a map of graded rings and thus defines a quasi-projective map. Thus, there is a unique quasi-projective map from W to $X \times_Z Y$ making the required diagrams commute.

The fact that the projection maps are proper follows from the fact that the maps from A and B to their images in $A \otimes_C B$ map onto the subrings of elements of degrees in $\mathbb{Z}^{k+m}$ and $\mathbb{Z}^{k+n}$, respectively. Hence, numbering the summands of $\mathbb{Z}^{k+m+n}$ correctly, the projection from $X \otimes_Z Y$ to X is induced by the composition of a surjection and the embedding of $(A \otimes_C B)_{[k+m]}$ into $A \otimes_C B$, and similarly for the projection to Y.

To prove statement (ii) we first note that the map induced on a product by ϕ and ψ is the composition of the maps induced by ϕ and 1_B and $1_{A'}$ and ψ, respectively. Thus, we may assume that one of the maps, which we take to be ψ, is the identity map.

We consider the case in which ϕ is a closed embedding and the case in which ϕ is projective separately. If ϕ is a closed embedding, then ϕ is locally defined by a surjective map of rings, and the local description of the product shows that $\phi \times 1_B$ is also locally described by a surjective map of rings and hence is proper. If ϕ is a projective map, we may assume that ϕ is the embedding of $A_{[k+m-1]}$ into $A_{[k+m]}$ for some m. The map induced by ϕ will then be a surjective map, which defines a closed embedding, followed by the inclusion of $(A \otimes_C B)_{[k+m+n-1]}$ into $(A \otimes_C B)_{[k+m+n]}$. Thus, the map induced by ϕ is proper. $\qquad\square$

8.7 The Hyperplane Sections on a Projective Scheme

The remainder of this chapter is devoted to intersection operations defined by divisors. In this section we discuss one of the simplest cases, the hyperplane sections.

Let $X = \mathrm{Proj}(A)$, where A is a $\mathbb{Z}^n$-graded ring, and let $i \in \mathbb{Z}^n$. We define a map on the Chow group $h_i : A_k(X) \to A_{k-1}(X)$ for each k as follows: let $\mathfrak{p}$ be a prime ideal in $\mathrm{Proj}(A)$ of dimension k, and let x and y be nonzero elements of $A/\mathfrak{p}$ such that x/y has degree i. The existence of such elements x and y follows from the fact that the set of degrees of homogeneous elements not in $\mathfrak{p}$ contains a positive cone. We then define

$$h_i([A/\mathfrak{p}]) = \mathrm{div}(\mathfrak{p}, x) - \mathrm{div}(\mathfrak{p}, y).$$

This map is extended to arbitrary cycles by linearity. The fact that it does define a map on the Chow group follows from the next theorem. In this theorem we make use of two main facts. The first is that we can find elements not in any finite set of prime ideals by Proposition 8.2.3. Thus, we can chose representatives x and y as in the definition outside any finite set of prime ideals. The second is that if two elements x, y generate an ideal of height 2, then the intersection by the elements x and y commutes. This follows from Proposition 5.2.11 on Koszul complexes, which states that the product of intersections is equal to the Euler characteristic of the Koszul complex. We will use this fact several times in the next few sections.

Theorem 8.7.1

(i) *The rational equivalence class of* $h_i([A/p])$ *does not depend on the choice of x and y.*

(ii) *The map h_i preserves rational equivalence.*

Proof Proof of (i): If x' and y' are homogeneous elements of A such that x/y and x'/y' have the same degree, then the degree of $xy'/x'y$ is zero. Thus, $\mathrm{div}(\mathfrak{p}, xy') - \mathrm{div}(\mathfrak{p}, x'y)$ is rationally equivalent to zero. But

$$\mathrm{div}(\mathfrak{p}, xy') - \mathrm{div}(\mathfrak{p}, x'y)$$
$$= \mathrm{div}(\mathfrak{p}, x) + \mathrm{div}(\mathfrak{p}, y') - \mathrm{div}(\mathfrak{p}, x') - \mathrm{div}(\mathfrak{p}, y)$$
$$= (\mathrm{div}(\mathfrak{p}, x) - \mathrm{div}(\mathfrak{p}, y)) - (\mathrm{div}(\mathfrak{p}, x') - \mathrm{div}(\mathfrak{p}, y')).$$

Thus, the cycles defined by the two choices are rationally equivalent.

Proof of (ii): We have to show that $h_i(\mathrm{div}(\mathfrak{q}, u/v))$ is rationally equivalent to zero if u and v are homogeneous elements of the same degree. With the correct choice of x and y, by Proposition 8.2.3 we can assume that the ideals $(x, u), (x, v), (y, u), (y, v)$ are ideals of codimension 2 in $A/\mathfrak{q}$. For each of these pairs we let $[x, u]$ be the class in codimension 2 obtained by intersecting with the two elements; its coefficient on a prime ideal $\mathfrak{p}$ of codimension 2 is the alternating sum of the homology of the Koszul complex on the two elements. We then have $h_i(\mathrm{div}(\mathfrak{q}, u/v)) = [x, u] - [x, v] - [y, u] + [y, v]$.

Now since these cycles have codimension 2, we know that $[x, u] = [u, x]$ and similarly for the other terms in the sum. Thus, we have

$$h_i(\mathrm{div}(\mathfrak{q}, u/v)) = [x, u] - [x, v] - [y, u] + [y, v]$$
$$= ([u, x] - [u, y]) - ([v, x] - [v, y]).$$

Written in this form, we see that $h_i(\mathrm{div}(\mathfrak{q}, u/v))$ is the same as the cycle obtained by adding together the results of applying $\mathrm{div}(u/v)$ to all the components of $h_i([A/\mathfrak{q}])$. $\qquad\square$

The argument of part (i) of the above theorem, using the fact that $\text{div}(\mathfrak{p}, xy) = \text{div}(\mathfrak{p}, x) + \text{div}(\mathfrak{p}, y)$, also shows that if i and j are elements of $\mathbb{Z}^n$, then $h_{i+j} = h_i + h_j$. In particular, in the classical projective case, h_i is just i times the usual hyperplane section.

The hyperplane sections commute with proper push-forward and flat pull-back of projective maps. More precisely, we have the following:

Proposition 8.7.2 *Let $\phi : X = \text{Proj}(B) \to Y = \text{Proj}(A)$ be a quasi-projective map, defined by a map $f : A \to B$ of graded rings and let $\eta : \mathbb{Z}^m \to \mathbb{Z}^n$ be the corresponding map on index groups.*

(i) *If ϕ is proper, the following diagram commutes for each j:*

$$
\begin{array}{ccc}
A_j(X) & \stackrel{h_{\eta(i)}}{\to} & A_{j-1}(X) \\
\phi_* \downarrow & & \downarrow \phi_* \\
A_j(Y) & \stackrel{h_i}{\to} & A_{j-1}(Y)
\end{array}
$$

(ii) *If ϕ is flat of relative dimension k and is defined on the open subset U of X, then the following diagram commutes for each j:*

$$
\begin{array}{ccc}
A_j(Y) & \stackrel{h_i}{\to} & A_{j-1}(Y) \\
\phi^* \downarrow & & \downarrow \phi^* \\
A_{j+k}(U) & \stackrel{h_{\eta(i)}}{\to} & A_{j+k-1}(U)
\end{array}
$$

Proof We first prove (i). If ϕ is a closed embedding the result is clear, so we assume that ϕ is projective. Let $\mathfrak{q}$ be an element of $\text{Proj}(B)$, and let $\mathfrak{p} = f^{-1}(\mathfrak{q})$. If $[k(\mathfrak{q}) : k(\mathfrak{p})]$ is infinite, both sides of the diagram give zero, so we assume that $[k(\mathfrak{p}) : k(\mathfrak{q})]$ is finite. Let x and y be elements of A not in $\mathfrak{p}$ such that x/y has degree i. Then $f(x)$ and $f(y)$ are not in $\mathfrak{q}$, and $f(x)/f(y)$ has degree $\eta(i)$, so we may use $f(x)/f(y)$ to compute $h_{\eta(i)}([B/\mathfrak{q}])$.

Let d be the dimension of $A/\mathfrak{p}$ and of $B/\mathfrak{q}$. We have

$$
h_i(\phi_*([B/\mathfrak{q}])) = h_i([k(\mathfrak{q}) : k(\mathfrak{p})][A/\mathfrak{p}])
$$

$$
= [k(\mathfrak{q}) : k(\mathfrak{p})](\text{div}(\mathfrak{p}, x) - \text{div}(\mathfrak{p}, y)).
$$

On the other hand, $h_{\eta(i)}([B/\mathfrak{q}]) = (\text{div}(\mathfrak{q}, f(x)) - \text{div}(\mathfrak{q}, f(y)))$. It suffices to show that $\phi_*(\text{div}(\mathfrak{q}, f(x))) = [k(\mathfrak{q}) : k(\mathfrak{p})][(A/\mathfrak{p})/x(A/\mathfrak{p})]_{d-1}$; the corresponding result for y will then follow, giving the commutativity of the above diagram. This formula follows immediately from Proposition 8.5.2.

To prove (ii), we must show that for $\mathfrak{p} \in \text{Proj}(A)$ and $x \notin \mathfrak{p}$ we have

$$
\phi^*(\text{div}(\mathfrak{p}, x)) = \sum_{\mathfrak{q}} \text{length}(B \otimes_A A/\mathfrak{p})_{\mathfrak{q}} \, \text{div}(\mathfrak{q}, f(x))
$$

where the sum is taken over all minimal primes $\mathfrak{q}$ in $\text{Proj}(B \otimes_A A/\mathfrak{p})$. The proof of this statement follows by showing that both sides of the equation are equal to $[B \otimes_A (A/\mathfrak{p}/xA/\mathfrak{p})]_{d+k-1}$ and follows the same steps as the proof that flat pull-back preserves rational equivalence in the affine case in Theorem 1.2.3.

$\square$

8.8 Divisors on Projective Schemes

In this section we define the intersection operation for an arbitrary divisor on a projective scheme. If A is a local ring, we defined intersection with a principal divisor (x) in Section 5.1 as follows: if $\mathfrak{p} \in \text{Spec}(A)$ is a prime ideal of dimension d, we defined $(x) \cap [A/\mathfrak{p}]$ to be $[(A/\mathfrak{p})/x(A/\mathfrak{p})]_{d-1}$ if $x \notin \mathfrak{p}$ and to be zero otherwise. The general definition is essentially the same in the first case but will not necessarily be zero in the second case.

Let A be a graded ring, and let $X = \text{Proj}(A)$.

Definition 8.8.1 *An effective divisor D on X is a closed subscheme defined by an ideal that is locally a principal ideal generated by an element that is not a zero-divisor.*

The main result of this section is to define an operator on the Chow group given by intersecting with a divisor D. For this purpose we first restrict to the case in which A is an integral domain. A divisor on $\text{Proj}(A)$ defines a cycle of codimension 1 as follows: let $\mathfrak{p}$ be a prime ideal of codimension 1, and let x be a special homogeneous element of A such that $x \notin \mathfrak{p}$. Then $\mathfrak{p}$ corresponds to a prime ideal of $A_{(x)}$. Suppose D is generated in $A_{(x)}$ by the element α. Then the coefficient is $\text{length}(A_{(\mathfrak{p})}/\alpha A_{(\mathfrak{p})})$. It is easy to see that this definition determines a well-defined cycle.

To intersect with a general cycle, the procedure is to restrict to the quotient modulo a prime ideal and use the construction of the previous paragraph. Let $\mathfrak{a}$ denote the ideal that defines D. If $\mathfrak{p}$ does not contain $\mathfrak{a}$, this construction is rather straightforward; we take the image of $\mathfrak{a}$ in $A/\mathfrak{p}$, which is still locally principal. The result is a well-defined cycle on $\text{Proj}(A/(\mathfrak{a} + \mathfrak{p}))$, so if we denote the closed subset $\text{Proj}(A/\mathfrak{p})$ by Z, we have a cycle in $Z_*(Z \cap D)$.

If $\mathfrak{p}$ does contain $\mathfrak{a}$, the situation is more complicated. The idea is to find another ideal that is isomorphic to $\mathfrak{a}$ but is not contained in $\mathfrak{p}$. The necessary results are contained in the following Proposition:

Proposition 8.8.1 *Let $\mathfrak{p}$ be a prime ideal of dimension d in $X = \text{Proj}(A)$, and let $\mathfrak{a}$ be an ideal of A that defines a divisor D. Then*

(i) *There exists a prime ideal $\mathfrak{q}$ contained in $\mathfrak{p}$ such that $\mathfrak{a}$ is not contained in $\mathfrak{q}$.*

(ii) *Let q be as in part (i), and let $B = A/q$. Then there exist nonzero homogeneous elements x and y in B with $x \notin \mathfrak{p}$ and such that $y\mathfrak{a} \subseteq xB$ but $y\mathfrak{a} \not\subseteq \mathfrak{p}$.*

(iii) *Let x and y be as in part (ii), and let i be the degree of x/y in $\mathbb{Z}^n$. Let $\mathfrak{b}$ be the image of the ideal $(y/x)\mathfrak{a}$ of B in $B/\mathfrak{p} = A/\mathfrak{p}$. Then the divisor*

$$[(A/\mathfrak{p})/\mathfrak{b}]_{d-1} - h_i([A/\mathfrak{p}])$$

does not depend on the choice of x and y up to rational equivalence in $A_*(A/\mathfrak{p})$.

(iv) *The divisor $[(A/\mathfrak{p})/\mathfrak{b}]_{d-1}$ does not depend on the choice of q up to rational equivalence in $A_*(A/\mathfrak{p})$.*

Proof The first statement follows from the fact that $\mathfrak{a}$ is locally generated by an element that is not a zero-divisor and so is not contained in any minimal prime ideal of A.

To prove the existence of x and y as in part (ii), we consider the local ring $B_{(\mathfrak{p})}$ at the prime $\mathfrak{p}$. The localization of $\mathfrak{a}$ is principal, say generated by an element z/w, where z and w are homogeneous elements of B of the same degree and $w \notin \mathfrak{p}$. Consider the fractional ideal $(w/z)\mathfrak{a}$ in the quotient field of B. Since (z/w) generates the localization of $\mathfrak{a}$ in $B_{(\mathfrak{p})}$, after localization at $\mathfrak{p}$ the fractional ideal $(w/z)\mathfrak{a}$ is equal to $B_{(\mathfrak{p})}$. Hence there is a homogeneous element u of B that is not in $\mathfrak{p}$ such that $u(w/z)\mathfrak{a}$ is contained in B but is not in contained in $\mathfrak{p}$. Then the elements $y = uw$ and $x = z$ satisfy the conditions of statement (ii).

To prove (iii), we assume that x' and y' are also homogeneous elements that satisfy the conditions of statement (ii), and let i' be the degree of x'/y'. We then have $(y/x)\mathfrak{a} = (y/x)(x'/y')(y'/x')\mathfrak{a}$. Hence, denoting the images of $(y/x)\mathfrak{a}$ and $(y'/x')\mathfrak{a}$ in $B/\mathfrak{p}$ by $\mathfrak{b}$ and $\mathfrak{b}'$, we have

$$[(A/\mathfrak{p})/\mathfrak{b}]_{d-1} - h_i([A/\mathfrak{p}]) = h_{i-i'}([A/\mathfrak{p}]) + [(A/\mathfrak{p})/\mathfrak{b}']_{d-1} - h_i([A/\mathfrak{p}])$$

$$= [(A/\mathfrak{p})/\mathfrak{b}']_{d-1} - h_{i'}([A/\mathfrak{p}])$$

up to rational equivalence.

Let q' be another prime ideal contained in $\mathfrak{p}$ with the property that $\mathfrak{a} \not\subseteq q'$. We consider $q \cap q'$. Again we localize at $\mathfrak{p}$ as in the proof of (ii), and the same argument shows that there exist elements x and y with the properties of statement (ii) with respect to the intersection $q \cap q'$. Thus, these elements work both for q and for q', and the resulting ideal $(y/x)\mathfrak{a}$ is the same. Using part (iii), we conclude that any choice of prime ideal q gives the same result up to rational equivalence. This proves (iv). □

Using Proposition 8.8.1, we define an intersection operation on the Chow group of X with values in the Chow group of D.

Definition 8.8.2 *Let D be a divisor on $X = \mathrm{Proj}(A)$ defined by a locally principal ideal $\mathfrak{a}$. Let $\mathfrak{p}$ be a prime ideal of A of dimension d. We define $D \cap [A/\mathfrak{p}]$ to be*

(i) *The class $[A/(\mathfrak{p} + \mathfrak{a})]_{d-1}$ if $\mathfrak{p} \not\supseteq \mathfrak{a}$.*
(ii) *The class $[A/(\mathfrak{p} + \mathfrak{b})]_{d-1}$, where $\mathfrak{b}$ is an ideal satisfying the conditions of Proposition 8.8.1, if $\mathfrak{p} \supseteq \mathfrak{a}$.*

If η is a cycle supported in Z, then it follows from Proposition 8.8.1 that $D \cap \eta$ is well-defined up to rational equivalence in $Z \cap D$.

If A is a nongraded ring and (x) is a principal divisor, and if $\mathfrak{p}$ is a prime ideal of A with $x \in \mathfrak{p}$, then $\mathfrak{b}$ can be chosen to be any principal ideal (y) not contained in $\mathfrak{p}$. Then $(x) \cap [A/\mathfrak{p}] = \mathrm{div}(\mathfrak{p}, y)$ is rationally equivalent to zero in $A_*(A/\mathfrak{p})$. Thus, this definition agrees with the definition in Section 5.1 in the affine case.

We show next that this operation commutes with proper push-forward and with flat pull-back of divisors.

Theorem 8.8.2 *Let $X = \mathrm{Proj}(B)$ and $Y = \mathrm{Proj}(A)$ be projective schemes, and let $\phi : X \to Y$ be a quasi-projective map. Let D be an effective divisor on Y. Then*

(i) *If ϕ is proper, and if $\phi^*(D)$ is a divisor on X, then*

$$\phi^*(D) \cap \eta = D \cap \phi_*(\eta)$$

for all $\eta \in A_*(X)$.
(ii) *If ϕ is flat of relative dimension k, then*

$$\phi^*(D) \cap \phi^*(\eta) = \phi^*(D \cap \eta)$$

for all $\eta \in A_*(Y)$.
(iii) *If i is an element of the index group of X, then $h_i(D \cap \eta) = D \cap (h_i(\eta))$.*

Proof We remark first that the condition that $\phi^*(D)$ is a divisor in X means that if $\mathfrak{a}$ is the ideal of A defining D, then the ideal generated by $\mathfrak{a}$ in B is locally generated by an element that is not a zero-divisor. In the second part of the theorem this condition is implied by the flatness of ϕ.

We prove the first statement. Let $\mathfrak{q}$ be a prime ideal in $\mathrm{Proj}(B)$, and let $\mathfrak{p}$ be its inverse image in A. We then have an injection $A/\mathfrak{p} \to B/\mathfrak{q}$. If $\mathfrak{a} \subseteq \mathfrak{p}$, we can replace $\mathfrak{a}$ as in Proposition 8.8.1 by an ideal that is not contained in $\mathfrak{p}$. The image of this ideal will not be contained in $\mathfrak{q}$, and, using the fact that intersection with hyperplane sections commutes with ϕ_*, we may thus reduce to the case in which $\mathfrak{a} \not\subseteq \mathfrak{p}$. We may then replace $\mathfrak{a}$ by its image in $A/\mathfrak{p}$ and reduce to the case in which $X = \mathrm{Proj}(B/\mathfrak{q})$ and $Y = \mathrm{Proj}(A/\mathfrak{p})$.

If $\dim(B/\mathfrak{q}) - \dim(A/\mathfrak{p}) \geq 1$, both sides of the equation are zero for dimension reasons. Assume that the difference of dimensions is 0. Let $\mathfrak{p}'$ be a prime ideal of codimension 1 that contains $\mathfrak{a}$. We can localize at $\mathfrak{p}'$, and the ideal $\mathfrak{a}$ will then become principal. We then use Proposition 8.5.2 to deduce that

$$\sum_{\mathfrak{q}'} [k(\mathfrak{q}') : k(\mathfrak{p}')]\left(\mathrm{length}\left(B_{(\mathfrak{q}')}/\mathfrak{a}_{(\mathfrak{q}')}\right)\right) = [k(\mathfrak{q}) : k(\mathfrak{p})]\left(\mathrm{length}\left(A_{(\mathfrak{p}')}/\mathfrak{a}_{(\mathfrak{p}')}\right)\right)$$

where the sum is taken over all $\mathfrak{q}'$ lying over $\mathfrak{p}'$. This proves the first part of the theorem.

To prove the second part, let $\eta = [A/\mathfrak{p}]$. We may again reduce to the case in which D is defined by an ideal $\mathfrak{a}$ that is not contained in $\mathfrak{p}$. We must then show that $[B/\mathfrak{p}/\mathfrak{a}(B/\mathfrak{p})]$ is the inverse image of $[A/\mathfrak{p}/\mathfrak{a}(A/\mathfrak{p})]$ under flat pull-back. This follows from taking a filtration with quotients of the form $A/\mathfrak{p}_i$, pulling back to X, and using that ϕ is flat of relative dimension k as in the proof of Theorem 1.2.3.

The third part of this theorem reduces to a local computation. As in the previous parts, the theorem can be reduced to the case in which $\mathfrak{a}$ is not contained in $\mathfrak{p}$. We may, as in the proof of Theorem 8.7.1, choose x and y so that x/y has degree i and the codimensions of $(x, \mathfrak{a})$ and $(y, \mathfrak{a})$ are 2. The result then follows again from the fact that in codimension 2 the intersection product commutes on the level of cycles as in the proof of Theorem 8.7.1. $\square$

If D and D' are effective divisors defined by ideals $\mathfrak{a}$ and $\mathfrak{a}'$, then we denote $D + D'$ the divisor defined by the product $\mathfrak{a}\mathfrak{a}'$. We note that a local computation, using the fact that $\mathrm{div}(\mathfrak{p}, xy) = \mathrm{div}(\mathfrak{p}, x) + \mathrm{div}(\mathfrak{p}, y)$, implies that $(D + D') \cap \eta = (D \cap \eta) + (D' \cap \eta)$ for all cycles η. The ideal $\mathfrak{a}\mathfrak{a}'$ is also locally principal generated by an element that is not a zero-divisor. We define a general divisor to be a quotient of two divisors. More formally, we define a general divisor to be an ordered pair (D, D'), thought of as the difference $D - D'$, where we identify (D, D') with (E, E') if $D + E' = D' + E$. The resulting intersection operator is the difference of the two; we have $(D - D') \cap \eta = D \cap \eta - D' \cap \eta$. The fact that $(D + D') \cap \eta = D \cap \eta + D' \cap \eta$ for effective divisors implies that this operation is well-defined.

8.9 Commutativity of Intersection with Divisors

In Chapter 5 we showed that if two principal divisors in $\mathrm{Spec}(A)$ intersect in codimension 2, then the intersection operators they define commute as maps on cycles and can be computed from the homology of a Koszul complex. We have used this fact to prove that the hyperplane sections are well-defined up to rational equivalence and to prove the commutativity of intersection with divisors with hyperplane sections. If the codimension is not 2, then in fact these intersection operators are not well-defined as cycles, so the situation is more complicated. In fact, we gave an example in the exercises in Chapter 5 where the simple definition given in Section 5.1 in the affine case does not commute as a map on cycles. The main result of this section is that they do commute modulo rational equivalence on the support of their intersection. This is a fundamental theorem that will be of crucial importance in the theory of local Chern characters.

Throughout we may assume that the ring involved is an integral domain. As mentioned above, we have shown that if the codimension of the intersection is 2 for principal divisors, then the intersection is given by the Euler characteristic of the Koszul complex. We first verify that this result holds for arbitrary divisors.

Proposition 8.9.1 *Suppose that D and D' are divisors on $X = \operatorname{Proj}(A)$, where A is an integral domain, and that the intersection of subschemes $D \cap D'$ has codimension 2 in X. Then for any prime ideal $\mathfrak{p}$ of codimension 2, the coefficient of $[A/\mathfrak{p}]$ in $D \cap D' \cap [X]$ is the Euler characteristic of the Koszul complex on the local generators of D and D' on $A_{(\mathfrak{p})}$.*

Proof If D and D' are defined by ideals $\mathfrak{a}$ and $\mathfrak{a}'$, then the hypothesis states that $\mathfrak{a} + \mathfrak{a}'$ has codimension 2 in A. Let $\mathfrak{p}$ be a minimal prime ideal over $\mathfrak{a} + \mathfrak{a}'$. Localizing at $\mathfrak{p}$, the ideals $\mathfrak{a}$ and $\mathfrak{a}'$ become principal; let the generators of $\mathfrak{a}_{(\mathfrak{p})}$ and $\mathfrak{a}'_{(\mathfrak{p})}$ be x and x', respectively. Then the theorem states that $(x) \cap (x') \cap [A]$ is the Euler characteristic of the Koszul complex $K_\bullet(x, x'; A)$, which follows from Theorem 5.2.12. $\qquad\square$

We next prove that intersection with divisors commutes up to rational equivalence in the case where the codimension is not 2. This proof is essentially the same as that in Fulton [17]. Let D and D' be divisors, defined by ideals $\mathfrak{a}$ and $\mathfrak{a}'$, respectively, such that the intersection of subschemes $D \cap D'$ has components of codimension 1. Let $\mathfrak{p}$ be one of the components of codimension 1. Then in the local ring $A_{(\mathfrak{p})}$, the quotients $A_{(\mathfrak{p})}/\mathfrak{a}_{(\mathfrak{p})}$ and $A_{(\mathfrak{p})}/\mathfrak{a}'_{(\mathfrak{p})}$ have finite length. Since there are finitely many prime ideals minimal over $\mathfrak{a} + \mathfrak{a}'$, we may take the maximal value of the products $\operatorname{length}(A_{(\mathfrak{p})}/(\mathfrak{a}_{(\mathfrak{p})}))\operatorname{length}(A_{(\mathfrak{p})}/(\mathfrak{a}'_{(\mathfrak{p})}))$ and obtain an integer. If all of the products are zero, then the intersection of D and D' has codimension 2. The proof of commutativity is by induction on the supremum of these products, which is reduced using the Rees ring of the ideal generated by $\mathfrak{a}$ and $\mathfrak{a}'$. We first describe the necessary properties of this ring in the case in which $\mathfrak{a}$ and $\mathfrak{a}'$ are principal for use in the local computations.

Let A be a local integral domain, and let x and y be nonzero elements of A. Let

$$R = R(x, y) = A \oplus (x, y) \oplus (x, y)^2 \oplus \cdots$$

be the Rees ring on (x, y) with the usual grading, and let $X = \mathrm{Proj}(R)$. Since x and y generate the irrelevant ideal of R, X is covered by U_x and U_y. The ring $R_{(x)}$ consists of quotients a/x^n, where a is in $(x, y)^n = (x^n, x^{n-1}y, \ldots, y^n)$. Writing $a = a_n x^n + a_{n-1}x^{n-1}y + \cdots + a_0 y^n$, we have

$$\frac{a}{x^n} = a_n + a_{n-1}\frac{y}{x} + \cdots + a_0 \left(\frac{y}{x}\right)^n.$$

From this expression it follows that $A_{(x)}$ is the subring $A[y/x]$ of the quotient field of A generated by y/x. Similarly, $A_{(y)}$ is the subring $A[x/y]$ of the quotient field.

Let D_x and $\overline{D}_x$ be the divisors defined by the ideals generated by x in degree 0 and $x \in (x, y)$ in degree 1, respectively. Let D_y and $\overline{D}_y$ be defined similarly. Let E be the divisor defined by the ideal (x, y) in degree 0. To show that E is a divisor, we consider the local situation. In the ring $A_{(x)} = A[y/x]$, the ideal (x, y) is generated by x, since $y = (y/x)x$, and hence is locally principal. Similarly, in $A_{(y)}$ the ideal of E is locally generated by y.

We next compute local generators for the ideals of D_x and $\overline{D}_x$. The ideal of D_x is generated by x globally, so it is locally generated by x in both $A_{(x)}$ and $A_{(y)}$. Since the ideal of $\overline{D}_x$ is generated by the element x in degree 1, it is generated by $x/x = 1$ in $A_{(x)}$ and by x/y in $A_{(y)}$. Similar results hold for y.

Combining these local computations, we see that we have the equality $D_x = \overline{D}_x + E$. In fact, we can check locally; in $A_{(x)}$, the ideal of $\overline{D}_x$ is generated by 1 and that of E is generated by x, while the ideal of D_x is generated by the product $1 \cdot x = x$. In $A_{(y)}$, the ideal of $\overline{D}_x$ is generated by x/y and that of E is generated by y; again the ideal of D_x is generated by the product $(x/y)y = x$. Similarly, $D_y = \overline{D}_y + E$.

Theorem 8.9.2 *Let $X = \mathrm{Proj}(A)$, where A is an integral domain, and let D and D' be divisors on X. Let Z be the intersection of subschemes $D \cap D'$. Then $D \cap D'([X])$ is rationally equivalent to $D' \cap D([X])$ in $Z_{d-2}(Z)$.*

Proof Let $\mathfrak{a}$ and $\mathfrak{a}'$ be ideals defining D and D'. Let $n(\mathfrak{a}, \mathfrak{a}')$ denote the supremum of the products

$$\left(\text{length}\left(A_{(\mathfrak{p})}/\mathfrak{a}_{(\mathfrak{p})}\right)\right)\left(\text{length}\left(A_{(\mathfrak{p})}/\mathfrak{a}'_{(\mathfrak{p})}\right)\right)$$

over all prime ideals in Z of codimension 1 in X. This product is zero for a given $\mathfrak{p}$ if and only if $\mathfrak{p}$ is not in the support of $A/(\mathfrak{a} + \mathfrak{a}')$. We prove the general result by induction on $n(\mathfrak{a}, \mathfrak{a}')$. If $n = 0$, the divisors D and D' meet in codimension at least 2, so the result is a restatement of Proposition 8.9.1. Let $Y = \text{Proj}(R(\mathfrak{a} + \mathfrak{a}'))$, where $R(\mathfrak{a} + \mathfrak{a}')$ is the Rees ring of the ideal $\mathfrak{a} + \mathfrak{a}'$, and let ϕ denote the map from Y to X. Then $\phi^*(D)$ is the divisor corresponding to the image of $\mathfrak{a}$ in $R(\mathfrak{a}+\mathfrak{a}')$ in degree 0. Let $\overline{D}$ denote the divisor corresponding to the ideal generated by the image of $\mathfrak{a}$ in $\mathfrak{a} + \mathfrak{a}'$ in degree 1 in $R(\mathfrak{a} + \mathfrak{a}')$, and define $\overline{D}'$ similarly. Let E be the divisor defined by the sum $\mathfrak{a} + \mathfrak{a}'$ in degree 0. Using the local computations above, we have

$$\phi^*(D) = \overline{D} + E \quad \text{and} \quad \phi^*(D') = \overline{D}' + E.$$

Furthermore, since D and D' in Y are the pull-backs of D and D' in X, and ϕ is a projective map, by Proposition 8.8.2 we have

$$\phi_*(\phi^*(D) \cap \phi^*(D')([Y])) = D \cap \phi_*(\phi^*(D')([Y])) = D \cap D' \cap \phi_*([Y])$$
$$= D \cap D' \cap [X]$$

and a similar formula when the order of D and D' is reversed. Hence, using the above expressions for $\phi^*(D)$ and $\phi^*(D')$, it suffices to prove that

$$(\overline{D} + E) \cap (\overline{D}' + E) \cap [Y] = (\overline{D}' + E) \cap (\overline{D} + E) \cap [Y].$$

These intersection products expand into four terms, and it suffices to show that we have commutativity for each term. Clearly $E \cap E \cap [Y] = E \cap E \cap [Y]$. The intersection of schemes $\overline{D} \cap \overline{D}'$ is empty, so we have $\overline{D} \cap \overline{D}' \cap [Y] = \overline{D}' \cap \overline{D} \cap [Y] = 0$. It remains to prove the commutativity for $\overline{D}$ and E and for $\overline{D}'$ and E; we prove the first case, since the second is the same.

To show that $\overline{D} \cap E \cap [Y] = E \cap \overline{D} \cap [Y]$ we use induction on the number $n(\mathfrak{a}, \mathfrak{a}')$ defined at the beginning of the proof. Denote the Rees ring $R(\mathfrak{a} + \mathfrak{a}')$ by B, and denote the ideals defining $\overline{D}$ and E by $\overline{\mathfrak{a}}$ and $\mathfrak{b}$, respectively. Let $\mathfrak{q}$ be a prime ideal of codimension 1 that is minimal over $\overline{\mathfrak{a}} + \mathfrak{b}$. The inverse image $\mathfrak{p}$ of $\mathfrak{q}$ in A must also have codimension 1, and because $\mathfrak{q}$ contains $\mathfrak{b}$, $\mathfrak{p}$ must contain $\mathfrak{a} + \mathfrak{a}'$. We then localize at $\mathfrak{p}$. We must show that

$$\left(\text{length}\left(B_{(\mathfrak{q})}/\overline{\mathfrak{a}}_{(\mathfrak{q})}\right)\right)\left(\text{length}\left(A_{(\mathfrak{q})}/\mathfrak{b}_{(\mathfrak{q})}\right)\right)$$

is strictly less than

$$\left(\text{length}\left(A_{(\mathfrak{p})}/\mathfrak{a}_{(\mathfrak{p})}\right)\right)\left(\text{length}\left(A_{(\mathfrak{p})}/\mathfrak{a}'_{(\mathfrak{p})}\right)\right).$$

We have an extension $A_{(\mathfrak{p})} \subseteq B_{(\mathfrak{q})} \subseteq K$, where K is the quotient field of $A_{(\mathfrak{p})}$. By the local computation above, we also have $\mathfrak{a}_{(\mathfrak{q})} = \overline{\mathfrak{a}}_{(\mathfrak{q})} \mathfrak{b}_{(\mathfrak{q})}$ and $\mathfrak{a}'_{(\mathfrak{q})} = \overline{\mathfrak{a}}'_{(\mathfrak{q})} \mathfrak{b}_{(\mathfrak{q})}$. Thus, since these ideals are principal, we have

$$\text{length}\left(A/\mathfrak{a}_{(\mathfrak{q})}\right) = \text{length}\left(A/\overline{\mathfrak{a}}_{(\mathfrak{q})}\right) + \text{length}\left(A/\mathfrak{b}_{(\mathfrak{q})}\right)$$

and

$$\text{length}\left(A/\mathfrak{a}'_{(\mathfrak{q})}\right) = \text{length}\left(A/\overline{\mathfrak{a}}'_{(\mathfrak{q})}\right) + \text{length}\left(A/\mathfrak{b}_{(\mathfrak{q})}\right).$$

Furthermore, since $\mathfrak{a}_{(\mathfrak{p})}$ and $\mathfrak{a}'_{(\mathfrak{p})}$ are principal ideals, it follows that

$$\text{length}\left(B_{(\mathfrak{q})}/\mathfrak{a}_{(\mathfrak{q})}\right) = \left(\text{rank}_{A_{(\mathfrak{p})}}\left(B_{(\mathfrak{q})}\right)\right)\left(\text{length}\left(A_{(\mathfrak{p})}/\mathfrak{a}_{(\mathfrak{p})}\right)\right)$$
$$= \text{length}\left(A_{(\mathfrak{p})}/\mathfrak{a}_{(\mathfrak{p})}\right)$$

and similarly for $\mathfrak{a}'$.

Combining these equalities, we have

$$\left(\text{length}\left(A/\mathfrak{a}_{(\mathfrak{q})}\right)\right)\left(\text{length}\left(A/\mathfrak{a}'_{(\mathfrak{q})}\right)\right)$$
$$= \left(\text{length}\left(A/\overline{\mathfrak{a}}_{(\mathfrak{q})}\right) + \text{length}\left(A/\mathfrak{b}_{(\mathfrak{q})}\right)\right)$$
$$\times \left(\left(\text{length}\left(A/\overline{\mathfrak{a}}'_{(\mathfrak{q})}\right) + \text{length}\left(A/\mathfrak{b}_{(\mathfrak{q})}\right)\right)\right).$$

This product expands into four terms, all of which are nonnegative and of which $(\text{length}(A/\mathfrak{b}_{(\mathfrak{q})}))^2$ is positive. Since one of the terms is

$$\left(\text{length}\left(A/\overline{\mathfrak{a}}_{(\mathfrak{q})}\right)\right)\left(\text{length}\left(A/\mathfrak{b}_{(\mathfrak{q})}\right)\right)$$

this proves the required inequality. $\qquad\qquad\qquad\qquad\qquad\qquad\square$

8.10 Bézout's Theorem

In Section 6.3 we discussed the multiplicity conjectures for graded modules and proved a formula for Euler characteristics of tensor products of complexes of free graded modules. In this section we show that a very similar result implies that the definition of intersection multiplicities using Euler characteristics satisfies a classic form of Bézout's theorem.

Let k be an algebraically closed field. Let $\mathbb{P}^n = \text{Proj}(k[X_0, \ldots, X_n])$ be projective space of dimension n over k; $k[X_0, \ldots, X_n]$ is given the usual $\mathbb{Z}$-grading. Let A denote the graded ring $k[X_0, \ldots, X_n]$. In Section 6.3 we were concerned with the Euler characteristic defined by the length of a graded module of finite length. For that purpose it was convenient to use the Hilbert polynomial whose value at n is the length of $\oplus_{i \leq n} M_n$ for a graded module M. Here we are interested in the Hilbert polynomial of a graded ring that defines a projective scheme of the form $\text{Proj}(A/\mathfrak{a})$, where $A/\mathfrak{a}$ defines a projective scheme of dimension zero. By Proposition 8.3.3 this condition implies that the Hilbert polynomial whose value at n is the length of $(A/\mathfrak{a})_n$ is constant, and its value is the length of $(A/\mathfrak{a})_{(x)}$ for any homogeneous element that is in no minimal prime ideal of $A/\mathfrak{a}$. For this reason we consider the Hilbert polynomial P_M such that $P_M(n) = \text{length}(M_n)$. The theory of Section 6.3 goes through in the same way, and to every bounded complex $F_\bullet$ of graded free modules there is a sequence of numbers c_k defined by

$$c_k = \frac{1}{k!} \sum_{i,j} (-1)^i n_{ij}^k$$

where the n_{ij} are the integers for which $F_i = \oplus_j A[n_{ij}]$. As in Theorem 6.3.1, we have

$$\sum_i (-1)^i P_{H_i(F_\bullet)} = \sum_k c_k P_A^{(k)}.$$

Theorem 8.10.1 (Bézout's Theorem) *Let Y and Z be closed subschemes of $\mathbb{P}^n$ of dimensions s and t, respectively, where $s + t = n$. Assume that Y and Z are defined by prime ideals $\mathfrak{p}$ and $\mathfrak{q}$ and that $Y \cap Z$ consists of a finite set of points. For each point p in the intersection, let $\mathfrak{m}_p$ be the prime ideal of $k[X_0, \dots, X_n]$ corresponding to p, and let $e_p(Y, Z) = \chi((A/\mathfrak{p})_{(\mathfrak{m}_p)}, (A/\mathfrak{q})_{(\mathfrak{m}_p)})$ over the regular local ring $k[X_0, \dots, X_n]_{(\mathfrak{m}_p)}$. Then*

$$\sum_p e_p(Y, Z) = \text{degree}(Y)\,\text{degree}(Z).$$

Proof For a graded A-module M, we denote $e(M)$ the degree of M, which is defined to be $d!$ times the leading coefficient of the Hilbert polynomial of M, where d is the dimension of M as defined in this chapter. In particular, we have $e(A) = 1$ since A is a polynomial ring. Let $F_\bullet$ and $G_\bullet$ be resolutions of $A/\mathfrak{p}$ and $A/\mathfrak{q}$ by bounded complexes of graded free modules. The assumption that Y and Z meet in a finite number of points implies that $A/\mathfrak{p} \otimes_A A/\mathfrak{q}$ has dimension 1, so that its Hilbert polynomial is constant; the same holds for $\text{Tor}_i^A(A/\mathfrak{p}, A/\mathfrak{q})$ for $i > 0$. For each i we let $e(\text{Tor}_i^A(A/\mathfrak{p}, A/\mathfrak{q}))$ be the constant value of the associated Hilbert polynomial. Following the proof of the statement (iii) of Theorem 6.3.2, we can conclude that

$$\sum_i (-1)^i e\big(\text{Tor}_i^A(A/\mathfrak{p}, A/\mathfrak{q})\big) = \frac{e(A/\mathfrak{p})e(A/\mathfrak{q})}{e(A)} = \text{degree}(Y)\,\text{degree}(Z).$$

To complete the proof it remains to show that for each i we have

$$e\big(\text{Tor}_i^A(A/\mathfrak{p}, A/\mathfrak{q})\big) = \sum_\mathfrak{m} \text{length}\big(\text{Tor}_i\big(A_{(\mathfrak{m})}/\mathfrak{p}_{(\mathfrak{m})}, A_{(\mathfrak{m})}/\mathfrak{q}_{(\mathfrak{m})}\big)\big)$$

where the sum is over all primes $\mathfrak{m}$ minimal over $\mathfrak{p} + \mathfrak{q}$. But it follows from Proposition 8.3.3 that the degree is the sum of the lengths of the localizations of $\text{Tor}_i^A(A/\mathfrak{p}, A/\mathfrak{q})$ at all such primes $\mathfrak{m}$. Thus, it follows from the exactness of localization that these are also the lengths of $\text{Tor}_i(A_{(\mathfrak{m})}/\mathfrak{p}_{(\mathfrak{m})}, A_{(\mathfrak{m})}/\mathfrak{q}_{(\mathfrak{m})})$, so this completes the proof. $\square$

Exercises

8.1   Prove that if U is an open subset of $X = \text{Proj}(A)$, then there is an exact sequence

$$A_*(X - U) \to A_*(X) \to A_*(U) \to 0.$$

8.2   Let A be a $\mathbb{Z}$-graded ring. Show that if A is flat over A_0 then $A_{(x)}$ is flat over A_0 for all $x \in A_1$. Show also that the converse does not hold.

8.3   Show that if C is a positive cone in $\mathbb{Z}^n$ defined by the integers $k_1, \ldots, k_n$, then $C - (k_1 k_2 \cdots k_n, k_2 \cdots k_n, \ldots, k_{n-1} k_n, k_n)$ is a positive cone with vertex at zero.

8.4   Let $f : A \to B$ be a map of graded rings. Show that $\{\mathfrak{p} \in \text{Proj}(B) \mid f^{-1}(\mathfrak{p}) \in \text{Proj}(A)\}$ is open. Give an example where this set is empty.

8.5   Let k be a field, and let ϕ be the map from $\text{Spec}(k[X])$ to $\text{Spec}(k)$ induced by the inclusion of k into $k[X]$. Show that the map ϕ_* from $Z_*(\text{Spec}(k[x]))$ to $Z_*(\text{Spec}(k))$ defined by sending the zero prime ideal to 0 and the cycle $[k[x]/(f(x))]$ to the cycle $(\dim_k(k[x]/(f(x))))[k]$ for all nonzero prime ideals $(f(x))$ does not preserve rational equivalence. Thus, the map defined for proper maps in Section 8.5 does not necessarily preserve rational equivalence for nonproper maps.

8.6   Show that a proper map is a closed map of topological spaces.

8.7   Let k be a field, and let the polynomial ring $k[S, T, U, V]$ be given a $\mathbb{Z}^2$ grading in which S and T have degree $(1, 0)$ and U and V have degree $(0, 1)$. Let $k[W, X, Y, Z]$ be given the usual $\mathbb{Z}$-grading.
      (a) Show that the map

$$f : k[W, X, Y, Z]/(XY - ZW) \to k[S, T, U, V]$$

      defined by sending W, X, Y, Z to SU, SV, TU, TV, respectively, defines a proper map.
      (b) Show that the map ϕ induced by f defines a bijection between the projective schemes $\text{Proj}(k[W, X, Y, Z]/(XY - ZW))$ and $\text{Proj}(k[S, T, U, V])$.

9

Chern Classes of Locally Free Sheaves

The purpose of this chapter is to define Chern classes of locally free sheaves on projective schemes. This theory will be used in the definition of local Chern characters of complexes in Chapter 11. We first define the concept of a sheaf on Proj(A), where A is a graded ring; we define a sheaf to be an equivalence class of graded A-modules, where two modules are equivalent when their restrictions to any open set of Proj(A) are the same (the precise definition is given in Section 1). This is not the usual definition of a sheaf, but the theory is essentially equivalent to that of quasi-coherent sheaves for projective schemes (see Hartshorne [21]). This definition suffices for our purposes, and it makes it easy to prove the important facts that we need. We are particularly interested in locally free sheaves, for which it is possible to define certain intersection classes on the Chow group of Proj(A), called Chern classes, which in some sense measure how far the module is from being free. We describe the theory of Chern classes in this chapter and prove the functorial properties necessary for later applications. In the last section we define the push-forward of a sheaf by a projective map, which will be used in the proof of the local Riemann-Roch formula in Chapter 12.

9.1 Sheaves

Let A be a graded ring, and let E and F be finitely generated graded modules. We define an equivalence relation on the set of such modules by saying that E is equivalent to F if there is a map f of graded modules from E to F such that the supports of the kernel and cokernel of f, which are sets of graded prime ideals, contain no elements of Proj(A). In other words, if $\mathfrak{p}$ is in the support of the kernel or cokernel of f, then the set of indices of elements not in $\mathfrak{p}$ does not contain a positive cone with vertex at zero. This condition is equivalent to the condition that, for all $\mathfrak{p} \in$ Proj(A), the map from $E_{\mathfrak{p}}$ to $F_{\mathfrak{p}}$ is an isomorphism,

177

which is in turn equivalent to the condition that, for all $\mathfrak{p} \in \mathrm{Proj}(A)$, the map from $E_{(\mathfrak{p})}$ to $F_{(\mathfrak{p})}$ is an isomorphism. This relation is reflexive and transitive but not symmetric, and we take the equivalence relation that it generates. A *sheaf* is defined to be an equivalence class of modules under this equivalence relation. We will usually denote the sheaf associated to a module E by the symbol $\mathcal{E}$. The sheaf on $X = \mathrm{Proj}(A)$ defined by the module A will be denoted $\mathcal{O}_X$.

We say that a sheaf $\mathcal{E}$ is *locally free* if there is a cover of $\mathrm{Proj}(A)$ consisting of special basic open sets U_x such that $E_{(x)}$ is a free $A_{(x)}$ module for all x. If E and F are modules representing the same sheaf, then $E_{(x)}$ is isomorphic to $F_{(x)}$ for all x, so this definition is independent of the module chosen to represent the sheaf. The sheaf $\mathcal{E}$ is locally free if and only if the localization $E_{(\mathfrak{p})}$ is a free $A_{(\mathfrak{p})}$ module for all $\mathfrak{p} \in \mathrm{Proj}(A)$.

The simplest locally free sheaves are the sheaves defined by the A-modules $A[i]$, where i is an element of $\mathbb{Z}^n$. The fact that these sheaves are locally free depends on the fact that, for every $\mathfrak{p}$ in $\mathrm{Proj}(A)$, the set of degrees of elements not in $\mathfrak{p}$ contains a positive cone so that there are elements s and t not in $\mathfrak{p}$ such that s/t has degree i. Thus, there is an isomorphism $g : A_{(\mathfrak{p})} \to (A[i])_{(\mathfrak{p})}$ defined by letting $g(a/u) = as/ut$. We denote the sheaf associated to $A[i]$ by $\mathcal{O}_X(i)$.

Let $\phi : \mathrm{Proj}(B) \to \mathrm{Proj}(A)$ be a projective map defined by a map f of graded rings from A to B. If E is a graded A-module, then $B \otimes_A E$ can be given a grading as follows: if η is the associated map on index groups, then the component of degree i is generated (as an Abelian group) by all $B_k \otimes E_j$ with $k + \eta(j) = i$. In this way $B \otimes_A E$ becomes a graded B-module.

Proposition 9.1.1 *Let $X = \mathrm{Proj}(B)$ and $Y = \mathrm{Proj}(A)$, and let $\phi : X \to Y$ be a proper map defined by a map f from A to B.*

(i) *If E and F are equivalent A-modules, then $B \otimes_A E$ and $B \otimes_A F$ are equivalent B-modules.*

(ii) *If E is an A-module that defines a locally free sheaf on Y, then $B \otimes_A E$ defines a locally free sheaf on X.*

Proof To prove the first part, we may assume that there is a map $g : E \to F$ such that the supports of the kernel and cokernel of g do not meet $\mathrm{Proj}(A)$. Let $\mathfrak{q} \in \mathrm{Proj}(B)$, and let $\mathfrak{p} = f^{-1}(\mathfrak{q})$; then $\mathfrak{p} \in \mathrm{Proj}(B)$ since the map ϕ is proper. Thus, the map induced by g from $E_{(\mathfrak{p})}$ to $F_{(\mathfrak{p})}$ is an isomorphism.

To complete the proof, we first show that we have an isomorphism

$$B_{(\mathfrak{q})} \otimes_{A_{(\mathfrak{p})}} E_{(\mathfrak{p})} \cong (B \otimes_A E)_{(\mathfrak{q})}.$$

To see this, we define a map from $B_{(q)} \otimes_{A_{(p)}} E_{(p)}$ to $(B \otimes_A E)_{(q)}$ by sending $(b/t) \otimes (e/s)$ to $(b \otimes e)/tf(s)$. If $E = A[i]$ for some i, then $A[i]_{(p)}$ is isomorphic to $A_{(p)}$, so this map is an isomorphism from the module $B_{(q)} \otimes_{A_{(p)}} A[i]_{(p)}$ to $B[i]_{(q)}$. The general case now follows from the fact that E has a finite presentation by sums of modules of the form $A[i]$ and the exactness of localization and of taking the component of degree 0.

Thus, since the map induced by g from $E_{(p)}$ to $F_{(p)}$ is an isomorphism, we have that

$$(B \otimes_A E)_{(q)} \cong B_{(q)} \otimes_{A_{(p)}} E_{(p)} \to B_{(q)} \otimes_{A_{(p)}} F_{(p)} \cong (B \otimes_A F)_{(q)}$$

is an isomorphism. Thus, $\phi^*(E)$ is equivalent to $\phi^*(F)$.

The second part of the theorem is proven similarly, using that fact that if $E_{(p)}$ is free then $(B \otimes_A E)_{(q)} \cong B_{(q)} \otimes_{A_{(p)}} E_{(p)}$ is free. $\square$

If E is a graded B module and if $\mathcal{E}$ denotes the associated sheaf on $Y = \mathrm{Proj}(B)$, and if $\phi : X \to Y$ is a proper map with $X = \mathrm{Proj}(A)$, we denote the sheaf on X defined by $A \otimes_B E$ by $\phi^*(\mathcal{E})$. Proposition 9.1.1 implies that $\phi^*(\mathcal{E})$ is well-defined.

The next proposition is useful in going between the local and global situations.

Proposition 9.1.2 *Let A be a graded ring, and let T be a multiplicative set that contains a unit special homogeneous element. Then the functor* **F** *that takes an A_T module to the component of degree 0, and the functor* **G** *that takes a graded $A_{(T)}$-module E to $A_T \otimes_{A_{(T)}} E$, define an equivalence of categories.*

Proof It follows from Proposition 8.2.5 that A_T is isomorphic to the ring $A_{(T)}[T_1, T_1^{-1}, \ldots, T_n, T_n^{-1}]$, where T_i is homogeneous of degree $e_i = (0, \ldots, 0, 1, 0, \ldots, 0)$. It is clear that the part of degree 0 of $A_T \otimes_{A_{(T)}} E$ is $A_{(T)} \otimes_{A_{(T)}} E = E$, so **FG** is the identity. On the other hand, if E' is a graded A_T-module, the map from $A_T E'_0 \otimes_{A_{(T)}} E'$ to E' defined by sending $T_1^{i_1} \cdots T_n^{i_n} \otimes e$ to $T_1^{i_1} \cdots T_n^{i_n} e$ is an isomorphism. Hence **F** and **G** define an equivalence of categories. $\square$

A map of sheaves is an equivalence class defined by a map of associated graded modules. Most constructions that are defined for modules carry over to sheaves. For example, the direct sum and tensor product of sheaves are defined by taking the direct sum and tensor product of modules that define them; it is straightforward to check that the construction does not depend on which modules are chosen up to equivalence. Similarly, the kernel and cokernel of a map of sheaves, as well as quotient sheaves, can be defined in this way. We can also define a functor Hom by defining $\mathrm{Hom}(\mathcal{E}, \mathcal{F})$ to be the module of homomorphisms from the underlying modules E and F. The module

$\text{Hom}(E, F)$ becomes a graded module by letting the component of degree n be the set of homomorphisms g such that $g(E_k) \subseteq F_{k+n}$ for all k. Using this definition we have $\text{Hom}(\mathcal{O}_X, \mathcal{E}) = \mathcal{E}$. A particular case of this construction is the dual $\mathcal{E}^* = \text{Hom}(\mathcal{E}, \mathcal{O}_X)$. If the constructions preserve the property of being free when applied to modules, they preserve the property of being locally free when applied to sheaves. Thus, the direct sum and tensor product of locally free sheaves are locally free, and if $\mathcal{E}$ and $\mathcal{F}$ are locally free sheaves, then $\text{Hom}(\mathcal{E}, \mathcal{F})$ is a locally free sheaf.

Let

$$0 \to E' \to E \to E'' \to 0$$

be a sequence of A-modules such that the resulting map from E' to E'' is zero. We say that the associated sequence of sheaves

$$0 \to \mathcal{E}' \to \mathcal{E} \to \mathcal{E}'' \to 0$$

is exact if the support of the homology of the sequence does not meet $\text{Proj}(A)$. If we have such an exact sequence and $\mathcal{E}'$ and $\mathcal{E}''$ are locally free, then $\mathcal{E}$ is locally free, and if $\mathcal{E}$ and $\mathcal{E}''$ are locally free then $\mathcal{E}'$ is locally free.

Two other constructions on sheaves that will be important are symmetric and exterior powers. If E is a graded A-module defining a sheaf $\mathcal{E}$, then the symmetric power $\text{Sym}^r(E)$ and the exterior power $\Lambda^r(E)$ are graded modules that define sheaves $\text{Sym}^r(\mathcal{E})$ and $\Lambda^r(\mathcal{E})$. If T is a multiplicative set of homogeneous elements containing a unit special homogeneous element, we have isomorphisms $(\text{Sym}^r(E))_{(T)} \cong \text{Sym}^r(E_{(T)})$ and $(\Lambda^r(E))_{(T)} \cong \Lambda^r(E_{(T)})$. We verify this fact in the case of the symmetric power, using subscripts to denote the ring over which the symmetric power is being taken. First, since taking symmetric powers commutes with localization, we have

$$\left(\text{Sym}^r_A(E)\right)_T \cong \text{Sym}^r_{A_T}(E_T).$$

On the other hand, since taking symmetric powers commutes with tensor products of rings, we have

$$\text{Sym}^r_{A_{(T)}}\left(E_{(T)}\right) \otimes_{A_{(T)}} A_T \cong \text{Sym}^r_{A_T}(E_T).$$

It now follows from Proposition 9.1.2 that $\text{Sym}^r_{A_{(T)}}(E_{(T)})$ is isomorphic to the component of $\text{Sym}^r_{A_T}(E_T)$ of degree 0, which by the previous isomorphism is isomorphic to $(\text{Sym}^r_A(E))_{(T)}$.

9.2 Chern Classes

In this section we define the Chern classes of a locally free sheaf over a projective scheme. The presentation here is essentially that given by Grothendieck

[20]. Fulton [17] gives another construction; we sketch a proof that these methods give the same result in the exercises. In fact, it is also possible to give the definition and construction of Chern classes via embeddings into Grassmannians directly without the kind of machinery we develop here; a description of this method in a topological setting can be found in Milnor and Stasheff [45]. However, it becomes much more difficult to prove some of the essential properties of Chern classes, such as the multiplicativity of Chern characters with respect to tensor products. These properties are essential for the applications to local algebra.

The Chern classes of a locally free sheaf $\mathcal{E}$ of rank r over X will be defined to be operators on the Chow group of X. For each integer $i = 1, \ldots, r$ we will define the Chern class $c_i(\mathcal{E})$ to be a map from $A_k(X)$ to $A_{k-i}(X)$ for each k.

To motivate the general construction, we consider two examples. In the first example, let A be a local ring, and let F be a free module. The symmetric algebra on F is isomorphic to a polynomial ring $A[T_1, \ldots, T_r]$, where r is the rank of F. Let $A[T_1, \ldots, T_r]$ have the usual $\mathbb{Z}$-grading, and let $X = \text{Proj}(A[T_1, \ldots, T_r])$. We may use the T_i to define the hyperplane section h_1, which we denote h, and we see that $h^r([A[T_1, \ldots, T_r])$ is the set of prime ideals in $\text{Proj}(A[T_i])$ containing $(T_1, \ldots, T_r)$, which is empty. Thus, $h^r([A[T_1, \ldots, T_r]) = 0$. On the other hand, in the situation where F defines a locally free sheaf but is not a free module, and where we carry out a similar construction for the symmetric algebra on F in place of the polynomial ring, the result will not be zero in general, but only a cycle that is supported in lower dimension than the dimension of A. The Chern classes will give expressions for these elements, and in a sense they measure how far a locally free sheaf is from being free.

As a second example, let A be a graded integral domain, let $X = \text{Proj}(A)$, and let $\mathcal{E}$ be a rank 1 locally free sheaf on X defined by a torsion free rank 1 module E. Since $\mathcal{E}$ has rank 1, there will be only one Chern class $c_1(\mathcal{E})$. Suppose that there exists a nonzero element $e \in E$ of degree 0. The Chern class of $\mathcal{E}$ applied to the class $[X]$ can be defined to be the class

$$c_1(\mathcal{E})([X]) = \sum_{\dim(\mathfrak{p})=\dim(A)-1} \text{length}(E/(e))_{(\mathfrak{p})}[A/\mathfrak{p}].$$

In other words, it is the class $[E/(e)]_{d-1}$ in dimension $d-1$ as defined in Chapter 1. Thus, if $\mathcal{E}$ has a nonzero element e of degree 0, the Chern class of $\mathcal{E}$ is defined to be the cycle of codimension 1 defined by dividing by the element e.

To complete the second example, we show how it can be constructed in terms of hyperplane sections on a symmetric algebra as in the first example. Let $\text{Sym}(E)$ be the symmetric algebra on the module E. We denote the projective scheme $\text{Proj}(\text{Sym}(E))$ by $P(\mathcal{E})$; it follows from the discussion in the previous section that $P(\mathcal{E})$ does not change if we replace E by an equivalent module. Note that since $\mathcal{E}$ is locally free of rank 1, we have $\dim(P(\mathcal{E})) = \dim(X) = d$. The element e in the component of $\text{Sym}(E)$ of degree 1 has degree $(0, 1)$, so it defines the hyperplane section $h_{(0,1)}$ in $P(\mathcal{E})$. Denote $h_{(0,1)}$ by h. We consider the cycle $h([P(\mathcal{E})])$, which, as we have just seen, is defined by the element e in degree 1, so it can be written as $[\text{Sym}(E)/e\,\text{Sym}(E)]_{d-1}$. One minimal prime ideal over e is the ideal generated by E, which contains all elements of degree (i, j) for $j > 0$ and so is not an element of $\text{Proj}(A)$ and is thus zero in $A_{d-1}(X)$. We claim that the other components are given by the pull-back of the cycle $[E/(e)]_{d-1}$ in X to $P(\mathcal{E})$. To see this, we may check locally on X, so we may assume that E is a free A-module. Then $P(\mathcal{E}) = \text{Proj}(A[T])$, and the element e corresponds to an element of $A[T]$ of the form aT. We note that, since $\dim(A[T]) = \dim(\text{Proj}(A[T])) + 1$, we are interested in the cycle defined by $A[T]/(aT)$ of dimension d rather than $d - 1$. However, in this case we have

$$[A[T]/(aT)]_d = [A[T]/(T)]_d + [A[T]/(a)]_d,$$

and $[A[T]/(a)]_d$ is the pull-back of $[A/aA]_{d-1}$ to $A[T]$, so the result is clear.

Let p denote the map from $P(\mathcal{E})$ to X. We note that p is a projective map and is also flat, since $\mathcal{E}$ is locally free. We can summarize the above computation by stating that $c_1(\mathcal{E})([X])$ is the class in $A_{d-1}(X)$ that satisfies the condition

$$p^*(c_1(\mathcal{E})([X])) = hp^*([X]).$$

The remainder of the chapter is a generalization of this construction to locally free sheaves of higher rank.

We wish to construct Chern classes $c_1(\mathcal{E}), \ldots, c_r(\mathcal{E})$ for a locally free sheaf $\mathcal{E}$ of rank r. As stated above, each $c_i(\mathcal{E})$ will be defined to be an operator on the Chow group of A of codimension i, so that for each k it defines an element $c_i(\mathcal{E})(\eta)$ in $A_{k-i}(X)$ for all $\eta \in A_k(X)$. We will also define $c_0(\mathcal{E})(\eta)$ in $A_k(X)$ by $c_0(\mathcal{E})(\eta) = \eta$. Chern classes will have certain naturality properties as well as other important properties that will be useful in their computation: for example, they commute with the operations we have defined up to this point and with each other.

Let $\mathcal{E}$ be a locally free sheaf over $X = \text{Proj}(A)$. Since we wish to define the action of the Chern classes of $\mathcal{E}$ on the Chow group, it suffices to define their action on a generator $[A/\mathfrak{p}]$; of course, we will then have to prove that the result is independent of rational equivalence. Replacing A with $A/\mathfrak{p}$ and $\mathcal{E}$ with its pull-back to $\text{Proj}(A/\mathfrak{p})$, we may assume that A is a domain. Let E be a module representing $\mathcal{E}$; since $\mathcal{E}$ is locally free, the torsion submodule of E is equivalent to the zero module, and we may assume that E is torsion-free. Let $\mathrm{Sym}(E)$ be the symmetric algebra on E. As in Chapter 8, we let the Rees algebra $R(E)$ be the quotient of $\mathrm{Sym}(E)$ by its torsion submodule. We note that since E is locally free, the map from $\mathrm{Sym}(E)$ to $R(E)$ is locally an isomorphism, so it induces an isomorphism from $\mathrm{Proj}(R(E))$ to $\mathrm{Proj}(\mathrm{Sym}(E))$. Let $P(\mathcal{E}) = \mathrm{Proj}(R(E)) = \mathrm{Proj}(\mathrm{Sym}(E))$. Let p denote the map from $P(\mathcal{E})$ to X; then p is a projective map; and since $\mathcal{E}$ is locally free, p is also a flat map. Let h denote the hyperplane section of $P(E)$ corresponding to the grading along the powers of $\mathcal{E}$; that is, let $h = h_i$ for $i = (0, \ldots, 0, 1)$. In the above example where $\mathcal{E}$ has rank 1, we derived the formula

$$hp^*([X]) = p^*(c_1(\mathcal{E})([X])).$$

The general formula is

$$h^r(p^*(\eta)) - h^{r-1}p^*(c_1(\mathcal{E})(\eta)) + \cdots + (-1)^r p^*(c_r(\mathcal{E})(\eta)) = 0.$$

To show that this really does define the classes $c_i(\mathcal{E})$, we must show that there are unique classes $\alpha_1, \ldots, \alpha_r$ with $\alpha_i \in A_{k-i}(X)$ such that

$$h^r(p^*(\eta)) - h^{r-1}p^*(\alpha_1) + \cdots + (-1)^r p^*\alpha_r = 0.$$

The results that make the machinery of Chern classes work are contained in the next few propositions.

Proposition 9.2.1 *Let $X = \mathrm{Proj}(A)$, $\mathcal{E}$, $P(\mathcal{E})$, p, and h be as above; and let r be the rank of $\mathcal{E}$. Let η be a cycle on X.*

(i) *If $k < r - 1$, then $p_*(h^k(p^*(\eta))) = 0$.*
(ii) *We have $p_*(h^{r-1}(p^*(\eta))) = \eta$ in the Chow group of X.*

Proof We may again assume that A is an integral domain and that η is the cycle $[X]$.

The first statement follows immediately from dimension considerations. Let d be the dimension of X. Then the dimension of $P(\mathcal{E})$ is $d + r - 1$, so the dimension of $h^k(p^*(\eta)) = h^k(p^*([X]))$ is $d + r - 1 - k$. If $k < r - 1$, then $d + r - 1 - k > d$, so the dimension of $h^k(p^*([X]))$ is greater than the dimension of X and $p_*(h^k(p^*([X]))) = 0$.

If $k = r - 1$, then the dimension argument of the preceding paragraph shows that $h^{r-1}(p^*([X]))$ is of the same dimension as X, so that its image under p_* is a multiple of $[X]$. Thus, we only need to compute the coefficient of $[X]$, and we can localize at the nonzero homogeneous elements of A and consider $A_{(0)} \otimes \mathrm{Sym}(E)$, where E is a module representing $\mathcal{E}$. Since $\mathcal{E}$ is locally free of rank r, $A_{(0)} \otimes \mathrm{Sym}(E)$ is a polynomial ring in r variables $T_1, \ldots, T_r$ over the field $A_{(0)}$, which we denote K. Now we may take the first $r - 1$ variables to define h^{r-1}; thus

$$h^{r-1}(p^*([X])) = h^{r-1}([\text{Proj}(K[T_1, \ldots, T_r])])$$

$$= [\text{Proj}(K[T_1, \ldots, T_r])/(T_1, \ldots, T_{r-1})]$$

$$= [\text{Proj}([K[T_r])].$$

Since the map from $\text{Proj}(K[T_r])$ to $\text{Spec}(K)$ is an isomorphism, we thus have $p_*([\text{Proj}(K[T_r])]) = [\text{Spec}(K)]$, and hence $p_*(h^{r-1}(p^*([X]))) = [X]$. □

Proposition 9.2.2 *With notation as above, the Chow group $A_*(P(\mathcal{E}))$ is generated by $p^*(A_*(X)), hp^*(A_*(X)), \ldots, h^{r-1}p^*(A_*(X))$.*

Proof The proof of this statement involves two steps, both of which use an exact sequence defined by an open subset. First, we choose an affine open set on which $\mathcal{E}$ is free and reduce to the case in which $\mathcal{E}$ is defined by a free module E over a ring. Then, in the case where E is free, we divide $\text{Proj}(\text{Sym}(E))$ into open sets and show that images of the Chow groups of the pieces correspond to the images of powers of h and that they generate the Chow group of $P(\mathcal{E})$.

If q is a graded prime ideal of $\text{Sym}(E)$, where E is a module representing $\mathcal{E}$, then the intersection of q with A is a prime ideal p of $\text{Proj}(A)$. Let $\mathcal{F}$ denote the pull-back of $\mathcal{E}$ to $\text{Proj}(A/\mathfrak{p})$. Then $P(\mathcal{F})$ can be embedded in $P(\mathcal{E})$, and, since q contains p, q is in the image of this embedding. Since this embedding preserves the action of h, if we can show that the theorem holds with X replaced by $\text{Proj}(A/\mathfrak{p})$ and $P(\mathcal{E})$ replaced by $P(\mathcal{F})$, then it will hold also for X and $\mathcal{E}$. Hence we may assume that A is an integral domain.

If A is a domain, there exists a special basic open set U_x of X on which E is free. Let $Y = X - U_x$. Then $Y = \text{Proj}(A/xA)$ is a projective scheme of lower dimension, while U_x is isomorphic to $\text{Spec}(A_{(x)})$ by Proposition 8.2.5. We have the diagram with exact rows:

$$
\begin{array}{ccccccc}
A_*(Y) & \to & A_*(X) & \to & A_*(U_x) & \to & 0 \\
\downarrow & & \downarrow & & \downarrow & & \\
A_*(P(\mathcal{E}_Y)) & \to & A_*(P(\mathcal{E})) & \to & A_*(P(\mathcal{E})_x) & \to & 0
\end{array}.
$$

Furthermore, the maps in the bottom row commute with h by Proposition 8.7.2. Thus, if $A_*(P(\mathcal{E}_Y))$ and $A_*(P(\mathcal{E})_x)$ are generated by the images of the first $r - 1$ powers of h composed with p^*, the same is true for $A_*(P(\mathcal{E}))$. By induction on the dimension, we may assume that the result is true for $A_*(P(\mathcal{E}_Y))$. We need to prove it for $A_*(P(\mathcal{E}_x))$; since we have assumed that $\mathcal{E}_x$ is defined by a free module, this reduces to the case in which E is a free module over a ring A and $X = \text{Spec}(A)$.

If E is a free module over A, the graded ring $\mathrm{Sym}(E)$ is a polynomial ring $A[T_1, \ldots, T_r]$, which we denote $A[T]$, and $P(\mathcal{E}) = \mathrm{Proj}(A[T])$. We will prove by induction on r that every element of $A_*(P(\mathcal{E}))$ is in the subgroup generated by $p^*(A_*(X))$, $hp^*(A_*(X))$, $\ldots$, $h^{r-1}p^*(A_*(X))$. If $r = 1$, the projection from $P(\mathcal{E})$ to X is an isomorphism, so that p^* is an isomorphism from $A_*(X)$ to $A_*(P(\mathcal{E}))$ and the result is immediate.

Suppose the theorem holds for $r - 1$. Let F be the free submodule of E of rank $r - 1$ generated by $T_1, \ldots, T_{r-1}$. There is a map from $\mathrm{Sym}(E)$ onto $\mathrm{Sym}(F)$ obtained by sending T_r to 0, and this map defines an embedding of $P(\mathcal{F})$ into $P(\mathcal{E})$. Denoting the projections of $P(\mathcal{F})$ and $P(\mathcal{E})$ to X by $p_{\mathcal{F}}$ and $p_{\mathcal{E}}$, and representing the hyperplane h on $P(\mathcal{E})$ by the divisor generated by T_r, we have that

$$p_{\mathcal{F}}^*(\eta) = hp_{\mathcal{E}}^*(\eta) \qquad (*)$$

for all cycles $\eta \in A_*(X)$.

Let U be the open subset U_{T_r} of $P(\mathcal{E})$. The complement of U is $P(\mathcal{F})$. Furthermore, it follows from Proposition 8.2.5 that U_{T_r} is isomorphic to $\mathrm{Spec}(A_{(T_r)}) = \mathrm{Spec}(A[T_1/T_r, \ldots, T_{r-1}/T_r])$. We have an exact sequence

$$A_*(P(\mathcal{F})) \to A_*(P(\mathcal{E})) \to A_*(U) \to 0.$$

Since $U = \mathrm{Spec}(A[T_1/T_r, \ldots, T_{r-1}/T_r])$ and $A[T_1/T_r, \ldots, T_{r-1}/T_r]$ is a polynomial ring, Theorem 1.2.6 implies that flat pull-back from X to U induces a surjection on Chow groups. Hence every element η of $A_*(P(\mathcal{E}))$ is the sum of an element of the form $p^*(\alpha_0)$ and an element β in the image of $A_*(P(\mathcal{F}))$. By induction on r, we can write

$$\beta = p_F^*(\alpha_1) + hp_F^*(\alpha_2) + \cdots + h^{r-2}p_F^*(\alpha_{r-1}).$$

Hence using $(*)$ we have

$$\eta = p_{\mathcal{E}}^*(\alpha_0) + \left(p_{\mathcal{F}}^*(\alpha_1) + hp_{\mathcal{F}}^*(\alpha_2) + \cdots + h^{r-2}p_{\mathcal{F}}^*(\alpha_{r-1})\right)$$
$$= p_{\mathcal{E}}^*(\alpha_0) + hp_{\mathcal{E}}^*(\alpha_1) + \cdots + h^{r-1}p_{\mathcal{E}}^*(\alpha_{r-1}).$$

This proves the Proposition. $\qquad\qquad\qquad\qquad\qquad\qquad\qquad\qquad\qquad\quad\square$

The next Proposition follows immediately from Proposition 9.2.2.

Proposition 9.2.3 *Again with the above notation, every element η of the Chow group $A_*(P(\mathcal{E}))$ can be written uniquely in the form*

$$\eta = h^{r-1}p^*(\alpha_1) + h^{r-2}p^*(\alpha_2) + \cdots + p^*(\alpha_r),$$

where $\alpha_1, \ldots, \alpha_r$ are elements of $A_*(X)$.

Proof The fact that η can be written in this form is Proposition 9.2.2. To prove the uniqueness, we may assume that $\eta = 0$ and show that $\alpha_i = 0$ for each i. Applying p_* to the above expression, Proposition 9.2.1 implies that we obtain α_1, so if $\eta = 0$ we have $\alpha_1 = 0$. We thus have

$$h^{r-2}p^*(\alpha_2) + \cdots + p^*(\alpha_r) = 0.$$

Applying h to this expression and then applying p_* as before, we obtain that $\alpha_2 = 0$. Continuing this process, we conclude that $\alpha_i = 0$ for all i. $\qquad\square$

We now have the necessary theorems to make the following definition.

Definition 9.2.1 *Let $\mathcal{E}$ be a locally free sheaf of rank r over $X = \mathrm{Proj}(A)$, and let p, h, and $P(\mathcal{E})$ be as above. Let $\eta \in Z_*(X)$. Then the Chern classes of $\mathcal{E}$ applied to η are the unique classes in $A_*(X)$ such that*

$$h^r(p^*(\eta)) - h^{r-1}p^*(c_1(\mathcal{E})(\eta)) + h^{r-2}p^*(c_2(\mathcal{E})(\eta)) - \cdots$$
$$\cdots + (-1)^r p^*(c_r(\mathcal{E})(\eta)) = 0.$$

We define the total Chern class of $\mathcal{E}$ to be $1 + c_1(\mathcal{E}) + \cdots + c_r(\mathcal{E})$.

We sometimes use an indeterminate t in the total Chern class and write it in the form $1 + c_1(\mathcal{E})t + \cdots + c_r(\mathcal{E})t^r$.

As mentioned above, we also define $c_0(\mathcal{E})$ by $c_0(\mathcal{E})(\eta) = \eta$. We can then write the above definition in the form

$$\sum_{i=0}^{r}(-1)^i h^{r-i}p^*(c_i(\mathcal{E})(\eta)) = 0.$$

We now verify formally the special construction of the Chern class of a rank 1 sheaf.

Proposition 9.2.4 *Let $X = \mathrm{Proj}(A)$ where A is an integral domain, and let $\mathcal{E}$ be a locally free sheaf on X of rank 1. Let d be the dimension of X. Let e be an element of a module E representing $\mathcal{E}$, and let x be a nonzero element of A such that e/x has degree $(0, \ldots, 0, 1)$. Then $c_1(\mathcal{E})([X])$ can be represented by the cycle $\sum n_{\mathfrak{p}}[A/\mathfrak{p}]$, where we have $n_{\mathfrak{p}} = \mathrm{length}(E_{(\mathfrak{p})}/(e)) - \mathrm{length}(E_{(\mathfrak{p})}/xE_{(\mathfrak{p})})$ for all graded prime ideals $\mathfrak{p}$ of dimension $d - 1$.*

Proof Since e/x has the correct degree, we may use it to define the hyperplane section h on $P(\mathcal{E})$. The definition of $c_1(\mathcal{E})$ states that the pull-back of $c_1(\mathcal{E})([X])$ to $P(\mathcal{E})$ is equal to $hp^*([X])$, where p is the projection from $P(\mathcal{E})$ to X. In this case we thus have

$$p^*(c_1(\mathcal{E})([X])) = [\mathrm{Sym}(E)/e\mathrm{Sym}(E)]_{d-1} - [\mathrm{Sym}(E)/x\mathrm{Sym}(E)]_{d-1}$$

where e is taken in degree 1.

To verify that this cycle is the one defined in the statement of the proposition, we may check it locally for each prime ideal $\mathfrak{p}$ of dimension $d - 1$. Let $\mathfrak{p}$ be such a prime ideal, and consider the localization $A_{(\mathfrak{p})}$. The localization $E_{(\mathfrak{p})}$ is free of rank 1, so its symmetric algebra is a polynomial ring $A_{(\mathfrak{p})}[T]$. We use e and x also to denote the images of e and x in the localizations. Since the element e is taken in degree 1, the corresponding element in the localization can be written eT. Thus, arguing as in the discussion preceding the definition of Chern classes, the right-hand side of the above equation is

$$\left[A_{(\mathfrak{p})}[T]/(eT)A_{(\mathfrak{p})}[T]\right]_{d-1} - \left[A_{(\mathfrak{p})}[T]/xA_{(\mathfrak{p})}[T]\right]_{d-1}$$
$$= \left[A_{(\mathfrak{p})}[T]/(e)A_{(\mathfrak{p})}[T]\right]_{d-1} - \left[A_{(\mathfrak{p})}[T]/(x)A_{(\mathfrak{p})}[T]\right]_{d-1}.$$

Since the latter cycle is precisely the pull-back of the cycle in the statement of the Proposition, this completes the proof.   □

There is an alternative definition of Chern classes using what are called Segre classes, which are defined by taking powers of h higher than $r - 1$ applied to the class of $p^*([A/\mathfrak{p}])$ and projecting back down to X (see Fulton [17]). We describe Segre classes and outline a proof that this definition gives the same result (it essentially follows formally from the definitions) in the exercises.

9.3 Functorial Properties of Chern Classes

In this section we show that Chern classes commute with the main operations of intersection theory. More specifically, they are compatible with proper push-forward, flat pull-back, and intersection with divisors. These properties follow formally from the corresponding properties of the hyperplane section proven in Chapter 8. We prove the compatibility with proper push-forward (the projection formula) in detail and leave the details of the other properties as exercises. At the end of this section we show that Chern classes are independent of rational equivalence, so they are defined as operators on the Chow group.

Proposition 9.3.1 (The Projection Formula) *Let $\phi : X \to Y$ be a proper map, and let $\mathcal{E}$ be a locally free sheaf on Y. Let η be an element of $Z_*(X)$. Then*

$$\phi_*(c_i(\phi^*(\mathcal{E}))(\eta)) = c_i(\mathcal{E})(\phi_*(\eta)).$$

Proof Let $X = \text{Proj}(B)$ and $Y = \text{Proj}(A)$, and assume that ϕ is defined by the map $f : A \to B$. If $\mathcal{E}$ is defined by the A-module E, then $\phi^*(\mathcal{E})$ is defined by the B-module $B \otimes_A E$. Let $\phi_{\mathcal{E}}$ be the induced map from $P(\phi^*(\mathcal{E}))$ to $P(\mathcal{E})$. We have a commutative diagram:

$$
\begin{array}{ccc}
P(\phi^*(\mathcal{E})) & \to & X \\
\downarrow & & \downarrow \\
P(\mathcal{E}) & \to & Y
\end{array} .
$$

We denote the hyperplane on $P(\mathcal{E})$ by h and the hyperplane on $P(\phi^*(\mathcal{E}))$ by $\tilde{h}$, and we denote the projections to Y and X, respectively, by p and $\tilde{p}$.

Using the definition of the Chern classes of $\phi^*(\mathcal{E})$, we have that

$$\sum_{i=0}^{r} (-1)^i \tilde{h}^{r-i} \tilde{p}^*(c_i(\phi^*(\mathcal{E}))(\eta)) = 0.$$

We next apply $(\phi_{\mathcal{E}})_*$ to this equation and use the fact that $(\phi_{\mathcal{E}})_* \tilde{h} = h(\phi_{\mathcal{E}})_*$ (Proposition 8.7.2) to obtain

$$\sum_{i=0}^{r} (-1)^i h^{r-i} (\phi_{\mathcal{E}})_* \tilde{p}^*(c_i(\phi^*(\mathcal{E}))(\eta)) = 0.$$

We claim that for all cycles η in $Z_*(X)$, we have $(\phi_{\mathcal{E}})_* \tilde{p}^*(\eta) = p^* \phi_*(\eta)$. We can assume that η is $[B/\mathfrak{q}]$, where $\mathfrak{q}$ is an element of X. Then $\phi_*([B/\mathfrak{q}]) = [k(\mathfrak{q}) : k(\mathfrak{p})][A/\mathfrak{p}]$, where $\mathfrak{p} = f^{-1}(\mathfrak{q})$, assuming that $[k(\mathfrak{q}) : k(\mathfrak{p})]$ is finite. To prove the equality we may localize and assume that $\mathcal{E}$ is defined by a free module E over the field $k(\mathfrak{p})$. If r is the rank of E, the equality $(\phi_E)_* \tilde{p}^*(\eta) = p^* \phi_*(\eta)$ then reduces to the fact that the degree of $k(\mathfrak{q})(T_1, \ldots, T_r)$ over $k(\mathfrak{p})(T_1, \ldots, T_r)$ is equal to the degree of $k(\mathfrak{q})$ over $k(\mathfrak{p})$, which is clear.

Hence, using the fact that $(\phi_E)_* \tilde{p}^* = p^* \phi_*$, we have

$$\sum_{i=0}^{r} (-1)^i h^{r-i} p^* \phi_*(c_i(\phi^*(\mathcal{E}))(\eta)) = 0.$$

The Chern classes of $\mathcal{E}$ applied to $\phi_*(\eta)$ are the unique classes in the Chow group of Y such that

$$\sum_{i=0}^{r} (-1)^i h^{r-i} p^*((c_i(\mathcal{E}))(\phi_*(\eta))) = 0$$

with $c_0(\mathcal{E})(\phi_*(\eta)) = \phi_*(\eta)$. We have $\phi_*(c_0(\phi^*(\mathcal{E}))(\eta)) = \phi_*(\eta)$. Thus, comparing the last equation with the previous one and using the uniqueness of Proposition 9.2.3, we must thus have

$$\phi_*(c_i(\phi^*(\mathcal{E}))(\eta)) = (c_i(\mathcal{E}))(\phi_*(\eta))$$

for $i = 1, \ldots, r$. This proves the Proposition. $\qquad\square$

Proposition 9.3.2 *Let D be a divisor on X and let η be an element of $Z_*(X)$. Then*

$$c_i(\mathcal{E}_D)(D \cap \eta) = D \cap (c_i(\mathcal{E})(\eta))$$

in $A_*(D)$, where $\mathcal{E}$ denotes the restriction of $\mathcal{E}$ to D.

Proof We may assume that D is an effective divisor and defines a closed subscheme of X, which we also denote D. Let $P(\mathcal{E}_D)$ be the projective scheme defined by $\mathrm{Sym}(\mathcal{E}_D)$, and let h_D and p_D denote the hyperplane on $P(\mathcal{E}_D)$ and the projection to D, respectively. The divisor D also defines a divisor on $P(\mathcal{E})$ that defines the subscheme $P(\mathcal{E}_D)$; we denote this divisor also by D. As in the proof of Proposition 9.3.1, we have a commutative product diagram

$$
\begin{array}{ccc}
P(\mathcal{E}_D) & \to & D \\
\downarrow & & \downarrow \\
P(\mathcal{E}) & \to & X
\end{array}.
$$

Let η be an element of $Z_*(X)$. By the definition of Chern classes, we have

$$\sum_{i=0}^{r} (-1)^i h^{r-i} p^*(c_i(\mathcal{E})(\eta)) = 0.$$

We intersect this element with D and use the facts that intersection with D commutes with h and with flat pull-back by p (Theorem 8.8.2) to obtain the equation

$$\sum_{i=0}^{r} (-1)^i h_D^{r-i} p_D^*(D \cap (c_i(\mathcal{E})(\eta))) = 0.$$

It now follows from the definition of $c_i(\mathcal{E}_D)(D \cap \eta)$ and the fact that $D \cap (c_0(\mathcal{E})(\eta)) = D \cap \eta$ that we have

$$D \cap c_i(\mathcal{E})(\eta) = c_i(\mathcal{E}_D)(D \cap \eta)$$

for $i = 0, \ldots, r$. $\qquad\square$

A special case of Proposition 9.3.2 is the following.

Proposition 9.3.3 *Let h_i be a hyperplane section on a projective scheme X. Then*

$$c_i(\mathcal{E})(h_i(\eta)) = h_i(c_i(\mathcal{E})(\eta)).$$

A similar argument shows the following.

Proposition 9.3.4 *Let $f : X \to Y$ be a flat map of relative dimension k, and let η be an element of $A_*(Y)$. Then $f^*(c_i(\mathcal{E})(\eta)) = c_i(f^*\mathcal{E})(f^*(\eta))$.*

Proof This proposition uses the compatibility of the hyperplane section with flat maps (Proposition 8.7.2) and an argument similar to that of Propositions 9.3.1 and 9.3.2.   □

Up to now we have defined Chern classes of cycles but have not shown that the result is independent of rational equivalence. This fact is now a consequence of Proposition 9.3.2.

Theorem 9.3.5 *Chern classes are independent of rational equivalence and thus define operations on the Chow group.*

Proof Let $X = \text{Proj}(A)$ be a projective scheme, and let $\mathcal{E}$ be a locally free sheaf on X. Let $\mathfrak{q}$ be a graded prime ideal of A of dimension $k + 1$, and let x and y be elements of A of the same degree that are not in $\mathfrak{q}$. Let $\eta = \text{div}(\mathfrak{q}, x/y)$ in $Z_k(X)$. By Proposition 9.3.2, for each i we have

$$c_i(\mathcal{E})\text{div}(\mathfrak{q}, x/y) = c_i(\mathcal{E})((x) \cap [A/\mathfrak{q}] - (y) \cap [A/\mathfrak{q}])$$

$$= (x) \cap (c_i(\mathcal{E})([A/\mathfrak{q}])) - (y) \cap (c_i(\mathcal{E})([A/\mathfrak{q}])).$$

By Proposition 1.2.2, this last element is rationally equivalent to zero. Hence $c_i(\mathcal{E})(\text{div}(\mathfrak{q}, x/y))$ is rationally equivalent to zero, so $c_i(\mathcal{E})$ preserves rational equivalence and defines an operator on the Chow group of X.   □

9.4 Composition of Chern Classes

In this section, we derive an expression for the composition of Chern classes. If $\mathcal{E}$ and $\mathcal{F}$ are locally free sheaves, we show that the composition $c_i(\mathcal{E})c_j(\mathcal{F})$ can be defined in terms of an equality similar to the equality used in the definition of Chern classes. A consequence is that Chern classes commute with each other, so that we have $c_i(\mathcal{E})c_j(\mathcal{F}) = c_j(\mathcal{F})c_i(\mathcal{E})$.

Theorem 9.4.1 *Let $\mathcal{E}$ and $\mathcal{F}$ be locally free sheaves over X of ranks r and s, respectively. Let η be an element of $A_k(X)$. Then the elements $c_i(\mathcal{E})c_j(\mathcal{F})(\eta)$ of the Chow group of X are the unique classes α_{ij} in $A_*(X)$, defined for $i = 0, \ldots, r$ and $j = 0, \ldots, s$, such that the following two conditions are satisfied:*

(i)

$$\sum_{i=0}^{r}\sum_{j=0}^{s}(-1)^{i+j}h_{\mathcal{E}}^{r-i}h_{\mathcal{F}}^{s-j}p^*(\alpha_{ij}) = 0$$

where p is the projection from $P(\mathcal{E}) \times P(\mathcal{F})$ to X and $h_\mathcal{E}$ and $h_\mathcal{F}$ are
the hyperplane sections on $P(\mathcal{E})$ and $P(\mathcal{F})$, respectively, pulled back to
$P(\mathcal{E}) \times P(\mathcal{F})$.

(ii) $\alpha_{i0} = c_i(\mathcal{E})(\eta)$ *for $i = 0, \ldots r$ and* $\alpha_{0j} = c_j(\mathcal{F})(\eta)$ *for $j = 0, \ldots, s$.*

Proof This theorem, like the functorial properties of Chern classes proven in
the previous section, follows formally from the definitions and the theorems
that make the definitions possible. Let $p_\mathcal{E}$ and $p_\mathcal{F}$ be the projections from $P(\mathcal{E})$
and $P(\mathcal{F})$ to X; we also use $p_\mathcal{E}$ and $p_\mathcal{F}$ to denote the corresponding projections
from $P(\mathcal{E}) \times P(\mathcal{F})$ to $P(\mathcal{F})$ and $P(\mathcal{E})$. Then the Chern classes $c_j(\mathcal{F})(\eta)$ are
defined by the properties that $c_0(\mathcal{F})(\eta) = \eta$ and that we have the equation

$$\sum_{j=0}^{s} h_\mathcal{F}^{s-j} p_\mathcal{F}^*(c_j(\mathcal{F})(\eta)) = 0.$$

This is an equation in the Chow group of $P(\mathcal{F})$. We next apply the Chern
classes of $p_\mathcal{F}^*(\mathcal{E})$ to each of the terms of this equation. The resulting classes
satisfy the condition

$$\sum_{i=0}^{r}\sum_{j=0}^{s} h_\mathcal{E}^{r-i} p_\mathcal{E}^*\left(c_i\left(p_\mathcal{F}^*(\mathcal{E})\right)\right) \left(h_\mathcal{F}^{s-j} p_\mathcal{F}^* c_j(\mathcal{F})(\eta)\right) = 0.$$

We next use the compatibilities of the various operations in this formula
to express this equation in the form of statement (i) of the theorem. More
precisely, we use the facts that $h_\mathcal{F}$ commutes with the Chern class $c_i(p_\mathcal{F}^*(\mathcal{E}))$
(Proposition 9.3.3), that $h_\mathcal{F}$ commutes with flat pull-back by $p_\mathcal{E}$ (Proposition
8.7.2), and that $c_i(\mathcal{E})$ commutes with flat pull-back by $p_\mathcal{F}^*$ (Proposition 9.3.4)
to derive the equation

$$\sum_{i=0}^{r}\sum_{j=0}^{s} h_\mathcal{E}^{r-i} h_\mathcal{F}^{s-j} p_\mathcal{E}^* p_\mathcal{F}^*(c_i(\mathcal{E}) c_j(\mathcal{F})(\eta)) = 0.$$

Since $p_\mathcal{E}^* p_\mathcal{F}^* = p^*$, this shows that the elements $\alpha_{ij} = c_i(\mathcal{E}) c_j(\mathcal{F})$ satisfy condition (i) of the theorem. That they satisfy condition (ii) follows from the fact
that $c_0(\mathcal{E})$ and $c_0(\mathcal{F})$ are identity maps.

The uniqueness follows from the uniqueness part of Proposition 9.2.3. In
fact, since the values of α_{ij} are determined by condition (ii) if i or j is zero, it
suffices to show that if we have elements α_{ij} in the Chow group of X such that

$$\sum_{i=0}^{r-1}\sum_{j=0}^{s-1} h_\mathcal{E}^{r-i} h_\mathcal{F}^{s-j} p^*(\alpha_{ij}) = 0$$

then all the α_{ij} must be zero. Given such α_{ij}, it follows from Proposition 9.2.3 that, for each i, we have $\sum_{j=0}^{s-1} h_{\mathcal{F}}^{s-j} p^*(\alpha_{ij}) = 0$, and then the same theorem implies that each $\alpha_{ij} = 0$.   □

An important corollary is that Chern classes commute with each other.

Corollary 9.4.2 *Let the notation be as in the above Theorem. Then*

$$c_i(\mathcal{E})c_j(\mathcal{F}) = c_j(\mathcal{F})c_i(\mathcal{E})$$

for all i and j.

Proof From Theorem 9.4.1, it suffices to show that we have

$$\sum_{i=0}^{r}\sum_{j=0}^{s}(-1)^{i+j} h_{\mathcal{E}}^{r-i} h_{\mathcal{F}}^{s-j} p^*(c_j(\mathcal{F})c_i(\mathcal{E})) = 0$$

and that we also have that $c_0(\mathcal{F})c_i(\mathcal{E}) = c_i(\mathcal{E})(\eta)$ for $i = 0, \ldots r$ and $c_j(\mathcal{F})c_0(\mathcal{E}) = c_j(\mathcal{F})(\eta)$ for $j = 0, \ldots, s$. The last two conditions are immediate since c_0 acts as the identity map. Theorem 9.4.1 implies that

$$\sum_{i=0}^{r}\sum_{j=0}^{s}(-1)^{i+j} h_{\mathcal{F}}^{s-j} h_{\mathcal{E}}^{r-i} p^*(c_j(\mathcal{F})c_i(\mathcal{E})) = 0.$$

Hence the desired result follows from the fact that $h_{\mathcal{F}}$ and $h_{\mathcal{E}}$ commute with each other.   □

9.5 The Whitney Sum Formula

One of the main tools in computing with Chern classes is the Whitney sum formula, which expresses the total Chern class of the middle term of a short exact sequence in terms of those of the outer terms. It states that if

$$0 \to \mathcal{E}' \to \mathcal{E} \to \mathcal{E}'' \to 0$$

is a short exact sequence of locally free sheaves over $X = \text{Proj}(A)$, then

$$c(\mathcal{E}) = c(\mathcal{E}')c(\mathcal{E}'').$$

We first represent the short exact sequence of sheaves by a sequence of modules

$$0 \to E' \to E \to E'' \to 0$$

where we can assume that the map from E to E'' is surjective by replacing E'' by the image of E if necessary. There is then a surjective map from $\text{Sym}(E)$ to

$\mathrm{Sym}(E'')$, which induces an imbedding of $P(\mathcal{E}'')$ into $P(\mathcal{E})$. We first show that we can define the class of $P(\mathcal{E}'')$ in the Chow group of $P(\mathcal{E})$ using the Chern classes of $\mathcal{E}'$.

Proposition 9.5.1 *Let*

$$0 \to \mathcal{E}' \to \mathcal{E} \to \mathcal{E}'' \to 0$$

be a short exact sequence of locally free sheaves of ranks r, $r + s$, *and* s, *respectively. Let* p *and* p'' *denote the projections from* $P(\mathcal{E})$ *and* $P(\mathcal{E}'')$ *to* X, *and let* h *denote the hyperplane on* $P(\mathcal{E})$. *Let* i *denote the embedding of* $P(\mathcal{E}'')$ *into* $P(\mathcal{E})$. *Then*

$$i_*(p''^*(\eta)) = \sum_{i=0}^{r} h^{r-i} p^*(c_i(\mathcal{E}')(\eta)).$$

Proof We assume that $X = \mathrm{Proj}(A)$ where A is a domain and $\eta = [X]$. Let d be the dimension of X. Let U be the complement of the image of $P(\mathcal{E}'')$ in $P(\mathcal{E})$. The image of $P(\mathcal{E}'')$ in $P(\mathcal{E})$ consists of those prime ideals of $\mathrm{Sym}(E)$ that define elements of $P(\mathcal{E})$ and that contain E', where E' and E are modules representing $\mathcal{E}'$ and $\mathcal{E}$. Thus, if $\mathfrak{p}$ is an element of $P(\mathcal{E})$ that is in U, the intersection of $\mathfrak{p}$ with $\mathrm{Sym}(\mathcal{E}')$ will not contain E', so it will be in $\mathrm{Proj}(\mathrm{Sym}(E')) = P(\mathcal{E}')$. Hence there is a map ϕ from U to $P(\mathcal{E}')$, and since $\mathrm{Sym}(E)$ is locally flat over $\mathrm{Sym}(E')$, ϕ is flat of relative dimension s.

Let p' denote the projection from $P(\mathcal{E}')$ to X, and let h' denote the hyperplane section on $P(\mathcal{E}')$. By the definition of Chern classes, we have

$$\sum_{i=0}^{r} h'^{r-i} p'^*(c_i(\mathcal{E}')(\eta)) = 0.$$

Pulling this equation back by ϕ, we have

$$\sum_{i=0}^{r} \phi^*(h'^{r-i} p'^*(c_i(\mathcal{E}')(\eta))) = 0.$$

Let ψ denote the embedding of U into $P(\mathcal{E})$. Then $p\psi = p'\phi$, and, since the map from $\mathrm{Sym}(E')$ to $\mathrm{Sym}(E)$ preserves degrees of elements, it also preserves the action of the corresponding hyperplane sections. Using these facts we can replace $\phi^* h'^{r-i} p'^*$ with $\psi^* h^{r-i} p^*$ and write

$$\psi^* \left(\sum_{i=0}^{r} (h^{r-i} p^*(c_i(\mathcal{E}')(\eta)) \right) = \sum_{i=0}^{r} \psi^*(h^{r-i} p^*(c_i(\mathcal{E}')(\eta))) = 0.$$

Thus, we have that $\sum_{i=0}^{r}(h^{r-i}p^*(c_i(\mathcal{E}')(\eta)))$ is zero when restricted to U. We also have that $i_*(p''^*(\eta)) = 0$ when restricted to U, since the image of $P(\mathcal{E}'')$ in $P(\mathcal{E})$ does not meet U.

Thus, we have shown that the two sides of the equation in the conclusion of the proposition are equal when restricted to U. Note that both sides of the equation are represented by cycles of dimension $d + s - 1$. We now consider the exact sequence

$$A_{d+s-1}(P(\mathcal{E}'')) \to A_{d+s-1}(P(\mathcal{E})) \to A_{d+s-1}(U) \to 0.$$

Since $P(\mathcal{E}'')$ itself has dimension $d + s - 1$, $A_{d+s-1}(P(\mathcal{E}''))$ is generated by the class of $P(\mathcal{E}'')$. Thus, to complete the proof it suffices to show that the coefficient of $[P(\mathcal{E}'')]$ in $= \sum_{i=0}^{r} h^{r-i}(p^*(c_i(\mathcal{E}')(\eta)))$ is equal to 1.

To prove the last statement we may localize and assume that E is a free module over a local ring A. In this case the symmetric algebra is a polynomial ring $A[T_1, \ldots, T_{r+s}]$, where we can choose $T_1, \ldots T_{r+s}$ in such a way that $T_1, \ldots, T_r$ generate E'. Since E' is free, $c_1(E'), \ldots, c_r(E')$ are zero, so the equation reduces to the statement that $h^r([P(E)]) = i_*([P(E'')])$. Using $T_1, \ldots, T_r$ to represent h^r, both sides of this equation are given by the class of $\mathrm{Proj}(A[T_1, \ldots, T_{r+s}]/(T_1, \ldots, T_r))$. Hence the coefficient of the image of $[P(\mathcal{E}'')]$ is 1, as was to be shown. $\square$

Theorem 9.5.2 (The Whitney Sum Formula) *Let*

$$0 \to \mathcal{E}' \to \mathcal{E} \to \mathcal{E}'' \to 0$$

be a short exact sequence of locally free sheaves. Then

$$c(\mathcal{E}) = c(\mathcal{E}')c(\mathcal{E}'').$$

Proof As usual, we may assume that A is an integral domain, and that we are computing the action of both sides of the equation on the class $[X]$. Let the ranks of $\mathcal{E}'$ and $\mathcal{E}''$ be r and s, respectively, so that the rank of $\mathcal{E}$ is $r + s$.

Let p'' be the projection from $P(\mathcal{E}'')$ to X, and let h'' denote the hyperplane section on $P(\mathcal{E}'')$. Let i denote the imbedding of $P(\mathcal{E}'')$ into $P(\mathcal{E})$. We have

$$\sum_{i=0}^{s} h''^{s-i} p''^*(c_i(\mathcal{E}''))(\eta) = 0.$$

Pushing this equation into $P(\mathcal{E})$ by i_*, we obtain

$$\sum_{i=0}^{s} i_*(h''^{s-i} p''^*(c_i(\mathcal{E}'')(\eta))) = \sum_{i=0}^{s} h^{s-i} i_*(p''^*(c_i(\mathcal{E}''))(\eta)) = 0.$$

We next apply Proposition 9.5.1 to get

$$\sum_{j=0}^{r}\sum_{i=0}^{s} h^{s-i}h^{r-j}(c_j(\mathcal{E}')c_i(\mathcal{E}''))(\eta) = 0.$$

This last expression can be written

$$\sum_{k=0}^{r+s}\sum_{i+j=k} h^{r+s-k}(c_j(\mathcal{E}')c_i(\mathcal{E}''))(\eta) = 0.$$

By the definition of Chern classes, we must thus have

$$c_k(\mathcal{E})(\eta) = \sum_{i+j=k}(c_j(\mathcal{E}')c_i(\mathcal{E}''))(\eta)$$

for all k. These equalities can be expressed in terms of the total Chern classes by the equation $c(\mathcal{E}) = c(\mathcal{E}')c(\mathcal{E}'')$, so this completes the proof. $\qquad\square$

9.6 The Splitting Principle

As we have seen in the discussion leading up to the definition of Chern classes, the situation for rank 1 sheaves is considerably simpler than for sheaves of higher rank. As a result, it is often easier to prove results for locally free sheaves of rank 1 than for general locally free sheaves. If the general case can be reduced to the rank 1 case by taking short exact sequences and using the Whitney sum formula, then it is useful to be able to find a filtration of a locally free sheaf with rank 1 quotients. In general this is not possible; however, it is always possible to construct a projective map so that the pull-back of the sheaf has such a filtration. Combining the above arguments with the projection formula then makes it possible to prove general results on Chern classes. The simplest way to construct a map with these properties is simply to pull-back to $P(\mathcal{E})$ itself.

Theorem 9.6.1 *Let $X = \mathrm{Proj}(A)$ be a projective scheme, where A is a graded integral domain. Let $\mathcal{E}$ be a locally free sheaf on X. Then there is a projective scheme Y together with a projective map $\phi : Y \to X$ such that*

(i) *$\phi^*(\mathcal{E})$ has a filtration*

$$0 = \mathcal{E}_0 \subseteq \mathcal{E}_1 \subseteq \cdots \subseteq \mathcal{E}_r = \phi^*(\mathcal{E})$$

   such that $\mathcal{E}_i/\mathcal{E}_{i-1}$ is a locally free sheaf of rank 1 for each i.

(ii) *The map induced by ϕ on cycles by proper push-forward is surjective.*

Proof It suffices to show that there exists a scheme Y together with a projective map ϕ to X such that the pull-back has a quotient that is locally free rank 1; the process can then be iterated to give the full result.

We let $Y = P(\mathcal{E})$, and let ϕ be the projection from Y to X. If $\mathcal{E}$ is defined by the module E, then $Y = \text{Proj}(\text{Sym}(E))$, and $\phi^*(\mathcal{E})$ is defined by the module $\text{Sym}(E) \otimes_A E$. We can map $\text{Sym}(E) \otimes_A E$ to $\text{Sym}(E)[1]$ by using the multiplication in $\text{Sym}(E)$. The image of this map consists of all elements of degree at least 0 (corresponding to elements of degree at least 1 in $\text{Sym}(E)$), so the map is surjective on $Y = \text{Proj}(\text{Sym}(E))$. Thus, $\phi^*(\mathcal{E})$ has a quotient that is a rank 1 locally free sheaf.

By Proposition 9.2.1, if η is a cycle on X, $\phi_*(h^{r-1}\phi^*(\eta)) = \eta$ in the Chow group of X. In fact, it also follows from that proof that specific divisors defining h can be chosen so that $\phi_*(h^{r-1}\phi^*(\eta)) = \eta$ as cycles on X. Thus, the map induced by ϕ on cycles is surjective. $\square$

Using the Whitney sum formula and the splitting principle makes it possible to prove many properties of Chern classes quite easily. Suppose that $\mathcal{E}$ has a filtration with quotients of rank 1 with Chern classes $\alpha_1, \ldots, \alpha_r$. The Whitney sum formula then implies that

$$c(\mathcal{E}) = \prod_{i=1}^{r}(1 + \alpha_i t).$$

In particular, the Chern classes are the elementary symmetric functions in the α_i. Since any symmetric function can be expressed as a polynomial in the elementary symmetric functions, a symmetric function of the α_i can be expressed as a polynomial in the Chern classes $c_i(\mathcal{E})$. Furthermore, any property that can be deduced from its expression in terms of the α_i holds when $\mathcal{E}$ has a filtration with rank 1 quotients by the Whitney sum formula, and thus it holds in general by Theorem 9.6.1. In fact, every cycle on X can be lifted to a cycle on $P(\mathcal{E})$, and any formula that holds on $P(\mathcal{E})$ also holds on X by the compatibility of Chern classes with proper push-forward. In the next section we apply these ideas to define Chern characters and prove their basic properties.

9.7 Chern Characters

Chern characters are defined in terms of Chern classes and, like Chern classes, are operators on the Chow group. However, the formulas for Chern characters involve rational numbers that are not integers, so it is necessary to use the tensor product of $A_*(X)$ with $\mathbb{Q}$. We call this group the *rational Chow group* of X, and, when it is necessary to specify that we are using the rational rather than the ordinary Chow group, denote it $A_*(X)_{\mathbb{Q}}$.

Chern characters are defined in such a way that it will follow formally from the theorems for Chern classes that Chern characters are additive on short exact sequences, and a simple computation for rank 1 sheaves (together with the splitting principle) will show that they are multiplicative on tensor products.

Let $\mathcal{E}$ be a locally free sheaf of rank r and assume that it has a filtration with quotients of rank 1 with Chern classes $\alpha_1, \ldots, \alpha_r$. Then we let the total Chern character of $\mathcal{E}$ be

$$\mathrm{ch}(\mathcal{E}) = \sum_{i=1}^{r} e^{\alpha_i}.$$

In this expression we use the power series representation of the exponential; thus, the term of degree j, which we denote $\mathrm{ch}_j(\mathcal{E})$, can be expressed by the formula

$$\mathrm{ch}_j(\mathcal{E})(\eta) = \sum_{i=1}^{r} \alpha_i^j(\eta)/j!.$$

For $j = 0$, $\mathrm{ch}_0(\mathcal{E})$ is the rank of $\mathcal{E}$. We note that for each j, $\mathrm{ch}_j(\mathcal{E})$ defines an operator on the rational Chow group that takes an element of $A_k(X)_{\mathbb{Q}}$ to $A_{k-j}(X)_{\mathbb{Q}}$ for each k.

In particular, the Chern character is a symmetric function in the α_i, so it can be expressed in terms of the Chern classes with rational coefficients (with denominators at most $j!$).

Before proving the main properties of Chern characters, we prove a result on Chern classes of tensor products of rank 1 sheaves.

Proposition 9.7.1 *Let $X = \mathrm{Proj}(A)$ and let $\mathcal{E}$ and $\mathcal{F}$ be locally free sheaves of rank 1 on X. Then*

$$c_1(\mathcal{E} \otimes \mathcal{F}) = c_1(\mathcal{E}) + c_1(\mathcal{F}).$$

Proof We may assume that A is an integral domain and that we are computing the action of the Chern classes on the class $[X]$. Let e and f be elements of modules E and F representing $\mathcal{E}$ and $\mathcal{F}$, and let x and y be nonzero elements of A such that e/x and f/y have degree $(0, \ldots, 0, 1)$. Then Proposition 9.2.4 implies that $c_1(\mathcal{E})([X])$ and $c_1(\mathcal{F})([X])$ can be represented by the cycles defined locally by intersecting with e/x and f/y, respectively. The element $(e \otimes f)/xy$ then has degree $(0, \ldots, 0, 1)$ in the symmetric algebra of $E \otimes F$ and can thus be used to compute $c_1(\mathcal{E} \otimes \mathcal{F})$.

Let $\mathfrak{p}$ be a prime ideal of codimension 1. The modules $E_{(\mathfrak{p})}$, $F_{(\mathfrak{p})}$, and $(E \otimes F)_{(\mathfrak{p})}$ are isomorphic to $A_{(\mathfrak{p})}$. Under these isomorphisms the element $(e \otimes f)/xy$ maps to the product of e/x and f/y. Thus, the coefficient of $[A/\mathfrak{p}]$ in this expression for $c_1(\mathcal{E} \otimes \mathcal{F})[X]$ is defined by the product of e/x and f/y, so it is the sum of the coefficients of $[A/\mathfrak{p}]$ in $c_1(\mathcal{E})[X]$ and $c_1(\mathcal{F})[X]$. $\qquad\square$

Theorem 9.7.2

(i) *Let*

$$0 \to \mathcal{E}' \to \mathcal{E} \to \mathcal{E}'' \to 0$$

be a short exact sequence of locally free sheaves over $X = \mathrm{Proj}(A)$. Then

$$\mathrm{ch}(\mathcal{E}) = \mathrm{ch}(\mathcal{E}') + \mathrm{ch}(\mathcal{E}'').$$

(ii) *Let $\mathcal{E}$ and $\mathcal{F}$ be locally free sheaves. Then*

$$\mathrm{ch}(\mathcal{E} \otimes \mathcal{F}) = \mathrm{ch}(\mathcal{E})\mathrm{ch}(\mathcal{F}).$$

(iii) *Let $\mathcal{E}^*$ be the dual $\mathrm{Hom}(\mathcal{E}, \mathcal{O}_X)$. Then $\mathrm{ch}_i(\mathcal{E}^*) = (-1)^i \mathrm{ch}_i(\mathcal{E})$ for each i.*

Proof Using the splitting principle, we may reduce each of these questions to the case in which the locally free sheaves have filtrations with quotients of rank 1. We carry out the details of this reduction for the second statement. Let η be an element of $A_*(X)_{\mathbb{Q}}$. By Theorem 9.6.1 there is a projective scheme Y and a projective map $\phi : Y \to X$ such that the pull-backs of $\mathcal{E}'$ and $\mathcal{E}''$ have filtrations with rank 1 quotients and such that ϕ_* is surjective on cycles. Let ρ be an element of $A_*(Y)_{\mathbb{Q}}$ such that $\phi_*(\rho) = \eta$. If the formula is known for the pull-back to Y, we have (using that tensor product commutes with pull-back of sheaves)

$$\mathrm{ch}(\phi^*(\mathcal{E} \otimes \mathcal{F}))(\rho) = \mathrm{ch}(\phi^*(\mathcal{E}))\mathrm{ch}(\phi^*(\mathcal{F}))(\rho).$$

Applying ϕ_* and using the projection formula, we obtain

$$\phi_*(\mathrm{ch}(\phi^*(\mathcal{E} \otimes \mathcal{F}))(\rho)) = \mathrm{ch}(\mathcal{E} \otimes \mathcal{F})(\phi_*(\rho)) = \mathrm{ch}(\mathcal{E} \otimes \mathcal{F})(\eta),$$

while

$$\phi_*(\mathrm{ch}(\phi^*(\mathcal{E}))\mathrm{ch}(\phi^*(\mathcal{F}))(\rho)) = \mathrm{ch}(\mathcal{E})(\phi_*(\mathrm{ch}(\phi^*(\mathcal{F}))(\rho)))$$

$$= \mathrm{ch}(\mathcal{E})(\mathrm{ch}(\mathcal{F}))(\eta).$$

Thus, we can conclude that $\mathrm{ch}(\mathcal{E} \otimes \mathcal{F})(\eta) = \mathrm{ch}(\mathcal{E})\mathrm{ch}(\mathcal{F})(\eta)$. By similar arguments, we may assume that $\mathcal{E}$, $\mathcal{E}'$, and $\mathcal{E}''$ have such filtrations to prove (i) and (iii).

To prove (i), we assume that $\mathcal{E}'$ and $\mathcal{E}''$ have filtrations with rank 1 quotients, and we denote the Chern classes of the quotients $\alpha_1, \ldots, \alpha_r$ and $\alpha_{r+1}, \ldots, \alpha_{r+s}$. Combining these filtrations gives a filtration of $\mathcal{E}$ with rank 1 quotients with Chern classes $\alpha_1, \ldots, \alpha_r, \alpha_{r+1}, \ldots, \alpha_{r+s}$. Thus, the Chern characters of $\mathcal{E}, \mathcal{E}'$, and $\mathcal{E}''$, respectively, are $\mathrm{ch}(\mathcal{E}) = \sum_{i=1}^{r+s} e^{\alpha_i}$, $\mathrm{ch}(\mathcal{E}) = \sum_{i=1}^{r} e^{\alpha_i}$, and $\mathrm{ch}(\mathcal{E}) = \sum_{i=r+1}^{s} e^{\alpha_i}$, and the first statement is clear.

For the second statement, let $0 = \mathcal{E}_0 \subset \mathcal{E}_1 \subset \cdots \subset \mathcal{E}_r = \mathcal{E}$ and $0 = \mathcal{F}_0 \subset \mathcal{F}_1 \subset \cdots \subset \mathcal{F}_s = \mathcal{F}$ be the filtrations on $\mathcal{E}$ and $\mathcal{F}$ with rank 1 locally free quotients. We take the filtration

$$0 = \mathcal{E}_0 \otimes \mathcal{F} \subset \mathcal{E}_1 \otimes \mathcal{F} \subset \cdots \subset \mathcal{E}_r \otimes \mathcal{F}.$$

The quotients in this filtration are $(\mathcal{E}_i/\mathcal{E}_{i-1}) \otimes \mathcal{F}$. Hence it follows from part (i) of the theorem that

$$\mathrm{ch}(\mathcal{E} \otimes \mathcal{F}) = \sum_{i=1}^{r} \mathrm{ch}((\mathcal{E}_i/\mathcal{E}_{i-1}) \otimes \mathcal{F}).$$

We now apply the same argument to each term of this sum, using the filtration on $\mathcal{F}$, to obtain

$$\mathrm{ch}(\mathcal{E} \otimes \mathcal{F}) = \sum_{i=1}^{r} \sum_{j=1}^{s} \mathrm{ch}((\mathcal{E}_i/\mathcal{E}_{i-1}) \otimes (\mathcal{F}_j/\mathcal{F}_{j-1})).$$

In addition, the additivity formula implies that we have the equalities $\mathrm{ch}(\mathcal{E}) = \sum_{i=1}^{r} \mathrm{ch}(\mathcal{E}_i/\mathcal{E}_{i-1})$ and $\mathrm{ch}(\mathcal{F}) = \sum_{i=j}^{r} \mathrm{ch}(\mathcal{F}_j/\mathcal{F}_{j-1})$. Suppose that the multiplicativity formula holds for rank 1 locally free sheaves. Then we have

$$\mathrm{ch}(\mathcal{E} \otimes \mathcal{F}) = \sum_{i=1}^{r} \sum_{j=1}^{s} \mathrm{ch}((\mathcal{E}_i/\mathcal{E}_{i-1}) \otimes (\mathcal{F}_j/\mathcal{F}_{j-1}))$$

$$= \sum_{i=1}^{r} \sum_{j=1}^{s} \mathrm{ch}(\mathcal{E}_i/\mathcal{E}_{i-1})\mathrm{ch}(\mathcal{F}_j/\mathcal{F}_{j-1})$$

$$= \left(\sum_{i=1}^{r} \mathrm{ch}(\mathcal{E}_i/\mathcal{E}_{i-1}) \right) \left(\sum_{j=1}^{s} \mathrm{ch}(\mathcal{F}_j/\mathcal{F}_{j-1}) \right) = \mathrm{ch}(\mathcal{E})\mathrm{ch}(\mathcal{F}).$$

Hence it suffices to prove the theorem in the case where $r = s = 1$.

Suppose now that $\mathcal{E}$ and $\mathcal{F}$ have rank 1. Let $\alpha = c_1(\mathcal{E})$ and $\beta = c_1(\mathcal{F})$. Theorem 9.4.1 implies that $c_1(\mathcal{E} \otimes \mathcal{F}) = \alpha + \beta$. Hence

$$\mathrm{ch}(\mathcal{E} \otimes \mathcal{F}) = e^{\alpha+\beta} = e^{\alpha} e^{\beta} = \mathrm{ch}(\mathcal{E})\mathrm{ch}(\mathcal{F}).$$

Thus, Chern characters are multiplicative on rank 1 sheaves, and this completes the proof of the theorem.

To prove statement (iii), we may again reduce to the case in which $\mathcal{E}$ has rank 1. In this case $\text{ch}(\mathcal{E}) = e^{\alpha}$ for $\alpha = c_i(\mathcal{E})$. In addition, since $\mathcal{E}$ has rank 1, we have $\mathcal{E} \otimes \mathcal{E}^* \cong \mathcal{O}_X$, and it thus follows from part (ii) that $\text{ch}(\mathcal{E})\text{ch}(\mathcal{E}^*) = 1$, so $\text{ch}(\mathcal{E}^*) = e^{-\alpha}$. Hence we have $\text{ch}_i(\mathcal{E}^*) = (-\alpha)^i/i! = (-1)^i\text{ch}_i(\mathcal{E})$ for each i.

$\square$

9.8 Push-Forward of Complexes of Sheaves

In the final section of this chapter we define the push-forward of a bounded complex of sheaves by a projective map. This topic has no direct connection with Chern classes, as it involves only sheaves and not the Chow group. However, this construction will be used in Chapter 12 in the proof of the Riemann-Roch theorem.

If $\phi : X \to Y$ is a projective map defined by the inclusion of a $\mathbb{Z}^n$-graded ring into a $\mathbb{Z}^{n+1}$-graded ring, we will define a map ϕ_* that takes a complex of sheaves on X to a complex of sheaves on Y. If X is a projective scheme defined by a $\mathbb{Z}$-graded ring over a field k, if ϕ is the map from X to $\text{Spec}(k)$, and if $\mathcal{M}$ is a sheaf on X, the homology of the complex $\phi_*(\mathcal{M})$ is what is usually known as the sheaf cohomology $H^i(X, \mathcal{M})$ (see, for example, [21]). However, for our purposes it is more convenient to define ϕ_* directly on complexes of sheaves.

We say that a sheaf is *coherent* if it is defined by a finitely generated module. One of the main properties of ϕ_* is that if $\mathcal{M}_\bullet$ is a bounded complex of coherent sheaves, then $\phi_*(\mathcal{M}_\bullet)$ is quasi-isomorphic to a bounded complex of coherent sheaves. In particular, its homology is coherent.

Let $Y = \text{Proj}(A)$, and let $X = \text{Proj}(B)$, where A is a $\mathbb{Z}^n$-graded ring and B is a $\mathbb{Z}^{n+1}$-graded ring, and where B is generated over A by elements $b_1, \ldots, b_k$. Let b_i have degree $(n_i, 1)$ with $n_i \in \mathbb{Z}^n$. We let $C^\bullet = C^\bullet(b_1, \ldots, b_k)$ be the complex defined as follows: for each $i \geq 1$, C^i is the product of the localizations $B_{b_{j_1} b_{j_2} \ldots b_{j_{i+1}}}$ for all $j_1 < j_2 < \cdots < j_{i+1}$. The boundary map d^i from C^i to C^{i+1} is defined on an element $b \in B_{b_{j_1} b_{j_2} \ldots b_{j_{i+1}}}$ by letting

$$d^i(b) = \sum_j (-1)^m \alpha_j(b)$$

where the sum is over all j not in $\{j_1, \ldots, j_{i+1}\}$, α_j is the embedding of $B_{b_{j_1} b_{j_2} \ldots b_{j_{i+1}}}$ into $B_{b_j b_{j_1} b_{j_2} \ldots b_{j_{i+1}}}$, and m is the number of j_i that are less than j. It is straightforward to verify that this defines a complex. Thus, $C^\bullet$ is a complex of graded modules that can be written

$$0 \to \prod_i B_{b_i} \to \prod_{i<j} B_{b_i b_j} \to \cdots \to B_{b_1 b_2 \ldots b_n} \to 0.$$

There is a map from B to C^0 defined by the natural map from B to the localization B_{b_i} in each component. We denote the complex obtained by adding B in degree -1 with this boundary map by $\overline{C}^\bullet$.

For computing with $C^\bullet(b_1, \ldots, b_k)$, it is useful to have a representation as a limit of Koszul complexes. For each integer $n > 0$, let $K_\bullet(n) = K_\bullet(b_1^n, \ldots, b_k^n)$ be the Koszul complex on $b_1^n, \ldots, b_k^n$. Let $\tilde{K}^i(n)$ denote the graded module $\mathrm{Hom}_B(K_{k-i-1}(n), B)$ for $i = 0, \ldots, k - 1$. Then $\tilde{K}^\bullet(n)$ is a complex with boundary maps induced from those of $K_\bullet(n)$. By Proposition 3.3.5, $\tilde{K}^\bullet(n)$ is isomorphic to the Koszul complex on $b_1^n, \ldots, b_k^n$ with degrees shifted by one and the resulting module in degree -1 removed. Furthermore, if $n' > n$, there is a map of complexes from $K_\bullet(n')$ to $K_\bullet(n)$ that takes the generator e_i of $K_1(n')$ to the $b_i^{n'-n}$ times the generator e_i of $K_1(n)$. This map induces a map from $\tilde{K}^\bullet(n)$ to $\tilde{K}^\bullet(n')$, that we denote $\alpha_{n',n}$.

Proposition 9.8.1 *With notation as above, the complex $C^\bullet(b_1, \ldots, b_k)$ is the direct limit over n of the complexes $\tilde{K}^\bullet(n)$ and maps $\alpha_{n,n'}$. The complex $\overline{C}^\bullet(b_1, \ldots, b_k)$ is the direct limit of the complexes*

$$\mathrm{Hom}(K_\bullet(b_1, \ldots, b_k), B)[1].$$

Proof Let $e_1, \ldots, e_k$ be a set of generators of $K_1(n) \cong B^k$. For each $i_1 < \cdots < i_s$, let $e_{i_1} \wedge \cdots \wedge e_{i_s}$ denote the corresponding basis element of $K_s(n)$, and let $(e_{i_1} \wedge \cdots \wedge e_{i_s})^*$ denote the element of the dual basis in $\mathrm{Hom}(K_s(n), B) = \tilde{K}^{k-s-1}$. We define a map β_n from $\tilde{K}^\bullet(n)$ to $C^\bullet$ by sending $(e_{i_1} \wedge \cdots \wedge e_{i_s})^*$ to $1/(b_{j_1}^n \ldots b_{j_{k-s}}^n) \in B_{b_{j_1} \ldots b_{j_{k-s}}}$, where $\{j_1, \ldots, j_{k-s}\}$ is the complement of $\{i_1, \ldots, i_s\}$. It is then straightforward to verify that β_n is a map of complexes and that if $n' > n$ we have $\beta_{n'} = \alpha_{n',n}\beta_n$. Thus, the maps β_n induce a map β on the direct limit. Since the localization $B_{b_{j_1} b_{j_2} \ldots b_{j_{k-s}}}$ is generated as a B-module by the elements $1/(b_{j_1}^n \cdots b_{j_{k-s}}^n)$, β is surjective. An element goes to zero if and only if it is annihilated by some power of $b_{j_1} \cdots b_{j_{k-s}}$, which implies that it is zero in the limit. Hence β is an isomorphism of complexes.

The proof of the second statement is the same, including the modules in degree -1. $\square$

It follows from Theorem 9.5.1 (and can also be proven directly) that permutations of the b_i define isomorphic complexes.

We now define the functor ϕ_*. Let $\mathcal{M}_\bullet$ be a bounded complex of coherent sheaves on X. We say that a homogeneous element of a graded B-module has degree k in the last component if it has degree $(k_1, \ldots, k_n, k)$ for some integers $k_1, \ldots, k_n$.

Definition 9.8.1 *We let $\phi_*(\mathcal{M}_\bullet)$ be the graded part of $\mathcal{M}_\bullet \otimes C_\bullet$ consisting of elements of degree 0 in the last component.*

One essential property of ϕ_* which has to be proven to make this definition valid is that it is independent of the choice of generators.

Proposition 9.8.2 *If $b_1, \ldots, b_k$ and $c_1, \ldots, c_m$ are two choices of generators for B over A, where the b_i and c_i have degree 1 in the last component, then*

(i) *The complexes*

$$C^\bullet(b_1, \ldots, b_k) \quad and \quad C^\bullet(c_1, \ldots, c_m)$$

are quasi-isomorphic.

(ii) *If $\mathcal{M}_\bullet$ is a bounded complex of sheaves, the complexes*

$$C^\bullet(b_1, \ldots, b_k) \otimes \mathcal{M}_\bullet \quad and \quad C^\bullet(c_1, \ldots, c_m) \otimes \mathcal{M}_\bullet$$

are quasi-isomorphic.

Proof It suffices to show that if we add one generator the resulting complexes are quasi-isomorphic. Let $b_1, \ldots, b_k$ be the original generators of B over A, and let y be the new generator. Let $C^\bullet = C^\bullet(b_1, \ldots, b_k)$ and $\tilde{C}^\bullet = C^\bullet(b_1, \ldots, b_k, y)$. Let $C_y^\bullet$ be the subcomplex of $\tilde{C}^\bullet$ generated by the factors of the form $B_{b_{j_1} \cdots b_{j_i} y}$. We have a short exact sequence of complexes

$$0 \to C_y^\bullet \to \tilde{C}^\bullet \to C^\bullet \to 0.$$

Thus, to complete the proof it suffices to show that $C_y^\bullet$ is exact. The complex $C_y^\bullet$ is isomorphic to the complex $\overline{C}^\bullet$ tensored with B_y with degree shifted by one. Thus, Proposition 9.5.1 implies that it is a direct limit of Koszul complexes $K_\bullet(b_1^n, \ldots, b_k^n)$ tensored with B_y. Now y is in the algebra generated by the b_i and all elements have degree 1 in the last component. Hence y is in the sub A-module generated by $b_1, \ldots, b_k$. Thus, the homology of the Koszul complex on $b_1^n, \ldots, b_k^n$ is annihilated by a power of y for all n, so this Koszul complex tensored with B_y is exact. Hence the direct limit, which is the complex $C_y^\bullet$, is exact. Thus, the above map from $C^\bullet(b_1, \ldots, b_k, y)$ to $C^\bullet(b_1, \ldots, b_k)$ is a quasi-isomorphism.

Hence the map from $\tilde{C}^\bullet$ to $C^\bullet$ is a quasi-isomorphism. Let $\mathcal{M}_\bullet$ be a bounded map of sheaves. To show that the induced map from $C^\bullet(b_1, \ldots, b_k) \otimes \mathcal{M}_\bullet$ to $C^\bullet(c_1, \ldots, c_m) \otimes \mathcal{M}_\bullet$ is a quasi-isomorphism, it suffices by the method of Proposition 3.2.4 to show that the map from $C^\bullet(b_1, \ldots, b_k) \otimes \mathcal{M}_i$ and $C^\bullet(c_1, \ldots, c_m) \otimes \mathcal{M}_i$ is a quasi-isomorphism for each i. Thus, we may prove the result for a single sheaf $\mathcal{M}$.

Consider the sequence

$$0 \to C_y^\bullet \otimes \mathcal{M} \to \tilde{C}^\bullet \otimes \mathcal{M} \to C^\bullet \otimes \mathcal{M} \to 0.$$

Since the original sequence is split in each degree, this sequence is still exact. Thus, it suffices to show that $C_y^\bullet \otimes \mathcal{M}$ is exact. The proof of this statement is the same as that of the proof that $C_y^\bullet$ is exact, replacing B_y by M_y throughout, where M is a graded module representing $\mathcal{M}$. $\qquad\square$

It follows from Theorem 9.5.2 that different choices of generators define the same functor ϕ_* up to quasi-isomorphism. We will prove various basic properties of ϕ_*, but first we compute the result of applying ϕ_* in an important example.

Theorem 9.8.3 *Let* $f : X \to Y$ *be a projective map as above, where* $Y = \mathrm{Proj}(A)$, $X = \mathrm{Proj}(B)$, *and where* $B = A[X_1, \ldots, X_k]$ *is a polynomial ring over* A. *Let* X_i *have degree* $(a_i, 1)$ *with* $a_i \in \mathbb{Z}^n$. *Let* $m = (m_1, m_2)$ *be an element of* $\mathbb{Z}^{n+1}$ *with* $m_1 \in \mathbb{Z}^n$ *and* $m_2 \in \mathbb{Z}$. *Then*

(i) *If* $m_2 \geq 0$, *then the homology of* $f_*(\mathcal{O}_X(m))$ *is zero except in degree zero, and*

$$H^0(f_*(\mathcal{O}_X(m))) = \oplus \mathcal{O}_Y \left(m_1 - \sum i_j a_j \right),$$

where the sum is taken over all k-*tuples* $i_1, \ldots, i_k$ *with* $i_1 + \cdots + i_k = m_2$ *and* $i_j \geq 0$.

(ii) *If* $-k < m_2 < 0$, *then the homology of* $f_*(\mathcal{O}_X(m))$ *is zero.*

(iii) *If* $m_2 \leq -k$, *then the homology of* $f_*(\mathcal{O}_X(m))$ *is zero except in degree* $k - 1$, *and*

$$H^{k-1}(f_*(\mathcal{O}_X(m))) = \oplus \mathcal{O}_Y \left(m_1 - \sum i_j a_j \right),$$

where the sum is taken over all k-*tuples* $i_1, \ldots, i_k$ *with* $i_1 + \cdots + i_k = m_2$ *and* $i_j < 0$.

Proof By Proposition 9.5.1, $\overline{C}^\bullet(X_1, \ldots, X_k)$ is a direct limit of Koszul complexes, and tensoring this complex with $\mathcal{O}_X(m)$ simply shifts the degrees of the modules by m. Since $X_1, \ldots, X_k$ form a regular sequence, this shows that $\overline{C}^\bullet(X_1, \ldots, X_k)$ is exact except in degree $k - 1$. Thus, the homology of $C^\bullet(X_1, \ldots, X_k)$ is isomorphic to $B[m]$ in degree 0, is a direct limit of quotients $B/(X_1^n, \ldots, X_k^n)[m]$ in degree $k - 1$, and is zero in every other degree.

To prove (i), we must find the component of $B[m]$ consisting of elements that have degree 0 in the last component. We note that an element of $B[m]$

has degree 0 if and only if the corresponding element of B has degree m. B is the polynomial ring $A[X_1, \ldots, X_k]$, and the degree in the last component of a monomial $a X_1^{i_1} \cdots X_k^{i_k}$ is $i_1 + \cdots + i_k$. Hence the part of $B[m] = B[(m_1, m_2)]$ of degree 0 in the last component is the direct sum of $A X_1^{i_1} \cdots X_k^{i_k}$ over all monomials with $i_1 + \cdots + i_k = m_2$.

We must now compute the degree shift that our assumptions impose on the coefficient a of $a X_1^{i_1} \cdots X_k^{i_k}$, where $i_1 + \cdots + i_k = m_2$. If a has degree $r \in \mathbb{Z}^n$, then the degree of $a X_1^{i_1} \cdots X_k^{i_k}$ in B is $(r, 0) + i_1(a_1, 1) + \cdots + i_k(a_k, 1) = (r + \sum i_j a_j, m_2)$. Thus, its degree in $B[(m_1, m_2)]$ is $(r + \sum i_j a_j - m_1, 0)$. Thus, the module of coefficients of $X_1^{i_1} \cdots X_k^{i_k}$ is A shifted by $m_1 - \sum i_j a_j$ or $A[m_1 - \sum i_j a_j]$. This proves statement (i).

It remains to compute the homology of $C^\bullet$ in degree $k - 1$. This homology is the quotient of the localization $B_{X_1 X_2 \cdots X_k}$ by the submodule generated by the localizations $B_{X_1 \cdots \hat{X}_i \cdots X_k}$ for $i = 1, \ldots, k$. A basis for the localization $B_{X_1 X_2 \cdots X_k}$ over A consists of all monomials $X_1^{i_1} \cdots X_k^{i_k}$ where the i_j can be any integers, positive or negative. The submodule generated by the localizations $B_{X_1 \cdots \hat{X}_i \cdots X_k}$ is generated by monomials $X_1^{i_1} \cdots X_k^{i_k}$ where $i_j \geq 0$ for at least one j. Thus, a basis for $H^{k-1}(C^\bullet)$ over A consists of monomials $X_1^{i_1} \cdots X_k^{i_k}$ with $i_j < 0$ for all j. The rest of the theorem now follows by the computation used in the proof of part (i). $\qquad\square$

Theorem 9.8.4 *Let the notation be as above.*

(i) *The functor ϕ_* takes short exact sequences of complexes to short exact sequences of complexes.*

(ii) *If $\mathcal{M}_\bullet$ and $\mathcal{N}_\bullet$ are bounded complexes that are quasi-isomorphic, then $\phi_*(\mathcal{M}_\bullet)$ is quasi-isomorphic to $\phi_*(\mathcal{N}_\bullet)$.*

(iii) *If $\mathcal{M}_\bullet$ is a bounded complex of coherent sheaves, then $\phi_*(\mathcal{M})$ is quasi-isomorphic to a bounded complex of coherent sheaves.*

Proof The first statement follows from the fact that the functor ϕ_* is defined by tensoring with localizations of B, which are flat over B. If $\mathcal{M}_\bullet$ is quasi-isomorphic to $\mathcal{N}_\bullet$, then flatness implies that $\mathcal{M}_\bullet \otimes C^i$ is quasi-isomorphic to $\mathcal{N}_\bullet \otimes C^i$ for each i, and the second statement follows from the argument used in Proposition 3.2.4 of Chapter 3.

To prove the third statement, we use the computation of Theorem 9.8.3. Using the results of Chapter 3, it suffices to show that if $\mathcal{M}$ is a coherent sheaf, then $\phi_*(\mathcal{M})$ has coherent homology. Let M be a finitely generated graded module representing $\mathcal{M}$. By taking a filtration of M and using part (i) of the

Theorem, we can reduce to the case where $M = B/\mathfrak{q}[n]$ for a graded prime ideal $\mathfrak{q}$ of B and some m in $\mathbb{Z}^{n+1}$.

If C is a graded homomorphic image of B for which M is a C-module, the localization of M at a set of generators of B is the same as the localization of M at the images of the generators of B in C, so we can replace B by C or C by B in the computation. This means, first of all, that since $\mathfrak{q}$ annihilates M we can replace B by $B/\mathfrak{q}$. Let $\mathfrak{p} = A \cap \mathfrak{q}$. Let $X_1, \ldots, X_k$ be indeterminates over $A/\mathfrak{p}$ of the correct degrees that map onto a set of generators of $B/\mathfrak{q}$ over $A/\mathfrak{p}$ that have degree 1 in the last component. The above remark implies that we may replace $B/\mathfrak{q}$ by $A/\mathfrak{p}[X_1, \ldots, X_n]$. Thus, we can assume that B is a polynomial algebra over $A/\mathfrak{p}$.

The remainder of the proof uses induction on the dimension of $A/\mathfrak{p}$. Let the dimension of $A/\mathfrak{p}$ be d, and suppose that the result is true for all modules with support of dimension less than d. Let S be the multiplicative set consisting of homogeneous elements in the complement of $\mathfrak{p}$ in A. Then $A_{(S)} = k(\mathfrak{p})$. Let $\tilde{B}_{(S)}$ denote the subring of elements in B_S of degrees $(0, \ldots, 0, i)$, and define $M_{(S)}$ similarly. Then $B_{(S)}$ is a polynomial ring over $A_{(S)} = k(\mathfrak{p})$, and $M_{(S)}$ is a finitely generated sum of modules of the form $B_{(S)}[k]$ for various $k \in \mathbb{Z}$. Clearing denominators by multiplying by certain nonzero homogeneous elements in $A/\mathfrak{p}$, we can find a finite complex $F_\bullet$ that maps to M and such that each F_i is a sum of B-modules of the form $B[m]$ for various $m \in \mathbb{Z}^{n+1}$. The complex $F_\bullet$ will not be exact, but its homology in degrees greater than zero will have support consisting of prime ideals that meet A in dimension strictly less than d. The module $B[m]$ defines the sheaf $\mathcal{O}_X(m)$ for each m, and by Theorem 9.8.3 $\phi_*(\mathcal{O}_X(m))$ has coherent homology for all m. Hence $\phi_*(F_\bullet)$ has coherent homology. Using induction on dimension and part (i) of the Theorem, we can then conclude that $\phi_*(\mathcal{M})$ has coherent homology as was to be shown. $\qquad\square$

Exercises

9.1 Show that the relation used to define the equivalence relation in the definition of sheaves is transitive.

9.2 Let $A = k[X, Y, Z]$ with the usual grading and let M be the A-module with three generators e_1, e_2, e_3 modulo the relation $Xe_1 + Ye_2 + Ze_3$.

(a) Show that the sheaf $\mathcal{E}$ defined by M is locally free.

(b) Show directly from the definition of Chern classes that the Chern classes of $\mathcal{E}$ are $1, h, h^2$, where h is the hyperplane section on Proj(A).

9.3 Let A and $\mathcal{E}$ be as in the previous problem. Write a short exact sequence involving $\mathcal{E}$ and sheaves of the form $\mathcal{O}(n)$ and compute the Chern classes of $\mathcal{E}$ from the Whitney sum formula.

9.4 (Definition of Chern classes using Segre classes.) Let $\mathcal{E}$ be a locally free sheaf of rank r over a projective scheme X. Let h be the hyperplane section on $P(\mathcal{E})$, and let p be the projection from $P(\mathcal{E})$ to X. We define the ith *Segre class* of $\mathcal{E}$ to be

$$s_i(\eta) = p_*(h^{i+r-1}(p^*(\eta)))$$

where η is an element of the Chow group of X.

(a) Show that $s_0(\eta) = \eta$.

(b) Let $s(\eta) = s_0(\eta) + s_1(\eta)t + \cdots + s_j(\eta)t^j + \cdots$. Show that $s(\mathcal{E})c(\mathcal{E})(\eta) = \eta$.

It follows that the total Segre class and the total Chern class are inverses in the ring of operators on the Chow group. This gives an alternative definition of Chern classes using the inverse of the Segre class.

9.5 Show that Chern classes commute with flat pull-back, completing the proof of Proposition 9.3.4.

9.6 Prove that the constructions of the direct sum and tensor product of modules really do preserve equivalence and thus define the corresponding constructions on sheaves.

10

The Grassmannian

In this chapter we develop the basic theory of the Grassmannian to the extent necessary for the definition of local Chern characters. The Grassmannian is classically defined as an algebraic variety whose points correspond to linear subspaces of a given dimension of a finitely generated vector space. This variety has an embedding into projective space called the Plücker embedding and thus defines a graded ring. Our point of view will be that this graded ring is the main object of study.

Let X be a projective scheme, and let $\mathcal{E}$ be a locally free sheaf on X. The Grassmannian we construct in this chapter will represent locally free subsheaves of $\mathcal{E}$ of a given rank r with locally free quotient. While in the classical case a subspace of rank r defines a point in the Grassmannian, in this case we consider there is a map π from the Grassmannian of $\mathcal{E}$ to X such that a locally free subsheaf of $\mathcal{E}$ defines a section of π. In fact, it can be shown that $G_r(\mathcal{F})$ represents the functor of subsheaves of rank r with these properties; we will not need this stronger result, so we refer to Kleiman [33] for details.

The other part of the construction is the definition of a canonical locally free subsheaf and quotient sheaf of the pullback of $\mathcal{E}$ to the Grassmannian. The Chern classes of these sheaves will be used to define local Chern characters of complexes in the next chapter.

In the first two sections we develop the local theory for a free submodule over a free module over a Noetherian ring. We then give the general definition and prove its important functorial properties. In the last section we describe how to compute the Chern classes of the canonical locally free sheaves using free resolutions.

10.1 Local Computations on the Grassmannian

We first briefly recall the geometric definition. Let n and r be positive integers with $r \leq n$, and let V be a vector space of dimension n over a field k. Let

$X(r, n)$ denote the set of r-dimensional subspaces of V. Then $X(r, n)$ can be made into a variety over k in a natural way. We describe below the graded ring that is associated to $X(r, n)$.

By choosing a basis for V we may identify V with k^n. With this identification, an r-dimensional subspace W can be specified an n-by-r matrix M of rank r whose columns form a basis of W. The rank of M will be r if and only if at least one of the r-by-r minors of M is nonzero. The Plücker embedding into projective space is defined by taking the coordinates of the point corresponding to the subspace spanned by the columns of M to be the set of all r-by-r minors of M. It is easy to see that if two matrices span the same subspace W, then the columns of the matrices define two bases of W, and there is an r-by-r matrix P that takes one basis to the other. Hence r-by-r minors of the two matrices are the same up to a constant multiple equal to the determinant of P. The converse is also true, although not so obvious. Since there are $\binom{n}{r}$ minors, the Plücker embedding maps the Grassmannian into projective space of dimension $\binom{n}{r} - 1$. Thus, there is a graded ring over k whose associated projective scheme is the Grassmannian.

We give a direct algebraic definition of this graded ring over a ring A. Let n and r be integers with $r \leq n$. Let $M = (X_{ij})$ be an n-by-r matrix whose entries are indeterminates over A. Let $\mathfrak{a}$ be the ideal of $A[X_{ij}]$ generated by the r-by-r minors of M. We index the minors by the sets $I = \{i_1, \ldots, i_r\}$, where we usually choose the ordering so that $i_1 < \cdots < i_r$ [although in some instances we will use this notation for arbitrary r-tuples $(i_1, \ldots, i_r)$]. We denote the set of such subsets I consisting of r elements by $S(r, n)$. For each $I \in S(r, n)$ we denote the corresponding minor of M by Δ_I. The subring of $A[X_{ij}]$ generated by the Δ_I is a graded ring that we denote $R_G(r, n)$, or simply R_G. We define the grading on this ring by letting Δ_I have degree 1. The Grassmannian $G(r, n)$, or $G_A(r, n)$ if it is necessary to specify the ring, is then defined to be $\mathrm{Proj}(R_G(r, n))$. The embedding of A into R_G defines a projective map π_G from $G(r, n)$ to $\mathrm{Spec}(A)$.

There are relations between the Δ_I, but locally the ring of the Grassmannian is a polynomial ring over A, and many computations with Grassmannians are carried out by using this local construction. Let $I \in S(r, n)$, and consider the open set U_{Δ_I} of $G(r, n)$ obtained by inverting Δ_I. The set U_{Δ_I} is then $\mathrm{Spec}((R_G)_{(\Delta_I)})$, where $(R_G)_{(\Delta_I)}$ is the part of the localization $(R_G)_{\Delta_I}$ of degree 0. Since the Δ_I generate $R_G(r, n)$ over A, the open sets U_{Δ_I} cover $G(r, n)$.

Proposition 10.1.1 *The ring $(R_G)_{(\Delta_I)}$ is a polynomial ring in $r(n-r)$ variables over A. More precisely, $(R_G)_{(\Delta_I)}$ is equal to the polynomial ring in the variables Δ_J / Δ_I, where J runs over all subsets of $S(r, n)$ having exactly $r - 1$ elements in common with I.*

Proof Let M be the matrix of indeterminates used in the definition of $G(r, n)$, and let M_I be the r-by-r submatrix of M consisting of the rows with indices in I; more generally, for any $J \in S(r, n)$ we denote the matrix of rows of M with indices in J by M_J. Let $\tilde{M} = MM_I^{-1}$ be the result of multiplying M on the right by the inverse of M_I. We note that since the determinant of M_I is Δ_I, the entries of M_I^{-1} lie in the localization of the polynomial ring $A[X_{ij}]$ at Δ_I. The matrix $\tilde{M}$ is another n-by-r matrix, and we claim that the r-by-r minors of $\tilde{M}$ generate $(R_G)_{(\Delta_I)}$. Let J be an element of $S(r, n)$. Then the matrix consisting of the rows of $\tilde{M}$ indexed by J is equal to $M_J M_I^{-1}$ and hence its determinant is Δ_J/Δ_I. Since R_G is generated over A by the elements Δ_J in degree 1, $(R_G)_{(\Delta_I)}$ is generated over A by the quotients Δ_J/Δ_I, so $(R_G)_{(\Delta_I)}$ is generated over A by the r-by-r minors of $\tilde{M}$.

We next describe $\tilde{M}$ more precisely. Since $\tilde{M} = MM_I^{-1}$ the entries in the rows corresponding to I form an identity matrix. Any minor of $\tilde{M}$ taken by choosing $r - 1$ rows from I and the other row not in I is equal, up to sign, to one of the entries in the other row. Thus, the other entries are equal to $\pm\Delta_J/\Delta_I$, where J is an element of $S(r, n)$ with exactly $r - 1$ elements in common with I. There are $r(n - r)$ such elements J of $S(r, n)$; it is also clear that there are $r(n - r)$ entries in $\tilde{M}$ outside the rows corresponding to elements of I. Thus, since $(R_G)_{(\Delta_I)}$ is generated over A by the minors of $\tilde{M}$, and each minor is a polynomial in the entries of $\tilde{M}$, these elements suffice to generate $(R_G)_{(\Delta_I)}$. They are also algebraically independent over A; to see this, it suffices to note that we can map X_{ij} to entries of an n-by-r matrix that consists of an identity matrix in the rows corresponding to I and a set of $r(n - r)$ elements that are algebraically independent over A; the images of the Δ_J/Δ_I map to algebraically independent elements and are hence themselves algebraically independent. $\qquad\square$

We next prove the basic fact that sections of the projective map π_G are defined by free submodules of F with locally free quotient. We recall that a *section* of π_G is a projective map ϕ from $\mathrm{Spec}(A)$ to G such that $\pi_G\phi$ is the identity map. In the next Proposition we use the fact that $\mathrm{Proj}(A[T]) \cong \mathrm{Spec}(A)$ when T is an indeterminate of degree 1.

Proposition 10.1.2 *Let F be a free module of rank n. Then every free submodule H of F of rank r with locally free quotient defines a section σ_H of π_G. The section σ_H is defined by a map of graded rings $f : R_G \to A[T]$ such that $f(\Delta_I) = \delta_I T$, where δ_I is the Ith minor of a matrix whose columns form a basis of H.*

Proof Let $N = (n_{ij})$ be an n-by-r matrix whose columns give a basis for H. Let S be an indeterminate, and let σ be the map from the polynomial ring $A[X_{ij}]$ to $A[S]$ that is the identity on A and that sends X_{ij} to $n_{ij}S$. This map induces a map from R_G to $A[S^r]$ that sends Δ_I to the Ith minor of N times S^r. Let $T = S^r$; we thus have a map f from R_G to $A[T]$ as in the statement of the Proposition. A different choice of basis for H will define a different map; denote the map defined by a different basis f'. Let u be the determinant of the associated change of basis matrix. Then u is a unit in A and we have $f(\Delta_I) = uf'(\Delta_I)$ for all I. Thus, $f(\Delta_I)/f(\Delta_J) = f'(\Delta_I)/f'(\Delta_J)$ for all I and J, so the induced map from $(R_G)_{(\Delta_I)}$ to $A[T]_{(T)}$ is uniquely defined by H.

To show that f defines a proper map we must show that f defines a closed embedding from $\mathrm{Proj}(A[T])$ to $\mathrm{Proj}(R_G)$. To show this, it suffices to show that for every prime ideal $\mathfrak{p}$ of A, $\mathfrak{p}A[T]$ does not contain the image of f in $A[T]$. This condition is equivalent to the fact that no prime ideal contains all the r-by-r minors of N, which is true since the quotient is locally free. Thus, H defines a section σ_H of π_G. $\qquad\square$

10.2 The Canonical Locally Free Sheaves

One of the most important properties of the Grassmannian is that it has a canonical locally free sheaf of rank r and one of rank $n - r$ that are, respectively, a subsheaf and a quotient of a free sheaf of rank n. Essentially, the subsheaf is generated by the columns of the matrix M whose minors generate R_G. However, the entries of this matrix lie in $A[X_{ij}]$ but not in R_G, and, in addition, the quotient defines a sheaf over $\mathrm{Proj}(A[X_{ij}])$, but this sheaf is not locally free. Let T denote the submodule of the free $A[X_{ij}]$ module $A[X_{ij}]^n$ that is generated by the columns of M, and let T_G denote the elements of T with entries in R_G; thus, T_G is a submodule of the free R_G-module R_G^n. We define the *canonical sheaves* of R_G to be the sheaves corresponding to the graded modules T_G and R_G^n/T_G. We denote the first sheaf $\mathcal{E}$ and the second $\mathcal{Q}$, or, when r and n need to be specified, $\mathcal{E}(r, n)$ and $\mathcal{Q}(r, n)$. There is thus a short exact sequence

$$0 \to \mathcal{E} \to \mathcal{F} \to \mathcal{Q} \to 0$$

where $\mathcal{F}$ is free of rank n.

Proposition 10.2.1 *The sheaves $\mathcal{E}(r, n)$ and $\mathcal{Q}(r, n)$ are locally free of ranks r and $n - r$, respectively.*

Proof To prove this statement we construct a set of elements in T_G and show that the quotient modulo the submodule generated by these elements is locally free of rank $n - r$. Since T is generated by r elements, the quotient must have rank at least $n - r$, so this will prove that these elements generate T_G locally, that Q is locally free of rank $n - r$, and thus that $\mathcal{E}$ is locally free of rank r.

Let $e_1, \ldots, e_n$ be a basis for the free module R_G^n. We define one element of T_G for each subset of $\{1, \ldots, n\}$ consisting of $r - 1$ elements. If K is a set $k_1, \ldots, k_{r-1}$ of $r - 1$ elements of $\{1, \ldots, n\}$, we define the element

$$w_K = \Delta_{(1,k_1,\ldots,k_{r-1})}e_1 + \Delta_{(2,k_1,\ldots,k_{r-1})}e_2 + \cdots + \Delta_{(n,k_1,\ldots,k_{r-1})}e_n.$$

We first show that w_K is in T_G; since the coefficients are in R_G, this amounts to showing that w_K is a linear combination of the columns of the matrix M. Let V be the r-by-1 column vector whose entry in the ith position is $(-1)^{i+1}$ times the $r - 1$-by-$r - 1$ minor obtained by taking the rows in K and omitting the ith column. Then the n-by-1 matrix MV is a linear combination of the columns of M and hence is in T. But the ith entry of MV is Δ_I, where $I = (i, k_1, \ldots, k_{r-1})$, so the entries of MV are precisely the coefficients of the above relation, which is thus in T_G.

We next show that the quotient of R_G^n divided by the submodule generated by the w_K is locally free of rank $n - r$. Let I be a set of r elements in $\{1, \ldots, n\}$, and consider the open set $U_{(\Delta_I)}$. We claim that the sheaf is free of rank $n - r$ when restricted to $U_{(\Delta_I)}$; more precisely, we claim that it is free on the set of e_k for $k \notin I$. Since we know that the rank of the quotient is at least $n - r$, it suffices to show that if $i \in I$, then e_i can be written as a linear combination of e_k for $k \notin I$ with coefficients in $(R_G)_{(\Delta_I)}$. Let K be the set of $r - 1$ elements obtained by deleting i from I, and consider the corresponding element w_K. The coefficient of e_j in w_K for $j \in I - \{i\}$ is zero, that of e_i is $\pm\Delta_I$, and those of the e_k with $k \notin I$ are $\Delta_{I'}$ for various I'. Hence in the ring $(G_R)_{(\Delta_I)}$ we can solve for e_i in terms of those e_k for which $k \notin I$. $\qquad\square$

Since $\mathcal{E}$ and Q are locally free sheaves, we can define their Chern classes and Chern characters. If $s = n - r$, we thus have Chern classes $c_1(\mathcal{E}), \ldots c_r(\mathcal{E})$ and $c_1(Q), \ldots c_s(Q)$, which define operators on the Chow group of $G(r, n)$, and, since we have an exact sequence

$$0 \to \mathcal{E} \to \mathcal{F} \to Q \to 0$$

where $\mathcal{F}$ is defined by a free module, the associated total Chern classes are inverses of each other. These operators will be used in the next chapter to define the local Chern characters of a complex.

10.3 The Grassmannian of a Locally Free Sheaf

Up to this point, we have constructed the Grassmannian $G(r, n)$ for a free module F of rank n over a ring A. The Grassmannian is a projective scheme together with a projective map π_G to $\mathrm{Spec}(A)$ such that free submodules of rank r of F with locally free quotient define sections of π_G. We have also defined a canonical subsheaf and quotient sheaf of a free module of rank n on $G(r, n)$. In this section we generalize these constructions to rank r locally free subsheaves of a locally free sheaf of rank n over a projective scheme X.

Let $\mathcal{F}$ be a locally free sheaf of rank n on X. We define the Grassmannian of locally free subsheaves of rank r of $\mathcal{F}$ with locally free quotient as a subscheme of $\mathrm{Proj}(\mathrm{Sym}(\Lambda^r(\mathcal{F}^*)))$, where $\mathrm{Sym}(\Lambda^r(\mathcal{F}^*))$ is the symmetric algebra of the rth exterior power of the dual of $\mathcal{F}$. Before carrying out this construction we prove a Lemma on exterior powers of locally free sheaves.

Lemma 10.3.1 *Let $\mathcal{F}$ be a locally free sheaf on a projective scheme X. There is an isomorphism*

$$\Lambda^r(\mathcal{F}^*) \to (\Lambda^r(\mathcal{F}))^*.$$

Proof We construct a map Ψ from $\Lambda^r(\mathcal{F}^*)$ to $(\Lambda^r(\mathcal{F}))^*$ by letting

$$\Psi\big(f_1^* \wedge \cdots \wedge f_r^*\big)(f_1 \wedge \cdots \wedge f_r)$$
$$= \sum_\sigma (-1)^{\mathrm{sign}(\sigma)} f_1^*\big(f_{\sigma(1)}\big) \cdot f_2^*\big(f_{\sigma(2)}\big) \cdots f_r^*\big(f_{\sigma(r)}\big)$$

where the sum is over all elements σ in the symmetric group S_r. Since this map is alternating in $f_1^*, \ldots, f_r^*$ and in $f_1, \ldots, f_r$, it is well-defined. To prove that it is an isomorphism of sheaves we may consider the local situation and assume that F is a free module over a ring. Let $f_1, \ldots, f_n$ be a basis for F. Then the set of $f_{i_1} \wedge \cdots \wedge f_{i_r}$ for integers $i_1 < \cdots < i_r$ between 1 and n forms a basis for $\Lambda^r(F)$. Let $f_1^*, \ldots, f_n^*$ be the dual basis of F^*. Let $j_1 < \cdots < j_r$ be integers between 1 and n, and consider the element $f_{j_1}^* \wedge \cdots \wedge f_{j_r}^*$ of $\Lambda^r(F^*)$; these elements form a basis of $\Lambda^r(F^*)$. Then $\Psi(f_{j_1}^* \wedge \cdots \wedge f_{j_r}^*)(f_{i_1} \wedge \cdots \wedge f_{i_r})$ is one if $i_k = j_k$ for all k and 0 otherwise. Hence the elements $\Psi(f_{j_1}^* \wedge \cdots \wedge f_{j_r}^*)$ form a dual basis to the basis described above for $\Lambda^r(F)$, and the map Ψ is an isomorphism. $\square$

We now define the ideal $\mathfrak{a}$ of $\mathrm{Sym}(\Lambda^r(\mathcal{F}^*))$, which defines the Grassmannian. For each λ of the form $f_1 \wedge \cdots \wedge f_r$ in $\Lambda^r(F)$, define the map $\phi_\lambda : \mathrm{Sym}(\Lambda^r(\mathcal{F}^*)) \to \mathcal{O}_X[T]$ by letting

$$\phi_\lambda(\eta) = \Psi(\eta)(\lambda)T.$$

Thus, ϕ_λ defines a map in degree 1, which extends to a map from $\mathrm{Sym}(\Lambda^r(\mathcal{F}^*))$ to $\mathcal{O}_X[T]$. Let $\mathfrak{a}$ be the intersection of the kernels of the maps ϕ_λ for all λ in the pull-back of $\Lambda^r(F)$ to any extension of $\mathcal{O}_X$. We let $R_G(r, \mathcal{F})$ denote $\mathrm{Sym}(\Lambda^r(\mathcal{F}^*))/\mathfrak{a}$. Let $G_r(\mathcal{F})$ denote $\mathrm{Proj}(R_G(r, \mathcal{F}))$, and let π or $\pi_{\mathcal{F}}$ denote the map from $G_r(\mathcal{F})$ to X.

An important fact is that this construction for locally free sheaves agrees with the original one for free modules, which allows us to carry out computations locally. Let F be a free module of rank n, and let $f_1, \ldots, f_n$ be a basis for F. Then a basis for $\Lambda^r(F^*)$ consists of the elements $f_{i_1}^* \wedge \cdots \wedge f_{i_r}^*$, where $i_1 < \cdots < i_r$ and where $f_1^*, \ldots, f_n^*$ is the dual basis of $f_1, \ldots, f_n$. We may define a map ψ from the symmetric algebra on $\Lambda^r(F^*)$ to the ring $R_G(r, n)$ defined in section one by sending $f_{i_1}^* \wedge \cdots \wedge f_{i_r}^*$ to Δ_I, where $I = (i_1, \ldots, i_r)$.

Proposition 10.3.2 *The map ψ as defined above induces an isomorphism between $R_G(r, F)$ and $R_G(r, n)$.*

Proof We must prove that the kernel of ψ is exactly $\mathfrak{a}$. To show this we examine the map ψ in detail.

To simplify notation, denote $f_{i_1}^* \wedge \cdots \wedge f_{i_r}^*$ by P_I where $I = (i_1, \ldots, i_r)$. Then $\mathrm{Sym}(\Lambda^r(F^*))$ is the polynomial ring over A in the variables P_I, and for each I we have $\psi(P_I) = \Delta_I$. Let $\lambda = e_{j_1} \wedge \cdots \wedge e_{j_r}$ be an element of $\Lambda^r(F)$. The elements e_i of F can be represented as the columns of an n-by-r matrix M with respect to the basis $f_1, \ldots, f_n$. For each I, let δ_I denote the minor of M of the submatrix with rows indexed by I. Then for each $I = (i_1, \ldots, i_r)$, $\phi_\lambda(P_I)$ is the coefficient of $f_{i_1} \wedge \cdots \wedge f_{i_r}$ in the expression of $e_{i_1} \wedge \cdots \wedge e_{i_r}$ in terms of the basis, and this coefficient is δ_I. Thus, if $g(P_I)$ is any polynomial with coefficients in A, we have $\phi_\lambda(g(P_I)) = g(\delta_I)$. Hence $g(P_I)$ is in $\mathfrak{a}$ if and only if $g(\delta_I) = 0$ for all n-by-r matrices M with coefficients in any extension of A.

In particular, we must have $g(\Delta_I) = 0$, which means that $\psi(\mathfrak{a}) = 0$. Conversely, if $f(\Delta_I) = 0$, since the X_{ij} are indeterminates, we can map X_{ij} to a_{ij} for any n-by-r matrix $M = (a_{ij})$ and conclude that $f(\delta_I) = 0$ where the δ_I are the minors of M, so that $f \in \mathfrak{a}$. $\square$

We next define the locally free sheaves in the general case. Let $\mathcal{E}_{\mathcal{F}}$ denote the submodule of $\pi^*(\mathcal{F})$ generated by all elements of the following form: let $f_1^*, \ldots, f_{r-1}^*$ be a set of elements of $\mathcal{F}^*$, let $e_1, \ldots, e_n$ be a set of n elements of $\mathcal{F}$, and let $e_1^*, \ldots, e_n^*$ be n elements of $\mathcal{F}^*$. Let $r_{ij} = e_i^*(e_j)$ for each i, j. Finally, let (μ_{ij}) be the adjoint of the matrix (r_{ij}). Then $\mathcal{E}_{\mathcal{F}}$ is generated by all elements of the form

$$\sum_{i,j} \mu_{ji} \left(e_i^* \wedge f_1^* \wedge \cdots \wedge f_{r-1}^* \right) \otimes e_j. \qquad (*)$$

Let $\mathcal{Q}_{\mathcal{F}}$ denote the quotient, so that we have a short exact sequence

$$0 \to \mathcal{E}_{\mathcal{F}} \to \pi^*(\mathcal{F}) \to \mathcal{Q}_{\mathcal{F}} \to 0.$$

We show that locally $\mathcal{E}_{\mathcal{F}}$ and $\mathcal{Q}_{\mathcal{F}}$ are the same locally free sheaves that we defined before. Let F be a free module, let $e_1, \ldots, e_n$ be a basis for F, and let $e_1^*, \ldots, e_n^*$ be the dual basis of F^*. For this particular choice of e_i and e_j^*, the matrix $(e_i^*(e_j))$ is the identity matrix, so its adjoint is also the identity matrix. We also choose $f_1^*, \ldots, f_{r-1}^*$ to be a set $e_{k_1}^*, \ldots, e_{k_{r-1}}^*$ chosen from the basis elements. Then the corresponding generator of $\mathcal{E}$ is

$$\sum_i e_i^* \wedge e_{k_1}^* \wedge \cdots \wedge e_{k_{r-1}}^* \otimes e_i \qquad (**)$$

which is mapped to

$$\sum \left(\Delta_{i,k_1,\ldots,k_{r-1}} \right) e_i$$

by the map ψ. Thus, we have the generators defined above in the local case. We now show that every element of the form $(*)$ is in the submodule generated by elements of the type $(**)$. Since the choice of $f_1^*, \ldots, f_{r-1}^*$ is independent of the other choices, it follows immediately from linearity that we can assume that $f_1^*, \ldots, f_{r-1}^*$ are equal to a set $e_{k_1}^*, \ldots, e_{k_{r-1}}^*$ of basis elements. To simplify the notation we let $e_{k_1}^* \wedge \cdots \wedge e_{k_{r-1}}^* = \mathbf{e}^*$. Let $g_1^*, \ldots, g_n^*$ and $g_1, \ldots, g_n$ be chosen arbitrarily. We can find α_{ij} and β_{ij} such that $g_i^* = \sum \alpha_{ik} e_k^*$ and $g_j = \sum \beta_{mj} e_m$ for all i and j. Let A and B denote the matrices (α_{ij}) and (β_{ij}), respectively. We then have

$$g_i^*(g_j) = \left(\sum_k \alpha_{ik} e_k^* \right) \left(\sum_m \beta_{mj} e_m \right) = \sum_k \alpha_{ik} \beta_{kj}.$$

Hence, using the above notation, we have $(r_{ij}) = AB$. Thus, $(\mu_{ij}) = \mathrm{adj}(AB) = \mathrm{adj}(B)\mathrm{adj}(A)$. Denote $\mathrm{adj}(B)$ by (ϵ_{ij}) and $\mathrm{adj}(A)$ by (γ_{ij}). Then, using the Kronecker delta notation and denoting the determinants of A and B by $|A|$ and $|B|$, we have

$$\sum_{i,j} \mu_{ji} g_i^* \wedge \mathbf{e}^* \otimes g_j = \sum_{i,j} \left(\sum_k \epsilon_{jk} \gamma_{ki} \right) \left(\sum_r \alpha_{ir} e_r^* \wedge \mathbf{e}^* \right) \otimes \left(\sum_s \beta_{sj} e_s \right)$$

$$= \sum_{i,j,k,r,s} \epsilon_{jk} \gamma_{ki} \alpha_{ir} \beta_{sj} \left(e_r^* \wedge \mathbf{e}^* \right) \otimes e_s$$

$$= \sum_{k,r,s} \left(\sum_i \gamma_{ki}\alpha_{ir} \right) \left(\sum_j \epsilon_{jk}\beta_{sj} \right) \left(e_r^* \wedge \mathbf{e}^* \right) \otimes e_s$$

$$= |A||B| \left(\sum_{k,r,s} \delta_{kr}\delta_{ks}(e_r \wedge \mathbf{e}^*) \otimes e_s \right)$$

$$= |A||B| \left(\sum_k \left(e_k^* \wedge \mathbf{e}^* \right) \otimes e_k \right).$$

Hence all of the generators can be expressed in terms of the generators defined in the previous section, so the two definitions agree. In particular, this shows that $\mathcal{E}$ is locally free of rank r and $\mathcal{Q}$ is locally free of rank $n - r$. We state this fact in the following theorem.

Theorem 10.3.3

(i) *The sheaves $\mathcal{E}_{\mathcal{F}}$ and $\mathcal{Q}_{\mathcal{F}}$ are locally free of ranks r and $n - r$, respectively.*

(ii) *If $\mathcal{H}$ is a locally free subsheaf of $\mathcal{F}$ of rank r with locally free quotient, there is a canonically defined section $\sigma_{\mathcal{H}}$ of $\pi_{\mathcal{F}}$ locally defined by the determinants of a matrix defining $\mathcal{H}$ as in Proposition 10.1.2. We have $\sigma_{\mathcal{H}}^*(\mathcal{E}_{\mathcal{F}}) \cong \mathcal{H}$.*

Proof The first statement follows from the local computations above.

To prove the second statement, let $\mathcal{H}$ be a locally free subsheaf of $\mathcal{F}$ of rank r. The map from $\mathcal{H}$ into $\mathcal{F}$ induces a map from $\mathcal{F}^*$ to $\mathcal{H}^*$ and hence a map from $\Lambda^r(\mathcal{F}^*)$ to $\Lambda^r(\mathcal{H}^*)$. Thus, we have a map from the symmetric algebra on $\Lambda^r(\mathcal{F}^*)$ to the symmetric algebra on $\Lambda^r(\mathcal{H}^*)$. Since $\Lambda^r(\mathcal{H}^*)$ is locally free of rank 1, the map from $\mathrm{Proj}(\mathrm{Sym}(\Lambda^r(\mathcal{H}^*)))$ to X is an isomorphism, so this map defines a section of $\pi_{\mathcal{F}}$. The fact that it is defined on all of X follows from the local computation of Proposition 10.1.2 together with the fact that this is the same as the map defined in that proposition, which we verify in the next paragraph.

Let F be a free module and let H be a submodule that is free of rank r with locally free quotient. Choose a basis $f_1, \ldots, f_n$ for F with dual basis $f_1^*, \ldots, f_n^*$. If $e_1, \ldots, e_r$ is a basis for H, the map from $\Lambda^r(\mathcal{F}^*)$ to $\Lambda^r(\mathcal{H}^*)$, expressed in terms of coordinates, takes the basis element $f_{i_1}^* \wedge \cdots \wedge f_{i_r}^*$ to the coefficient of $f_{i_1} \wedge \cdots \wedge f_{i_r}$ in the expression for $e_1 \wedge \cdots \wedge e_r$. If $e_1, \ldots, e_r$ are defined by the columns of an n-by-r matrix M with respect to the basis $f_1, \ldots, f_r$, this coefficient is the determinant of the minor of the submatrix of M consisting of the rows indexed by $i_1, \ldots, i_r$. Hence this is the map defined in Proposition 10.1.2. As before, we denote the section defined by $\mathcal{H}$ by the notation $\sigma_\mathcal{H}$.

To prove the last statement, we may again consider the local situation, and we describe the pull-back of the canonical subsheaf $\mathcal{E}$ explicitly. We assume that $X = \text{Spec}(A)$ for a local ring A. Assume that the section σ_H is defined by a free submodule H of F as in the previous paragraph, and that H is generated by the columns of a matrix M. If I is a subset of $\{1, \dots, n\}$, let f denote the map from $(R_G)_{\Delta_I}$ to A_{δ_I}, where δ_I is the minor of M taken from the rows indexed by I. Then $f(\Delta_J/\Delta_I) = \delta_J/\delta_I$. As shown in the proof of Proposition 10.2.1, the canonical sheaf $\mathcal{E}$ is generated locally by the elements

$$\sum \left(\Delta_{(i,i_1,\dots,i_{r-1})}/\Delta_I \right) e_i$$

where $i_1, \dots, i_{r-1}$ are $r - 1$ elements of I. The image of $\mathcal{E}$ under the map f is thus generated by

$$\sum \left(\delta_{(i,i_1,\dots,i_{r-1})}/\delta_I \right) e_i.$$

Since H is free of rank r and generated by the columns of M, these elements generate H locally. Thus, $\sigma_\mathcal{H}^*(\mathcal{E}) = \mathcal{H}$. $\qquad\square$

10.4 Functorial Properties of the Grassmannian

In this section we prove two of the basic functorial properties of the Grassmannian. The first of these is concerned with the pull-back of a sheaf by a map of projective schemes, and the second deals with direct sums. If $\mathcal{F}$ and $\mathcal{F}'$ are locally free sheaves of ranks n and n', respectively, and if $\mathcal{H}$ and $\mathcal{H}'$ are locally free subsheaves of ranks r and r', then $\mathcal{H} \oplus \mathcal{H}'$ is a locally free subsheaf of rank $r + r'$. We show that there is a map defined on Grassmannians from $G_r(\mathcal{F}) \times G_{r'}(\mathcal{F}')$ to $G_{r+r'}(\mathcal{F} \oplus \mathcal{F}')$ that represents this construction. Similarly, the map that sends a subsheaf to its pull-back by a map defines a map on Grassmannians over the corresponding projective schemes.

Let $\phi : X \to Y$ be a map of projective schemes, and let $\mathcal{F}$ be a locally free sheaf of rank n on Y. Then the mapping that sends $\mathcal{H}$ to $\phi^*(\mathcal{H})$ defines a map from locally free subsheaves of rank r on Y to locally free subsheaves of rank r on X.

Proposition 10.4.1 *There is a natural isomorphism ψ from $G_r(\phi^*\mathcal{F})$ to $G_r(\mathcal{F})$ $\times_Y X$ such that*

(i) *For any rank r locally free subsheaf $\mathcal{H}$ of $\mathcal{F}$ on Y we have*

$$\psi \sigma_{\psi^*(\mathcal{H})} = \sigma_\mathcal{H} \times X.$$

(ii) *Let p be the projection from $G_r(\phi^*(\mathcal{F}))$ to $G_r(\mathcal{F})$. Then $p^*(\mathcal{E}_\mathcal{F}) = \mathcal{E}_{\phi^*(\mathcal{F})}$ and $p^*(\mathcal{Q}_\mathcal{F}) = \mathcal{Q}_{\phi^*(\mathcal{F})}$.*

Proof Let $X = \mathrm{Proj}(B)$ and $Y = \mathrm{Proj}(A)$, and let ϕ be defined by a map of rings that we denote f. Let F be a module such that the associated sheaf is $\mathcal{F}$. Then $\phi^*(\mathcal{F})$ is defined by $B \otimes_A F$ and, since $\mathcal{F}$ is locally free, the natural map from $B \otimes (\Lambda^r(F^*))$ to $\Lambda^r(B \otimes F^*))$ defines an isomorphism of sheaves $h : \phi^*(\Lambda^r(\mathcal{F}^*)) \to (\Lambda^r(\phi^*(\mathcal{F})))^*$. This isomorphism in turn induces an isomorphism on symmetric algebras and, again using the local computation, we can deduce that we have an isomorphism between $B \otimes_A R_G(r, \mathcal{F})$ and $R_G(r, \phi^*(\mathcal{F}))$. This isomorphism induces the required isomorphism ψ from $G_r(\phi^*\mathcal{F})$ to $G_r(\mathcal{F}) \times_Y X$.

The fact that ψ has the required commutativity with sections follows from the commutativity of the following diagram:

$$
\begin{array}{ccc}
\phi^*(\Lambda^r(\mathcal{F}^*)) & \cong & \Lambda^r((\phi^*(\mathcal{F}))^*) \\
\downarrow & & \downarrow \\
\phi^*(\Lambda^r(\mathcal{H}^*)) & \cong & \Lambda^r((\phi^*(\mathcal{H}))^*)
\end{array} .
$$

together with the fact that the sections are induced by the vertical maps in this diagram.

To see the last statement, we note first that a generator of $\mathcal{E}_\mathcal{F}$ of the form described in the definition is mapped by the tensor product to a generator of the same form for $\mathcal{E}_{\phi^*(\mathcal{F})}$, so that $\phi^*(\mathcal{E}_\mathcal{F}) \subseteq \mathcal{E}_{\phi^*(\mathcal{F})}$. On the other hand, both sheaves are locally free subsheaves of $\phi^*(\mathcal{F})$ of rank r with locally free quotient, so they must be equal. Hence the quotients $\phi^*(\mathcal{Q}_\mathcal{F})$ and $\mathcal{Q}_{\phi^*(\mathcal{F})}$ are also equal. $\qquad\square$

We next define a map from $G_r(\mathcal{F}) \times G_{r'}(\mathcal{F}')$ to $G_{r+r'}(\mathcal{F} \oplus \mathcal{F}')$ defined by taking direct sums. This formula will be combined in the next chapter with the additivity formulas for Chern classes to derive additivity formulas for local Chern characters of short exact sequences of complexes.

Let $X = \mathrm{Proj}(A)$ be a projective scheme, and let $\mathcal{F}$ and $\mathcal{F}'$ be locally free sheaves on X of ranks n and n', respectively. For integers r and r' with $1 \le r \le n$ and $1 \le r' \le n'$ we have the product $G_r(\mathcal{F}) \times G_{r'}(\mathcal{F}') = \mathrm{Proj}(R_G(r, \mathcal{F}) \otimes R_G(r', \mathcal{F}'))$. We note first that if $\mathcal{H}$ and $\mathcal{H}'$ are locally free submodules of the locally free sheaves $\mathcal{F}$ and $\mathcal{F}'$ of ranks r and r, respectively, there are sections $\sigma_\mathcal{H}$ of $\pi_{G_r(\mathcal{F})}$ and $\sigma_{\mathcal{H}'}$ of $\pi_{G_{r'}(\mathcal{F}')}$. The product of these two maps over X defines a section of $G_r(\mathcal{F}) \times G_{r'}(\mathcal{F}')$, which we denote $\sigma_\mathcal{H} \times \sigma_{\mathcal{H}'}$. We will show that under the direct sum map this section will correspond to the section $\sigma_{\mathcal{H} \oplus \mathcal{H}'}$ defined by the direct sum.

The map defining the direct sum can be defined most easily by defining a dual map. We have a map

$$\rho : \Lambda^r(\mathcal{F}) \otimes \Lambda^{r'}(\mathcal{F}') \to \Lambda^{r+r'}(\mathcal{F} \oplus \mathcal{F}')$$

that sends $(f_1 \wedge \cdots \wedge f_r) \otimes (f'_1 \wedge \cdots \wedge f'_{r'})$ to $(f_1, 0) \wedge \cdots \wedge (f_r, 0) \wedge (0, f'_1)$ $\wedge \cdots \wedge (0, f'_{r'})$. It is clear that ρ is alternating in the f_i and the f'_i, so it is well defined. Thus, ρ defines a dual map ρ^* from $(\Lambda^{r+r'}(\mathcal{F} \oplus \mathcal{F}'))^*$ to $(\Lambda^r(\mathcal{F}) \otimes \Lambda^{r'}(\mathcal{F}'))^*$. Using Lemma 10.3.1 and the fact that taking duals commutes with tensor product, we thus have a map from $\Lambda^{r+r'}((\mathcal{F} \oplus \mathcal{F}')^*)$ to $\Lambda^r(\mathcal{F}^*) \otimes \Lambda^{r'}(\mathcal{F}'^*)$, which includes a map on symmetric algebras from $\mathrm{Sym}((\Lambda^{r+r'}(\mathcal{F} \oplus \mathcal{F}'))^*)$ to $\mathrm{Sym}(\Lambda^r(\mathcal{F}^*)) \otimes \mathrm{Sym}(\Lambda^{r'}(\mathcal{F}'^*))$ in which $\Lambda^{r+r'}(\mathcal{F} \oplus \mathcal{F}'))^*$ maps to the subalgebra of $\mathrm{Sym}(\Lambda^r(\mathcal{F}^*)) \otimes \mathrm{Sym}(\Lambda^{r'}(\mathcal{F}'^*))$ generated by $\Lambda^r(\mathcal{F}^*) \otimes \Lambda^{r'}(\mathcal{F}'^*)$. Thus, we have a map from $G_r(\mathcal{F}) \times G_{r'}(\mathcal{F}')$ to $G_{r+r'}(\mathcal{F} \oplus \mathcal{F}')$, which we denote $\psi_\oplus$.

Proposition 10.4.2 *The map $\psi_\oplus$ is a closed embedding from $G_r(\mathcal{F}) \times G_{r'}(\mathcal{F}')$ to $G_{r+r'}(\mathcal{F} \oplus \mathcal{F}')$.*

Proof Let f denote the map on symmetric algebras that induces $\psi_\oplus$. By definition, f sends an element of degree i to an element of degree (i, i). (We consider here only the degrees of grading of the symmetric algebra, the degrees of A, where the underlying projective scheme X is $\mathrm{Proj}(A)$, are preserved by this map.) Thus, the map on indices η is defined by $\eta(i) = (i, i)$, and we have a quasi-projective map.

To see that it is a closed embedding, we show that locally the map is surjective. First take an open affine subset of X on which $\mathcal{F}$ and $\mathcal{F}'$ are defined by free modules F and F'. Let $f_1, \ldots, f_n$ and $f'_1, \ldots, f'_{n'}$ be bases of F and F', and let f_i^* and $f_i'^*$ denote the associated dual bases. Then the set $\{(f_1^*, 0), \ldots, (f_n^*, 0), (0, f_1'^*), \ldots, (0, f_{n'}'^*)\}$ is a basis for $(F \oplus F')^*$, and their wedge powers thus form a basis for $\Lambda^{r+r'}((F \oplus F')^*)$. The image under ρ^* of a basis element in $\Lambda^{r+r'}((F \oplus F')^*)$ will be zero unless there are r factors of the form $(f_i^*, 0)$ and r' of the form $(0, f_i'^*)$, and the element

$$\left(f_{i_1}^*, 0\right) \wedge \cdots \wedge \left(f_{i_r}^*, 0\right) \wedge \left(0, f_{j_1}'^*\right) \wedge \cdots \wedge \left(0, f_{j_{r'}}'^*\right)$$

is sent to

$$f_{i_1}^* \wedge \cdots \wedge f_{i_r}^* \otimes f_{j_1}'^* \wedge \cdots \wedge f_{j_{r'}}'^*.$$

In terms of the Δ_I, the map is thus defined by sending Δ_I, where I is a subset of $r + r'$ elements of $\{1, \ldots, n + n'\}$, to zero unless $I = J \cup K$ with J consisting of r elements of $\{1, \ldots, n\}$ and K consisting of r' elements of

$\{n + 1, \ldots, n + n'\}$, in which case it is sent to $\Delta_J \otimes \Delta_K$ (where we identify a subset of $\{n + 1, \ldots, n + n'\}$ with the corresponding subset of $\{1, \ldots, n'\}$). Since the ring of the product $G_r(\mathcal{F}) \times G_{r'}(\mathcal{F}')$ is generated locally by elements of the form $(\Delta_J/\Delta_{J'}) \otimes (\Delta_K/\Delta_{K'})$, which is the image of $\Delta_{J \cup K}/\Delta_{J' \cup K'}$, it is thus clear that the map defining $\psi_\oplus$ is locally surjective and defines a closed embedding. $\qquad\square$

Proposition 10.4.3 *Let $\mathcal{F}$ and $\mathcal{F}'$ be locally free sheaves of ranks n and n', respectively, and let $\mathcal{H} \subseteq \mathcal{F}$ and $\mathcal{H}' \subseteq \mathcal{F}'$ be locally free subsheaves with locally free quotients of ranks r and r'. We then have*

$$\sigma_{\mathcal{H} \oplus \mathcal{H}'} = \psi_\oplus(\sigma_\mathcal{H} \times \sigma_{\mathcal{H}'}).$$

Proof The sections σ are defined by the maps from the exterior powers of $\mathcal{F}$, $\mathcal{F}'$, and $\mathcal{F} \oplus \mathcal{F}'$ to the exterior powers of $\mathcal{H}, \mathcal{H}'$, and $\mathcal{H} \oplus \mathcal{H}'$. Furthermore, the product of $\sigma_\mathcal{H}$ and $\sigma_{\mathcal{H}'}$ is then defined by the induced map on the tensor product of symmetric algebras. Hence the result follows from the commutativity of the diagram

$$
\begin{array}{ccc}
\mathrm{Sym}(\Lambda^{r+r'}(\mathcal{F} \oplus \mathcal{F}')^*) & \longrightarrow & \mathrm{Sym}(\Lambda^r(\mathcal{F}^*)) \otimes \mathrm{Sym}(\Lambda^{r'}(\mathcal{F}')^*) \\
\downarrow & & \downarrow \\
\mathrm{Sym}(\Lambda^{r+r'}(\mathcal{H} \oplus \mathcal{H}')^*) & \longrightarrow & \mathrm{Sym}(\Lambda^r(\mathcal{H}^*)) \otimes \mathrm{Sym}(\Lambda^{r'}(\mathcal{H}')^*)
\end{array}.
$$

$\qquad\square$

To complete the description of the direct sum map we have to show the relation between the canonical sheaves on $G_r(\mathcal{F}) \times G_{r'}(\mathcal{F}')$ and the canonical sheaves on $G_{r+r'}(\mathcal{F} \oplus \mathcal{F}')$.

Proposition 10.4.4 *Let $\mathcal{E}, \mathcal{E}'$, and $\mathcal{E}''$ be the canonical subsheaves and let $\mathcal{Q}, \mathcal{Q}'$, and $\mathcal{Q}''$ be the canonical quotient sheaves of $G_r(\mathcal{F}), G_{r'}(\mathcal{F}')$, and $G_{r+r'}(\mathcal{F} \oplus \mathcal{F}')$, respectively. Let p_1 and p_2 be the projections from $G_r(\mathcal{F}) \times G_{r'}(\mathcal{F}')$ onto the factors. Then*

$$\psi_\oplus^*(\mathcal{E}'') = p_1^*(\mathcal{E}) \oplus p_2^*(\mathcal{E}')$$

and

$$\psi_\oplus^*(\mathcal{Q}'') = p_1^*(\mathcal{Q}) \oplus p_2^*(\mathcal{Q}').$$

Proof It suffices to prove the first statement, since the second then follows from the short exact sequence used to define $\mathcal{Q}_\mathcal{F}$. To check that the sheaves $\psi_\oplus^*(\mathcal{E}'')$ and $p_1^*(\mathcal{E}) \oplus p_2^*(\mathcal{E}')$ are the same, it suffices to show that they are the same locally. Thus, we may assume that $\mathcal{F}$ and $\mathcal{F}'$ are defined by free modules F and F', and we may restrict our attention to generators of $\mathcal{E}_{F \oplus F'}$ defined by a basis of $F \oplus F'$ and a corresponding dual basis [generators of type (∗∗) in the definition]. Let $f_1, \ldots, f_n$ and $f'_1, \ldots, f'_{n'}$ be bases of F and F' with corresponding dual bases $f_1^*, \ldots, f_n^*$ and $f_1'^*, \ldots, f_{n'}'^*$. We consider generators of $\mathcal{E}_{F \oplus F'}$ that are not mapped to zero by the map induced by ρ^*, where ρ^* is the map from $\Lambda^{r+r'}(F^* \oplus F'^*)$ to $\Lambda^r(F^*) \otimes \Lambda^{r'}(F'^*)$ used in the definition of $\psi_\oplus$. These elements will be either of the form

$$\sum_{i=1}^{n} \left(f_i^*, 0 \right) \wedge \left(f_{j_1}^*, 0 \right) \wedge \cdots \wedge \left(f_{j_{r-1}}^*, 0 \right) \wedge \left(0, f_{k_1}'^* \right)$$

$$\wedge \cdots \wedge \left(0, f_{k_{r'}}'^* \right) \otimes (f_i, 0) + \sum_{i=1}^{n'} \left(0, f_i'^* \right) \wedge \left(f_{j_1}^*, 0 \right)$$

$$\wedge \cdots \wedge \left(f_{j_{r-1}}^*, 0 \right) \wedge \left(0, f_{k_1}'^* \right) \wedge \cdots \wedge \left(0, f_{k_{r'}}'^* \right) \otimes \left(0, f_i' \right)$$

or of a similar form in which r elements are chosen from F^* and $r' - 1$ are chosen from F'^*. We consider the image under ρ^* of the above element. The second term is mapped to zero. The first term is mapped to

$$\sum_{i=1}^{n} f_i^* \wedge f_{j_1}^* \wedge \cdots \wedge f_{j_{r-1}}^* \otimes f_{k_1}'^* \wedge \cdots \wedge f_{k_{r'}}'^* \otimes (f_i, 0).$$

Since $f_{k_1}'^* \wedge \cdots \wedge f_{k_{r'}}'^*$ is a general element of $\Lambda^{r'}((F')^*)$, the above elements form a set of generators for $p_1^*(\mathcal{E}_F)$. A similar computation shows that the images of generators of $\mathcal{E}_{F \oplus F'}$ of the second form generate $p_2^*(\mathcal{E}_{F'})$. Hence we have

$$\phi_\oplus^*(\mathcal{E}'') = p_1^*(\mathcal{E}) \oplus p_2^*(\mathcal{E}')$$

as was to be shown.   □

10.5 Computation of Chern Classes via Free Resolutions

In this section we return to the situation of Section 1, in which we have a free submodule of a free module, and we describe a method for computing the Chern classes of the associated canonical quotient sheaf. Let r be an integer with $1 \le r \le n$, and let M be an n-by-r matrix whose entries are indeterminates. Let $R_G = R_G(r, n)$ be the ring generated by the maximal minors Δ_I of M as in Section 1. We show that the Chern classes of the canonical sheaf $\mathcal{Q}$ of the Grassmannian $G_A(r, n)$ can be computed by means of a resolution of the ideal of R_G defined by minors of certain submatrices of M. In particular, they can be computed directly from R_G without pulling back cycles to the scheme $P(\mathcal{Q})$ as is necessary in the general definition of Chern classes.

To prove this result we need some facts about ideals generated by determinants of a generic matrix. Let A be a ring, and let m and n be integers with $m \geq n$. Let $A[X_{ij}]$ be a polynomial ring over A in mn variables, and let (X_{ij}) be an m-by-n matrix with entries X_{ij}. Let $\mathfrak{a}(m, n)$ denote the ideal generated by the maximal (n-by-n) minors of (X_{ij}). Then:

(i) If A is an integral domain, then $\mathfrak{a}(m, n)$ is a prime ideal.

(ii) $\mathfrak{a}(m, n)$ is a perfect ideal of height $m - n + 1$. In particular, if A is Cohen-Macaulay, then $A[X_{ij}]/\mathfrak{a}(m, n)$ is Cohen-Macaulay.

Proofs of these facts can be found in Bruns and Vetter [8, 2.B].

We will need one special result about generic determinantal ideals.

Proposition 10.5.1 *Let A be an integral domain, and let X_{ij} and $\mathfrak{a}(m, n)$ be as above. Let $B = A/\mathfrak{a}(m, n)$. Let $T_1, \ldots, T_m$ be indeterminates, and let r_i be the image of $X_{1i}T_1 + \cdots + X_{mi}T_m$ in $B[T_1, \ldots, T_m]$ for $i = 1, \ldots, n$. Then*

(i) *The elements $r_1, \ldots, r_{n-1}$ form a regular sequence.*

(ii) *The relations between $r_1, \ldots, r_n$ are generated by the relations $s_{ij} = (0, \ldots, r_j, 0, \ldots, -r_i, \ldots, 0)$ and the elements*

$$t_{I,j} = (\Delta_{I,1}, -\Delta_{I,2}, \Delta_{I,3}, \ldots, (-1)^{n+1}\Delta_{I,n}),$$

where $\Delta_{I,j}$ denotes the determinant of the $n - 1$-by-$n - 1$ submatrix of (X_{ij}) consisting of the rows indexed by I and all columns except the jth column.

Proof It is clear that the elements s_{ij} are relations between $r_1, \ldots, r_n$. The fact that the $t_{I,j}$ are relations comes from the cofactor expansion of the n-by-n minors of (X_{ij}) together with the fact that we are taking the quotient modulo $\mathfrak{a}(m, n)$.

We prove the converse by induction on n. If $n = 1$, there are no $n - 1$-by-$n - 1$ submatrices, and the X_{i1} are themselves in $\mathfrak{a}(m, n)$, so $r_1 = 0$ and the result should be interpreted to say that the annihilator of r_1 is generated by 1. Assume that $n > 1$. Then $X_{ij} \notin \mathfrak{a}(m, n)$ for all i and j, and since $\mathfrak{a}(m, n)$ is a prime ideal, the r_i are not zero-divisors. We consider the localization obtained by inverting X_{11}. In the quotient $B[T_1, \ldots, T_m]_{X_{11}}/(r_1)$ we can solve for T_1 and express the remaining $r_2, \ldots, r_n$ as polynomials in $T_2, \ldots, T_m$. Let $\bar{r}_i$ denote the image of r_i in $B[T_2, \ldots, T_m]_{X_{11}}/(r_1)$ expressed as a linear polynomial in $T_2, \ldots, T_m$. Let N be the $m - 1$-by-$n - 1$ matrix of the coefficients of $\bar{r}_2, \ldots, \bar{r}_n$. Then the entries in N are still algebraically independent over A. Furthermore, it is an exercise in linear algebra to show that a t-by-t minor of

N is, up to a power of X_{11}, equal to the $t + 1$-by-$t + 1$ minor of M obtained by adding the first row and column of M. By induction, we can conclude that $\bar{r}_2, \ldots, \bar{r}_{n-1}$ form a regular sequence and that the relations between $\bar{r}_2, \ldots, \bar{r}_n$ are generated by the images of the s_{ij} in $B[T_2, \ldots, T_n]_{X_{11}}/(r_1)$ together with the elements

$$\bar{t}_{\tilde{I},j} = (-\Delta_{\tilde{I},2}, \Delta_{\tilde{I},3}, \ldots, (-1)^{n+1}\Delta_{\tilde{I},n}),$$

where $\Delta_{\tilde{I},j}$ denotes the determinant of the $n - 2$-by-$n - 2$ submatrix of N consisting of the rows in $\tilde{I}$ and all columns except the jth column (where we number the rows of N by 2 through m). If we let I be the set of rows of (X_{ij}) obtained by adding the first row to $\tilde{I}$, this element is the image of

$$(\Delta_{I,1}, -\Delta_{I,2}, \Delta_{I,3}, \ldots, (-1)^{n+1}\Delta_{I,n}).$$

Hence we have shown that the relations given in the statement of the proposition generate the module of relations between the r_i after localization by X_{11}. Similarly, they form a set of generators after localization by X_{ij} for any i and j.

To complete the proof, we use an argument on depth. Suppose that $r_1, \ldots, r_{n-1}$ did not form a regular sequence. Then the annihilator of r_j modulo $r_1, \ldots, r_{j-1}$ would be strictly larger than $(r_1, \ldots, r_{j-1})$ for some j; let j be the lowest such integer. Let $\mathfrak{c}$ be the image of $((r_1, \ldots, r_{j-1}) : r_j)$ in $B[T_1, \ldots, T_m]/(r_1, \ldots, r_{j-1})$. Our assumptions imply that $\mathfrak{c} \neq 0$. The above argument shows that $\mathfrak{c}$ is annihilated by a power of X_{ij} for all i and j. Let $\mathfrak{p}$ be an associated prime ideal of the submodule $\mathfrak{c}$ of $B[T_1, \ldots, T_m]/(r_1, \ldots, r_{j-1})$. Then $\mathfrak{p}$ must contain all the X_{ij}. Since the X_{ij} form a regular sequence over $A[X_{ij}]$ and $\mathfrak{a}(m, n)$ is a perfect ideal of height $m - n + 1$, the depth of $B[X_{ij}]_{\mathfrak{p}}$ is at least $mn - (m - n + 1)$. By the minimality of j, $r_1, \ldots, r_{j-1}$ form a regular sequence, and $j - 1 < n - 1$, so the depth of $(B[T_1, \ldots, T_m]/(r_1, \ldots, r_{j-1}))_{\mathfrak{p}}$ is greater than $mn - (m - n + 1) - (n - 1) = m(n - 1) > 0$. However, we have chosen $\mathfrak{p}$ so that the depth of $(B[T_1, \ldots, T_m]/(r_1, \ldots, r_{j-1}))_{\mathfrak{p}}$ is zero, so this is a contradiction. Thus, $r_1, \ldots, r_{n-1}$ form a regular sequence.

To show that the relations between $r_1, \ldots, r_n$ are as stated, it suffices to show that the annihilator of r_n in $B[T_1, \ldots, T_m]/(r_1, \ldots, r_{n-1})$ is generated by the s_{in} and the $n - 1$-by-$n - 1$ minors of the matrix formed by the last $n - 1$ columns of the matrix (X_{ij}). Let $\mathfrak{b}$ denote the ideal generated by these minors. Arguing as in the previous paragraph, it suffices to show that for all prime ideals $\mathfrak{p}$ containing the X_{ij}, the depth of $(B/\mathfrak{b}[T_1, \ldots, T_m]/(r_1, \ldots, r_{n-1}))_{\mathfrak{p}}$ is greater than zero. Since $\mathfrak{b}$ is a perfect ideal of height $m - n + 2$ and $r_1, \ldots, r_{n-1}$ form a regular sequence, this depth is at least $mn - (m - n + 2) - (n - 1) = mn - m - 1$. If $mn - m - 1 = 0$, then $m(n - 1) = 1$, so $m = 1$ and $n = 2$. Since we are assuming that $m \geq n$, this is impossible. Thus, the annihilator of r_n is as described, so this completes the proof. ☐

The next proposition follows from a similar argument.

Proposition 10.5.2 *Let A be an integral domain, and let X_{ij} and $\mathfrak{a}(m, n)$ be as above. Let $k \geq 1$ be an integer, and let Y_{ij}, $i = 1, \ldots, k$ and $j = 1, \ldots, n$ be another set of indeterminates. We define B to be $A[X_{ij}, Y_{ij}]/\mathfrak{a}(m, n)$. Let $S_1, \ldots, S_k$ and $T_1, \ldots, T_m$ be indeterminates, and let r_i be the image of $Y_{1i}S_1 + \cdots + Y_{ki}S_k + X_{1i}T_1 + \cdots + X_{mi}T_m$ in $B[S_1, \ldots, S_k, T_1, \ldots, T_m]$ for each i. Then the elements $r_1, \ldots, r_n$ form a regular sequence.*

Proof The proof is the same as that of Proposition 10.5.1, using induction and localizing at the X_{ij}, except for the verification when $n = 1$. In this case we must verify that $Y_{11}S_1 + \cdots + Y_{k1}S_k + X_{11}T_1 + \cdots + X_{m1}T_m$ is not a zero-divisor. However, this fact follows immediately from the fact that Y_{11} is not a zero-divisor in B. ☐

We now return to the construction of Chern classes. For a given integer i with $1 \leq i \leq n - r + 1$, let K_i be a subset of $\{1, \ldots, n\}$ consisting of $r + i - 1$ elements, and let $D(K_i)$ denote the set of all Δ_I for which I is contained in K_i. If $i = 1$, $D(K_i)$ will consist of a single determinant, while if $i = n - r + 1$, $D(K_i)$ will be the set of all of the Δ_I. Let $\mathfrak{a}_{K_i}$ be the ideal of R_G generated by the elements of $D(K_i)$. For $i = 0$, we let K_i be the empty set and let $\mathfrak{a}_{K_i}$ be the zero ideal.

In later constructions we will use the following local description of the ideal $\mathfrak{a}_{K_i}$. Let I be a subset of $\{1, \ldots, n\}$ consisting of r elements. As we computed in the proof of Proposition 10.1.1, the ideal defined in $(R_G)_{(\Delta_I)}$ by $\mathfrak{a}_{K_i}$ is generated by the maximal minors of the submatrix N_i of the matrix $M M_I^{-1}$ consisting of rows indexed by elements of K_i. If $I \subseteq K_i$, this ideal contains 1 and is trivial. If I is not a subset of K_i, let t be the number of elements in $I \cap K_i$. Then the ideal of maximal minors of N_i is generated by the maximal minors of the $r + i - 1 - s$-by-$r - s$ submatrix of N_i consisting of rows that are not in I and columns that do not contain a 1 (so their entries in the rows indexed by $I \cap K_i$ are zero). In particular, this ideal is a perfect ideal of codimension i. Thus, $\mathfrak{a}_{K_i}$ locally generates a perfect ideal of codimension i of $R_G(r, n)$.

Let $F_\bullet^i$ be a free graded resolution of $R_G/\mathfrak{a}_{K_i}$; that is, a resolution by sums of modules of the type $R_G(r, n)[k]$ for various integers k. The resolution $F_\bullet^i$ will not be finite in general, but, since the ideal $\mathfrak{a}_{K_i}$ is locally a perfect ideal of codimension i, the resolution is locally finite, so that for every I the complex

$(F^i_\bullet)_{(\Delta_l)}$ is quasi-isomorphic to a bounded complex. Let $\mathfrak{p}$ be a prime ideal in $\mathrm{Proj}(R_G)$; we will discuss the action of the Chern classes of the canonical quotient sheaf of $G_A(r, n)$ on the class $[R_G/\mathfrak{p}]$. Denote the dimension of $R_G/\mathfrak{p} \otimes R_G/\mathfrak{a}_{K_i}$ by d. Then the homology $H_k(R_G/\mathfrak{p} \otimes F^i_\bullet)$ of $R_G/\mathfrak{p} \otimes F^i_\bullet$ has dimension at most d for all k, and for all but finitely many k the support of $H_k(R_G/\mathfrak{p} \otimes F^i_\bullet)$ contains no prime ideals in $\mathrm{Proj}(R_G)$. Let

$$\chi_d\big(F^i_\bullet \otimes R_G/\mathfrak{p}\big) = \sum_i (-1)^i \big[H_i\big(F^i_\bullet \otimes R_G/\mathfrak{p}\big)\big]_d$$

denote the corresponding cycle of dimension d.

The main theorem of this section states that the Chern classes of the canonical quotient sheaf on $G(r, n)$ can be computed using the cycles $\chi_d(F^i_\bullet \otimes R_G/\mathfrak{p})$. Let Q be the graded R_G-module that defines the canonical quotient sheaf $\mathcal{Q}$ as described in Section 2, and let $e_1, \ldots, e_n$ denote the set of generators of Q described there. Denote the rank of Q by $s = n - r$. Let $\mathrm{Sym}_G(Q)$ denote the symmetric algebra of Q, and let p denote the projective map from $\mathrm{Proj}(\mathrm{Sym}_G(Q)) = P(\mathcal{Q})$ to $\mathrm{Proj}(R_G) = G(r, n)$. We denote the hyperplane section on $\mathrm{Proj}(\mathrm{Sym}_G(Q))$ by h.

Theorem 10.5.3 *With notation as above, let $\mathfrak{p}$ be a prime ideal in $G_A(r, n)$, and let $[R_G/\mathfrak{p}]$ be the associated cycle in $Z_*(G(r, n))$. Let d denote the dimension of $R_G/\mathfrak{p}$. For each integer i between 1 and s, let K_i be a subset of $\{1, \ldots, n\}$ with $r + i - 1$ elements such that $K_1 \subseteq K_2 \subseteq \cdots \subseteq K_s$. In addition, suppose that for each $i = 0, \ldots, s$ we have*

(i) *The dimension of $R_G/\mathfrak{p} \otimes R_G/\mathfrak{a}_{K_i}$ is at most $d - i$.*
(ii) *The dimension of $R_G/\mathfrak{p} \otimes \mathrm{Sym}_G(Q)/(\mathfrak{a}_{K_i} + (e_1, \ldots, e_{s-i}))$ is at most $d-1$, and the cycle $[R_G/\mathfrak{p} \otimes \mathrm{Sym}_G(Q)/(e_1, \ldots, e_{s+1})]_{d-1}$ is zero.*

Then

$$c_i(\mathcal{Q})([R_G/\mathfrak{p}]) = \chi_{d-i}(F_\bullet \otimes R_G/\mathfrak{p})$$

for each i.

Proof Recall that the Chern classes of $\mathcal{Q}$ applied to a cycle η are defined by the formula

$$h^s p^*(\eta) - h^{s-1} p^*(c_1(\mathcal{Q})(\eta)) + \cdots + (-1)^s p^*(c_s(\mathcal{Q})(\eta)) = 0.$$

We wish to show that, under the above hypotheses, $c_i(\mathcal{Q})([R_G/\mathfrak{p}]) = \chi_{d-i}(F^i_\bullet \otimes R_G/\mathfrak{p})$ for $i = 1, \ldots, s$. For each i, let η_i denote the cycle $\chi_{d-i}(F^i_\bullet \otimes R_G/\mathfrak{p})$; let $\eta_0 = [R_G/\mathfrak{p}]$. We will show that there are cycles $\beta_{-1}, \beta_0, \ldots, \beta_s$ in

$A_*(P(\mathcal{Q}))$ such that

(i) $\beta_{-1} = \beta_s = 0$.
(ii) $\beta_i + \beta_{i-1} = h^{s-i} p^*(\eta_i)$ for $i = 0, \ldots, s$.

Assume that we have defined cycles β_i with the above properties. If we substitute the expressions for $h^{s-i} p^*(\eta_i)$ in terms of the β_j in the above formula for Chern classes, we get

$$h^s p^*(\eta_0) - h^{s-1} p^*(\eta_1) + \cdots + (-1)^s p^*(\eta_s)$$

$$= (\beta_{-1} + \beta_0) - (\beta_0 + \beta_1) + \cdots + (-1)^s (\beta_{s-1} + \beta_s)$$

$$= \beta_{-1} + (-1)^s \beta_s = 0.$$

Thus, using the uniqueness of the expression used in the definition of Chern classes (Proposition 9.2.3), this shows that the classes η_j are the Chern classes of $\mathcal{Q}$ applied to $[R_G/\mathfrak{p}]$ as claimed.

Renumbering if necessary, we take K_i to be the set of the last $r + i - 1$ elements of the set $\{1, \ldots, n\}$; that is, $K = \{s-i+2, \ldots, n\}$. The intersections with the hyperplane h^{s-i} can be computed by taking $s - i$ elements of Q, which we take to be $e_1, \ldots, e_{s-i}$. The hypotheses imply that $e_1, \ldots, e_{s-i}$ intersect the cycles $p^*(\eta_i)$ in the correct codimension; we discuss this point in more detail below. We first construct ideals that will define the cycles β_i in the case where $\mathfrak{p} = 0$. Let $\mathfrak{c}_i$ be the ideal of $\mathrm{Sym}_G(Q)$ generated by $e_1, \ldots, e_{s-i}$ and $\mathfrak{a}_{K_i}$, and let $\mathfrak{b}_i$ be the ideal of $\mathrm{Sym}_G(Q)$ generated by $e_1, \ldots, e_{s-i}$ and $\mathfrak{a}_{K_{i+1}}$. Let $\mathfrak{b}_{-1}$ be the ideal generated by $e_1, \ldots, e_{s+1}$. We will show below that the cycle defined by $\mathrm{Sym}_G(Q)/\mathfrak{c}_i$ is a representative of $h^{s-i} p^*(\eta_i)$ in the case where $\mathfrak{p} = 0$.

Lemma 10.5.4 *For each* $i = 0, \ldots, s$, *there is a short exact sequence*

$$0 \to \mathrm{Sym}_G(Q)/\mathfrak{b}_i \overset{e_{s-i+1}}{\to} \mathrm{Sym}_G(Q)/\mathfrak{c}_i \to \mathrm{Sym}_G(Q)/\mathfrak{b}_{i-1} \to 0.$$

Proof Since $\mathfrak{b}_{i-1}$ is generated by e_{s-i+1} and $\mathfrak{c}_i$, the sequence is clearly exact except possibly at $\mathrm{Sym}_G(Q)/\mathfrak{b}_i$. To prove exactness at $\mathrm{Sym}_G(Q)/\mathfrak{b}_i$ we use a local computation. Fix a subset I of $\{1, \ldots, n\}$, and consider the restriction of the module Q to $U_{(\Delta_I)}$. The restriction $Q_{(\Delta_I)}$ is a free module of rank s, and it can be defined as the quotient of a free module on $e_1, \ldots, e_n$ modulo the relations defined by the r columns in the matrix $\tilde{M} = MM_I^{-1}$ defined in the local description of the Grassmannian in Proposition 10.1.1. Recall that the matrix $\tilde{M}$ has an identity matrix in the rows corresponding to I and the other entries are of the form Δ_J/Δ_I for all J having exactly $r - 1$ elements in common with I. Furthermore, the ring $(R_G)_{(\Delta_I)}$ is a polynomial ring over A in the Δ_J/Δ_I.

There are two cases, depending on whether $s - i + 1$ is an element of the set I or not. In either case, the quotient $\text{Sym}_G(Q)/\mathfrak{b}_i$ is a polynomial ring on $e_{s-i+1}, \ldots, e_n$ modulo the ideal generated by the linear combinations of $e_{s-i+1}, \ldots, e_n$ defined by the columns in last $r + i + 1$ rows of the matrix $\tilde{M}$ and by the maximal minors of the submatrix of $\tilde{M}$ consisting of the last $r + i$ rows. To simplify notation, we denote the Δ_J / Δ_I entries of this matrix by a_{ij}.

If any of the integers $i = s - i + 2, \ldots, n$ are in I, we may solve for the corresponding generator e_i using the equation of Proposition 10.2.1. Each such operation removes the corresponding row and column from the matrix. Since this involves removing an identity matrix, the ideal of maximal minors is unchanged. We are left with one of the two following situations, the first when $s - i + 1$ is in I, and the second when it is not:

$$\begin{pmatrix} 1 & 0 & \cdots & 0 \\ a_{21} & a_{22} & \cdots & a_{2n} \\ \vdots & \vdots & & \vdots \\ a_{m1} & a_{m2} & \cdots & a_{mn} \end{pmatrix} \text{ or } \begin{pmatrix} a_{11} & a_{12} & \cdots & a_{1n} \\ a_{21} & a_{22} & \cdots & a_{2n} \\ \vdots & \vdots & & \vdots \\ a_{m1} & a_{m2} & \cdots & a_{mn} \end{pmatrix}.$$

We recall that the a_{ij} are indeterminates. In either case, we denote the matrix by N and the submatrix of N consisting of rows 2 through m by N_2. We denote the ideal of maximal minors of N_2 by $\mathfrak{b}$. Let $T_1, \ldots, T_m$ be indeterminates. Let $\mathfrak{a}$ be the ideal generated by $\mathfrak{b}$ and $a_{1j}T_1 + \cdots + a_{mj}T_m$ (where we use a_{1j} to denote entries in the first row also in the first case) for $j = 1, \ldots, n$. It must be shown that in $A[a_{ij}, T_k]/\mathfrak{a}$, the annihilator of T_1 is the ideal of maximal minors of N. In case one, we can solve for T_1 and replace it with $a_{21}T_2 + \cdots + a_{m1}T_m$. For $i = 1, \ldots, n$ let $r_i = a_{2i}T_2 + \cdots + a_{mi}T_m$. We must show that the annihilator of r_1 modulo the ideal generated by $r_2, \ldots, r_n$ and $\mathfrak{b}$ is equal to the ideal of maximal minors of N. By Proposition 10.5.1, the relations between the r_i modulo $\mathfrak{b}$ are generated by the relations of the form $(0, \ldots, -r_j, \ldots, r_i, \ldots, 0)$ together with relations

$$(\delta_{I,1}, -\delta_{I,2}, \delta_{I,3}, \ldots, (-1)^{n+1}\delta_{I,n})$$

where $\delta_{I,j}$ denotes the determinant of the $n - 1$-by-$n - 1$ submatrix of N_2 consisting of the rows indexed by I and all columns except the jth column. In particular, the coefficient of r_1 will be in the ideal generated by $r_2, \ldots, r_n$ and by the maximal ($n - 1$-by-$n - 1$) minors of the submatrix of N_2 obtained by omitting the first column. Since $r_2, \ldots, r_n$ are in $\mathfrak{a}$, they are zero in the quotient. On the other hand, the ideal generated by the maximal minors of the submatrix of N_2 obtained by omitting the first row is also the ideal generated by maximal minors of N, so this completes the proof in case one.

We now consider the second case. Suppose $f(T_i)$ is in the annihilator of T_1 modulo the ideal generated by the $a_{1j}T_1 + \cdots + a_{mj}T_m$ and $\mathfrak{b}$. Then there exist polynomials $g_j(T_i)$ in $(A/\mathfrak{b})[T_1, \ldots, T_m]$ such that

$$f(T_i)T_1 = \sum_{j=1}^{n} g_j(T_i)(a_{1j}T_1 + \cdots + a_{mj}T_m).$$

Suppose $g_j(T_i) = h_j(T_2, \ldots, T_m) + T_1 k_j(T_i)$. We can move the term $T_1 k_j(T_i)$ $(a_{1j}T_1 + \cdots + a_{mj}T_m)$ to the other side of the equation and replace $f(T_i)$ by $f(T_i) - k_j(T_i)(a_{1j}T_1 + \cdots + a_{mj}T_m)$, which has the same image in the quotient modulo $\mathfrak{a}$. Thus, we may assume that the g_j are polynomials in $T_2, \ldots, T_m$. We then must have

$$\sum_{j=1}^{n} g_j(T_2, \ldots, T_m)(a_{2j}T_2 + \cdots + a_{mj}T_m) = 0,$$

since all other terms in the equation contain a factor of T_1. For each i, let $r_i = a_{2i}T_2 + \cdots + a_{mi}T_m$. By Proposition 10.5.1, $(g_1, \ldots, g_n)$ is in the module generated by elements of the form $(0, \ldots, -r_j, \ldots, r_i, \ldots, 0)$ together with elements of the form

$$(\delta_{I,1}, -\delta_{I,2}, \delta_{I,3}, \ldots, (-1)^{n+1}\delta_{d,n}),$$

where the $\delta_{I,j}$ are $n - 1$-by-$n - 1$ minors of N_2 as above. Assume first that $(g_1, \ldots, g_n)$ is of the form $(0, \ldots, -r_j, \ldots, r_i, \ldots, 0)$. Then we have the following equation in $(A/\mathfrak{b})[T_1, \ldots, T_m]$:

$$f(T_i)T_1 = -r_j(a_{1i}T_1 + \cdots + a_{mi}T_m) + r_i(a_{1j}T_1 + \cdots + a_{mj}T_m)$$
$$= (-a_{1i}r_j + a_{1j}r_i)T_1.$$

Since this is an equation in $(A/\mathfrak{b})[T_1, \ldots, T_m]$, this implies that $f(T_i) = (-a_{1i}r_j + a_{1j}r_i)$. Now taking the quotient modulo $\mathfrak{a}$, we have $r_j \cong -a_j T_1$ and similiarly for r_i, so

$$f(T_i) = -a_{1i}(-a_{1j}T_1) + a_{1j}(-a_{1i}T_1) = 0.$$

Assume now that $(g_1, \ldots, g_n)$ is of the second form. Then, by an argument similar to the one in the previous paragraph, we conclude that

$$f(T_i) = a_{11}\delta_{I,1} - a_{12}\delta_{I,2} + \cdots + (-1)^{n+1}a_{1n}\delta_{I,n}$$

is the maximal minor of N corresponding to a submatrix containing the first row. Hence the annihilator of T_1 is generated by these minors together with the ideal $\mathfrak{b}$, and this ideal is the ideal generated by the maximal minors of N. $\square$

We can now complete the proof of the main theorem. Let $G_\bullet^i$ and $H_\bullet^i$ be free graded resolutions for $R_G(Q)/\mathfrak{b}_i$ and $R_G(Q)/\mathfrak{c}_i$, respectively. The hypotheses on dimension imply that $(R_G(G)/\mathfrak{b}_i) \otimes_{R_G} (R_G/\mathfrak{p})$ and $(R_G(G)/\mathfrak{c}_i) \otimes_{R_G} (R_G/\mathfrak{p})$

have dimension at most $d - 1$ for all i. The exact sequence of Lemma 10.5.4 thus implies that we have

$$\chi_{d-1}\left(G_\bullet^i \otimes R_G/\mathfrak{p}\right) + \chi_{d-1}\left(G_\bullet^{i-1} \otimes R_G/\mathfrak{p}\right) = \chi_{d-1}\left(H_\bullet^i \otimes R_G/\mathfrak{p}\right)$$

for $i = 0, \ldots, s$. We let $\beta_i = \chi_{d-1}(G_\bullet^{i+1} \otimes R_G/\mathfrak{p})$ for each $i = 0, \ldots, s$. For $i = s$, we have that $\mathfrak{b}_s$ contains all the Δ_I in R_G, so $\beta_s = 0$. For $i = -1$, the ideal $\mathfrak{b}_{-1}$ is generated by $e_1, \ldots, e_{s+1}$, and the hypothesis of the theorem states that the associated cycle of dimension $d - 1$ is zero, which implies that $\beta_{-1} = 0$. To complete the proof it remains to show that we have

$$\chi_{d-1}\left(H_\bullet^i \otimes R_G/\mathfrak{p}\right) = h^{s-i} p^*(\eta_i).$$

To prove this statement, we first show that $e_1, \ldots, e_{s-i}$ form a regular sequence on $R_G(Q)/\mathfrak{a}_{K_i}$. We must prove that e_k is not a zero-divisor modulo the ideal generated by $e_1, \ldots, e_{k-1}$ and $\mathfrak{a}_{K_i}$ for all $k \leq s - i$. We compute locally as in the proof of Lemma 10.5.4, and we have two cases just as in that proof, depending on whether k is in I. The difference is that there are now $s - k + 1$ rows between the kth row and row $s - i + 2$, and $s - k + 1 > 0$. Suppose first that $k \in I$, so that there is a 1 in the kth row. Then the corresponding relation identifies e_k with a linear combination of the e_i for $i > k$. There are now two subcases to consider. If all the rows between k and $s - i + 1$ contain a 1 (so all of $k, \ldots, s - i + 1$ are in I), then the e_j in these rows are identified with linear combinations of the e_j in the last $r + i - 1$ rows as in Lemma 10.5.4. We use the notation of that Lemma. In this case, because of the extra identifications, we are considering relations between fewer than n columns, and it follows from the first part of Proposition 10.5.1 that all relations are of the form $(0, \ldots, -r_j, \ldots, r_i, \ldots, 0)$. Hence, arguing as in Lemma 10.5.4, we can conclude that e_k is not a zero-divisor. In the other subcase, where not all the rows between k and $s - i + 1$ are in I, there is at least one row remaining after the identifications, and the result follows from a similar argument using Proposition 10.5.2.

We now consider the other case, where the kth row consists of indeterminates. In this case the argument is similar to that in the second case of Lemma 10.5.4, except that, as in case one in the previous paragraph, it follows either from Proposition 10.5.1 or from Proposition 10.5.2 that the only relations between the elements $r_i = a_{2i} T_2 + \cdots + a_{mi} T_m$ are of the form $(0, \ldots, -r_j, \ldots, r_i, \ldots, 0)$. Hence, following the argument of Lemma 10.5.4, we can conclude that $f(T_i) = 0$, so e_k is not a zero-divisor.

Thus, $e_1, \ldots, e_{s-i}$ form a regular sequence on $\mathrm{Sym}_G(Q)/\mathfrak{a}_{K_i}$, so a resolution for $\mathrm{Sym}_G(Q)/\mathfrak{c}_i$ can be found by tensoring a graded free resolution for $\mathrm{Sym}_G(Q)/\mathfrak{a}_{K_i}$ with the Koszul complex $K_\bullet(e_1, \ldots, e_{s-i})$. Let $F_\bullet^i$ be a graded free resolution of $R_G/\mathfrak{a}_{K_i}$ as in the statement of the theorem. Then, since $\operatorname{Sym}_G(Q)$ is locally flat over R_G, $\operatorname{Sym}_G(Q) \otimes F_\bullet^i$ is a free graded resolution of $\operatorname{Sym}_G(Q)/\mathfrak{a}_{K_i}$, and

$$K_\bullet(e_1, \ldots, e_{s-i}) \otimes \operatorname{Sym}_G(Q) \otimes F_\bullet^i$$

is thus a graded free resolution of $\operatorname{Sym}_G(Q)/\mathfrak{c}_i$ and so may be used for $H_\bullet^i$. However, $\chi_{d-i+s-1}(\operatorname{Sym}_G(Q) \otimes (F_\bullet^i \otimes R_G/\mathfrak{p}))$ is the cycle $\mathfrak{p}^*(\eta_i)$, and thus $\chi_{d-1}(H_\bullet^i \otimes R_G/\mathfrak{p}) = \chi_{d-1}(K_\bullet(e_1, \ldots, e_{s-i}) \otimes \operatorname{Sym}_G(Q) \otimes F_\bullet^i \otimes R_G/\mathfrak{p})$ represents the cycle class $h^{s-i}(p^*(\eta_i))$ as claimed. Hence this completes the proof. $\square$

We remark that the only really essential assumption on dimension for computing $c_i(Q)([R_G/\mathfrak{p}])$ is that the codimension of $R_G/\mathfrak{a}_{K_i} \otimes R_G/\mathfrak{p}$ must be i. While the other equalities on dimension do not follow from this one, it is possible to find a change of variables that does not change $\mathfrak{a}_{K_i}$ and for which the other dimension conditions also hold. On the other hand, the condition that the codimension of $R_G/\mathfrak{a}_i \otimes R_G/\mathfrak{p}$ is i is clearly necessary to assure that the computation gives a result of the correct dimension. The fact that one can adjust the other variables to produce a situation satisfying the conditions of the theorem follows from a generic transversality result of Kleiman [34].

If the quotient $R_G/\mathfrak{p}$ is also Cohen-Macaulay, then the computation is easier.

Corollary 10.5.5 *If $R_G/\mathfrak{p}$ is locally Cohen-Macaulay, then*

$$c_i(Q)([R_G/\mathfrak{p}]) = [R_G/\mathfrak{a}_{K_i} \otimes R_G/\mathfrak{p}]_{d-i}.$$

Proof This follows from the fact that the ideal defining $\mathfrak{a}_{K_i}$ is locally generated by the minors of a matrix of indeterminates and so is a perfect ideal of codimension i. Hence if $R_G/\mathfrak{p}$ is Cohen-Macaulay, the homology in degrees greater than zero in the complex $F_\bullet^i \otimes R_G/\mathfrak{p}$ vanishes, so we have

$$\chi_{d-i}\left(F_\bullet^i \otimes R_G/\mathfrak{p}\right) = [R_G/\mathfrak{a}_{K_i} \otimes R_G/\mathfrak{p}]_{d-i}.$$

$\square$

Exercises

10.1 Let k be a field. Show that a graded homomorphism from $R_k(r, n)$ to k that is the identity on k determines a subspace of k^n of dimension r.

10.2 Show that in the ring $R_G(2, 4)$, the ideal of relations between the Δ_I is generated by the relation

$$\Delta_{12}\Delta_{34} - \Delta_{13}\Delta_{24} + \Delta_{14}\Delta_{23}.$$

10.3   Verify the statements about determinants used in the proof of Proposition 10.5.1.

10.4   Verify explicitly the induction step from $n = 1$ to $n = 2$ in Proposition 10.5.1.

10.5   Let $R_G = R_G(2, 4)$. Letting $\mathfrak{p}$ be the prime ideal generated by Δ_{23}, show that the dimension condition in the statement of Theorem 10.5.3 is satisfied for $i = 1$ but not for $i = 2$, but that it can be fixed by a change of variables that leaves K_1 fixed.

10.6   Show that if $\mathfrak{p} = 0$ in Theorem 10.5.3, then the hypotheses on dimension hold.

11
Local Chern Characters

This chapter is devoted to developing the theory of local Chern characters of complexes. The main aim is to prove the necessary results for the applications to intersection multiplicities in Chapter 13. However, we also point out relations between local Chern characters and multiplicities of ideals and describe connections with other topics discussed in this book.

We first treat matrices with support at the maximal ideal of a local ring, where it makes sense to define Chern classes of an individual matrix and where these classes can be interpreted as numbers rather than classes in a Chow group (Roberts [57]). In the second section we show that this definition generalizes that of the multiplicity of an m-primary ideal.

However, for the important applications of the theory it is necessary to give a more general definition of the Chern character of a complex of locally free sheaves with given support, which we do in the following section. Essentially, this is simply the alternating sum of Chern characters of the maps in the complex, but it is more convenient to carry out the construction for all the maps at once.

The remainder of this chapter and most of the following one are devoted to proving the important properties of these invariants that will be used in the applications. These include additivity and multiplicativity formulas and the local Riemann-Roch formula, which relates local Chern characters to Euler characteristics. These theorems use a splitting principle for complexes that reduces the questions to the case of a complex of length one of a particularly simple type; however, the reduction requires the use of general projective schemes even to prove the results for the case of complexes of free modules over a local ring.

Local Chern characters were originally defined in a paper of Baum, Fulton, and Macpherson [5], in which they were used to prove a version of the Riemann-Roch theorem for singular varieties. A more general version, from which many of the ideas of this chapter are taken, can be found in Fulton [17]. Local

Chern characters were defined using étale cohomology by Iversen [30], and an algebraic construction of the component of codimension one was given by Foxby [16].

11.1 Chern Classes of m-primary Matrices

Let A be a local integral domain of dimension d. In this section we define Chern classes associated to a matrix supported at the maximal ideal $\mathfrak{m}$ of A.

Let M be a matrix with entries in A. Since A is an integral domain, M has a well-defined rank, which we denote r. For each positive integer i we let J_i denote the ideal of i-by-i minors of M. In this section we define the *support* of M to mean the support of A/J_r; another way of saying this is that the primes $\mathfrak{p}$ not in the support of M are those for which, after localizing at $\mathfrak{p}$, M can be put in the form $\left(\begin{smallmatrix} I & 0 \\ 0 & 0 \end{smallmatrix}\right)$ with respect to some bases, where I is an identity matrix. We say that M is $\mathfrak{m}$-*primary* if its support consists of the maximal ideal of A.

Let M be an m-by-n $\mathfrak{m}$-primary matrix. The construction of Chern classes uses an auxiliary matrix that we denote $\tilde{M}$. Let s be an indeterminate. We consider the $m+n$-by-n matrix with coefficients in $A[s]$, which is written in block form as

$$\begin{pmatrix} sI \\ M \end{pmatrix}$$

where I is an n-by-n identity matrix. Let J denote the ideal of maximal minors of $\tilde{M}$. We note that a maximal minor of $\tilde{M}$ will be of the form $s^{n-j}\Delta$, where Δ is a j-by-j minor of M. Hence J is the sum of the ideals $s^{n-j}J_j$, where j runs from 0 to the rank of M, and thus J is an ideal of $A[s]$ that contains information about the minors of all sizes of M.

We next take the Rees ring $R(J)$ of the ideal J, which is a graded ring over the ring $A[s]$. Since J is generated by the maximal minors of an $m+n$-by-n matrix, the generators of this Rees ring over $A[s]$ satisfy the relations of the Grassmannian $G_{A[s]}(n, m+n)$ of n-planes in $m+n$ space over $A[s]$, and hence we have an embedding of $\mathrm{Proj}(R(J))$ into this Grassmannian. From the results of the previous chapter we may apply the Chern classes of this Grassmannian to the cycle given by the image of $\mathrm{Proj}(R(J))$; however, the definition we actually want requires a preliminary step.

We first divide by the ideal of $R(J)$ generated by s (in degree 0), and denote the quotient Q.

Lemma 11.1.1 *Let J_i denote the ideal of i-by-i minors of the matrix M as above. The graded component Q_1 of Q in degree 1 is then isomorphic as an*

A-module to the direct sum

$$A/J_1 \oplus J_1/J_2 \oplus \cdots \oplus J_{r-1}/J_r \oplus J_r.$$

More generally, for each $k \geq 1$ there is a decreasing sequence of ideals

$$A = K_0 \supseteq K_1 \supseteq \cdots \supseteq K_{kr-1} \supseteq K_{kr} = J_r^k$$

such that Q_k is isomorphic as an A-module to the direct sum

$$K_0/K_1 \oplus K_1/K_2 \oplus \cdots \oplus K_{kr-1}/K_{kr} \oplus K_{kr}.$$

Proof As noted above, the ideal J is generated by the union of the sub-A-modules $s^{n-j}J_j$ for j between 0 and r. Hence, for each $k \geq 1$, the ideal J^k is generated by all $s^{nk-j}K_j$ for j between 0 and kr, where K_j is the sum of all products $J_{i_1}J_{i_2}\cdots J_{i_t}$ with $i_1 + i_2 + \cdots + i_t = j$. Since the J_i form a decreasing sequence of ideals, the K_i do as well, and since $J_i = 0$ for $i > r$ we have $K_{kr} = J_r^k$. Since the first statement of the lemma is a special case of the second, it suffices to prove the second statement.

We define a map f from the direct sum $\oplus_{j=0}^{kr}K_j$ of the ideals K_j of A to J^k by letting $f(k) = s^{nk-j}k$ for $k \in K_j$. Since the powers of s are linearly independent over A it is clear that this map is injective. Let ϕ denote the map induced by f from $\oplus_{j=0}^{kr}K_j$ to J^k/sJ^k. Since J contains s^n, J^k contains s^{nk}, and thus sJ^k contains s^{nk+1}. Hence J^k/sJ^k is generated by the sub-A-modules $s^{nk-j}K_j$ for $j = 0, \ldots, nk$, and ϕ is surjective.

We next prove that the kernel of ϕ is the direct sum $\oplus_{j=0}^{kr}K_{j+1}$, where the embedding is defined as the direct sum of the inclusions of K_{j+1} into K_j and where we let $K_{kr+1} = 0$. If $(a_0, a_1, \ldots, a_{kr})$ is mapped to zero by ϕ, then

$$s^{nk}a_0 + s^{nk-1}a_1 + \cdots a_{kr} = s(s^{nk}b_0 + s^{nk-1}b_1 + \cdots b_{kr})$$

for some $b_i \in K_i$. Since this is an equality of polynomials with coefficients in A, we must have $s^{nk-i}a_i = s^{nk-i}b_{i+1}$ and hence $a_i \in K_{i+1}$ for each i. In particular, a_{kr} must equal zero. Conversely, if $a_i = b_{i+1} \in K_{i+1}$ for each i, $s^{nk}a_0 + s^{nk-1}a_1 + \cdots a_{kr}$ will be an element of sJ^k. Thus, the kernel of ϕ is $\oplus K_{j+1}$, giving the required isomorphism. $\square$

We next examine several further properties of the ring Q. First, Q is a graded ring with $Q_0 = A[s]/(s) = A$. The ring A has dimension d, so $A[s]$ has dimension $d + 1$ and the Rees ring $R(J)$ has dimension $d + 2$. Since Q is obtained by dividing $R(J)$ by a principal ideal, Q has dimension $d + 1$. Furthermore, if we let q be the ideal of Q generated by $s^n, s^{n-1}J_1, \ldots s^{n-r+1}J_{r-1}$, it follows from Lemma 11.1.1 that the quotient on dividing by q is the Rees ring of J_r

over A. Since this Rees ring has dimension $d + 1$, q is a minimal prime ideal of Q, and $\text{Proj}(Q/\mathfrak{q})$ is one component of $\text{Proj}(Q)$.

However, it is the other components that are of interest. (In Section 3 we will define the Chern character of a complex, and we will show that the contributions given by the component corresponding to q will be zero in the alternating sum used in the definition.) Let $\mathfrak{a}$ be an ideal that defines the other components; for example, $\mathfrak{a}$ can be taken to be the set of elements annihilated by a power of q. We could also take $\mathfrak{a}$ to be the ideal of Q generated by J_r, since J_r annihilates q. Note that since the ideals J_i are all $\mathfrak{m}$-primary, using the structure of Q_1 as described in Proposition 9.1.1, we see that q is annihilated by a power of $\mathfrak{m}$, so that $\mathfrak{a}$ contracts to an $\mathfrak{m}$-primary ideal of A. Let η be the cycle $[Q/\mathfrak{a}]_d$. Then η is a sum of prime ideals of Q that contract to $\mathfrak{m}$. Since this cycle is contained in the Grassmannian $G_A(n, m + n)$ and is supported at $\mathfrak{m}$, it is contained in the Grassmannian $G_{A/\mathfrak{m}}(n, m + n)$ over $A/\mathfrak{m}$. Thus, we can apply the Chern classes of the canonical quotient sheaf as defined in the previous chapter. Let $\mathcal{Q}$ be the canonical quotient sheaf on $G_A(n, m + n)$, and denote the Chern class $c_i(\mathcal{Q})$ by c_i. Since η has dimension d, if $i_1, \ldots i_k$ are positive integers whose sum is d, the product of Chern classes $c_{i_1} \ldots c_{i_k}(\eta)$ will produce a cycle of dimension zero, which we then project to $\text{Spec}(A/\mathfrak{m})$. For each choice of $i_1, \ldots, i_k$ there is thus an integer $c_{i_1} \cdots c_{i_k}(\eta)$ such that this projection is equal to $c_{i_1} \cdots c_{i_k}(\eta)[A/\mathfrak{m}]$ in $A_0(\text{Spec}(A/\mathfrak{m}))$; we call these integers, as $i_1, \ldots i_k$ runs over sequences of positive integers with sum equal to d, the *local Chern classes* of the matrix M. Using the formula for the Chern character, we may similarly define a rational number $\text{ch}_d(M)$ (with denominator dividing $d!$), which we call the *local Chern character* of M.

We conclude this section with the simplest nontrivial example, in which A is a domain of dimension 1, x is a nonzero element of A, and M is the 1-by-1 matrix with entry x. In this case $\tilde{M}$ is the matrix $\binom{s}{x}$, and J, the ideal of 1-by-1 minors of $\tilde{M}$, is generated by s and x. Thus, $R(J)$ is the Rees ring on (s, x) and can be represented as the quotient $A[s][Y, Z]/(xY - sZ)$ by mapping Y to s and Z to x in degree 1. The ring Q is now obtained by dividing by s, and this quotient is $A[Y, Z]/(xY)$. Thus, Q has two minimal prime ideals. The first is generated by Y, and noting that Y maps to s in degree 1 in the Rees ring, we see that this is the ideal that we denoted q. The second is generated by x; thus, the component of interest corresponds to the image $(A/xA)[Y, Z]$. In this case the Grassmannian is just projective space $\mathbb{P}^1$, and the first Chern class of the canonical quotient sheaf is the usual hyperplane section; dividing by the hyperplane and projecting to $\text{Spec}(A/\mathfrak{m})$ we obtain the length of A/xA times the class $[A/\mathfrak{m}]$. Hence the local Chern class $c_1(M)$, which is the same as the local Chern character $\text{ch}_1(M)$ in this case, is equal to the length of A/xA. We note that this number is also the Euler characteristic of the complex

$$\cdots \to 0 \to A \xrightarrow{x} A \to 0 \to \cdots$$

defined by the matrix M.

11.2 An Example: Multiplicities of m-primary Ideals

A simple example of the Chern character of a matrix is given by the case of the multiplicity of an m-primary ideal $\mathfrak{a}$. We recall that the multiplicity of $\mathfrak{a}$ is equal to $d!$ times the leading coefficient of the Hilbert-Samuel polynomial $P_\mathfrak{a}$, the polynomial such that $P_\mathfrak{a}(n)$ is the length of $A/\mathfrak{a}^n$ for large n. We show in this section that the multiplicity of $\mathfrak{a}$ is equal to the local Chern character $\mathrm{ch}_d(M)$ up to a factor of $d!$, where M is a 1-by-n matrix whose entries form a set of generators for $\mathfrak{a}$.

We first compute $c_1^d(M)$ in a more general case.

Theorem 11.2.1 *Let M be a matrix of rank r, and suppose that J_r, the ideal of r-by-r minors of M, is* m-*primary. Then $(c_1^d(M))$ is equal to the multiplicity of J_r.*

Proof Consider the graded ring $Q/\mathfrak{a}$ defined as in the previous section, where we take $\mathfrak{a}$ to be the ideal of Q generated by J_r. It follows from Theorem 10.5.3 of Chapter 10 that $c_1(Q)$ on a cycle in the Grassmannian can be computed by taking the intersection with the divisor generated by one of the determinants that generate the graded ring of the Grassmannian. Since these determinants have degree 1 in the ring of the Grassmannian, these also define the intersection with the hyperplane section. Hence the push-forward of $c_1^d(Q)([Q/\mathfrak{a}]_d)$ to $A_0(\mathrm{Spec}(A/\mathfrak{m}))$ is the degree of the graded ring $Q/\mathfrak{a}$.

In Proposition 9.1.1 we showed that the component of degree k of the ring Q is isomorphic as an A-module to the sum of K_j/K_{j+1} where the K_j form a decreasing sequence of m-primary ideals with $K_0 = A$, $K_{rk} = J_r^n$, and $K_{rk+1} = 0$. Since J_r annihilates K_j/K_{j+1} for all $j < kr$ and we are letting $\mathfrak{a}$ be the ideal generated by J_r, we have

$$(Q/\mathfrak{a})_k \cong A/K_1 \oplus K_1/K_2 \oplus \cdots \oplus K_{rk-1}/J_r^k \oplus J_r^k/j_r^{k+1}.$$

Thus, the length of this graded piece is equal to the length of A/J_r^{k+1} for each k. Since the polynomial whose value at k is the length of A/J_r^{k+1} in high degrees has leading coefficient equal to $1/d!$ times the multiplicity of J_r, and the polynomial whose value at k is the length of $(Q/\mathfrak{a})_k$ has leading coefficient $1/d!$ times the degree of $Q/\mathfrak{a}$, this proves the result. $\square$

Now we consider the special case in which the matrix M is a row vector whose entries generate the ideal $\mathfrak{a}$. The matrix $\tilde{M}$ is then an n-by-n identity matrix times s followed by a row with entries generators of $\mathfrak{a}$. Now recall that by Theorem 10.5.3 of Chapter 10 the second Chern class of the canonical quotient sheaf, applied to the cycle defined by a matrix, can be computed by intersecting with the ideal of minors of a sufficiently generic n-by-$n + 1$ submatrix. But the whole matrix $\tilde{M}$ is n by $n + 1$, so its minors generate the entire irrelevant ideal of elements of positive degree, and hence this intersection is zero. For c_i with $i > 2$ the same argument holds. Thus $\mathrm{ch}_d(M) = c_1^d/d!$, and the above theorem then implies that $\mathrm{ch}_d(M)$ is equal to $1/d!$ times the multiplicity of $\mathfrak{a}$.

11.3 Cycles Associated to a Complex of Locally Free Sheaves

Let $\mathcal{F}_\bullet$ be a bounded complex of locally free sheaves over a projective scheme X. In the next section we will define the local Chern character of $\mathcal{F}_\bullet$, extending the definition of the local Chern character of a matrix defined in Section 1. The basic idea is to take an alternating sum of Chern characters associated to the maps in $\mathcal{F}_\bullet$. However, to assure that it is well-defined and commutes with the basic constructions of intersection theory, a more careful definition is required. This is essentially the construction described in Fulton [17, Section 18.1]. One of the main steps in this process is to define cycles in a certain product of Grassmannians associated to $\mathcal{F}_\bullet$, which we do in this section.

The *support* of $\mathcal{F}_\bullet$ is defined to be the closed subset of X consisting of those points at which $\mathcal{F}_\bullet$ is not exact. In the case in which $X = \mathrm{Spec}(A)$, the support is thus the set of prime ideals where the localization is not split exact. We note that the support of each map in the complex as defined in Section 1 is contained in the support of the complex as defined here but that, if we consider a map from one module to another as a complex, its support as defined here is usually larger than the support as a map as defined in Section 1.

For each i, let d_i denote the boundary map from $\mathcal{F}_i$ to $\mathcal{F}_{i-1}$ and let f_i denote the rank of $\mathcal{F}_i$. We assume that there are nonnegative integers r_i such that the rank of $\mathcal{F}_i$ is equal to $r_i + r_{i-1}$ and that $r_i = 0$ for $i \ll 0$; in particular, $\mathcal{F}_i$ is assumed to have constant rank. If $\mathcal{F}_\bullet$ is exact, then r_i is the rank of d_{i+1}. To simplify notation, we assume that the nonzero sheaves in the complex $\mathcal{F}_\bullet$ have degrees between 0 and n; the general case can then be obtained simply by shifting degrees.

We assume in this section that $X = \mathrm{Proj}(A)$, where A is a domain. Let d denote the dimension of X. Let s be an indeterminate of degree zero, and let $X[s]$ and $X[s, s^{-1}]$ denote $\mathrm{Proj}(A[s])$ and $\mathrm{Proj}(A[s, s^{-1}])$, respectively. The embedding of A into $A[s]$ induces a map of schemes from $X[s]$ to X that we denote ψ, and $X[s, s^{-1}]$ is an open subset of $X[s]$.

We first pull $\mathcal{F}_\bullet$ back to $X[s]$ by ψ and denote $\psi^*(\mathcal{F}_\bullet)$ by $\tilde{\mathcal{F}}_\bullet$. For each i from 0 to n, let $G_i[s]$ denote $G_{f_i}(\tilde{\mathcal{F}}_i \oplus \tilde{\mathcal{F}}_{i-1})$, the Grassmannian of rank f_i subsheaves of $\tilde{\mathcal{F}}_i \oplus \tilde{\mathcal{F}}_{i-1}$ over $X[s]$. We similarly let G_i denote $G_{f_i}(\mathcal{F}_i \oplus \mathcal{F}_{i-1})$. Let $G_{\mathcal{F}}$ and $G_{\mathcal{F}}[s]$ denote the products $\prod_{i=0}^{n+1} G_i$ and $\prod_{i=0}^{n+1} G_i[s]$, respectively. For each i, let $\mathcal{E}_i$ be the subsheaf of $\tilde{\mathcal{F}}_i \oplus \tilde{\mathcal{F}}_{i-1}$ defined by

$$\mathcal{E}_i = \{(sf, d_i(f)) \mid f \in \tilde{\mathcal{F}}_i\}.$$

Then $\mathcal{E}_i$ is a subsheaf of $\tilde{\mathcal{F}}_i \oplus \tilde{\mathcal{F}}_{i-1}$. If we restrict $\mathcal{E}_i$ to $G_i[s, s^{-1}]$ we can invert s, and $\mathcal{E}_i$ is the graph of the map d_i/s. Hence the quotient is isomorphic to $\tilde{\mathcal{F}}_{i-1}$ and is thus locally free. Thus, we have a section $\sigma_{\mathcal{E}_i}$ of the Grassmannian $G_i[s, s^{-1}]$, which defines an embedding of $X[s, s^{-1}]$ into $G_i[s, s^{-1}]$. Taking the product of these sections for all i defines an embedding of $X[s, s^{-1}]$ into $\prod_{i=0}^{n+1} G_{f_i}(\tilde{\mathcal{F}}_i \oplus \tilde{\mathcal{F}}_{i-1}) = G_{\mathcal{F}}[s, s^{-1}]$. We let $\alpha(\mathcal{F})$ denote the cycle defined by the closure of the image of this embedding in $G_{\mathcal{F}}[s]$. Thus, $\alpha(\mathcal{F})$ is a cycle of dimension $d + 1$ in $G_{\mathcal{F}}[s]$.

We next define a second cycle in $G_{\mathcal{F}}[s]$, which we call the "trivial cycle." Let $r_0, \ldots, r_n$ be integers such that $r_i + r_{i-1} = f_i$ for $i = 0, \ldots, n$ (letting $r_{-1} = 0$). For each i, we let H_i denote $G_{r_i}(\mathcal{F}_i)$, the Grassmannian of rank r_i subsheaves of $\mathcal{F}_i$. The direct sum map defined in Chapter 10 defines an embedding from $H_i \oplus H_{i-1}$ to $G_{f_i}(\mathcal{F}_i \oplus \mathcal{F}_{i-1}) = G_i$, and these together thus define a closed embedding of $\prod_{i=0}^n H_i$ into $G_{\mathcal{F}}$. We denote $\prod_{i=0}^n H_i$ by $H_{\mathcal{F}}$. We assume first that Z, the support of $\mathcal{F}_\bullet$, is not equal to X. On the open set $X - Z$, the kernel of d_i is a locally free subsheaf of F_i with locally free quotient, so that it defines a section from $X - Z$ to the restriction of H_i to $X - Z$. Combining these sections, we have a section of $H_{\mathcal{F}}$ over $X - Z$. We define the trivial cycle to be the cycle defined by the closure of the image of this section in $G_{\mathcal{F}}$, pulled back to $G_{\mathcal{F}}[s]$. We denote this cycle $\beta(\mathcal{F})$. Like $\alpha(\mathcal{F})$, $\beta(\mathcal{F})$ is a cycle of dimension $d + 1$ on $G_{\mathcal{F}}[s]$. We assumed above that Z is not equal to X; if $Z = X$, we let $\beta(\mathcal{F}) = 0$.

The trivial cycle has two main properties. Let (s) denote the principal divisor with ideal generated by s. Then the first property is that

$$(s) \cap \alpha(\mathcal{F}) = (s) \cap \beta(\mathcal{F})$$

if and only if $\mathcal{F}_\bullet$ is exact. The second main property is that the alternating sum of Chern characters that we define below vanishes on any cycle with support in the image of $H_{\mathcal{F}}$. We prove these results below; we first describe the cycles $\alpha(\mathcal{F})$ and $\beta(\mathcal{F})$ more explicitly in the case in which $\mathcal{F}_\bullet$ is a complex of free modules.

In this situation we use the specific representation of the ring of the Grass-mannian of a free module as defined in the first section of Chapter 10. That is, we assume that we have chosen bases for the free modules in $F_{\bullet}$ and for each i, the ring R_{G_i} is generated by the elements Δ_I, where I is a choice of f_i rows out of $f_i + f_{i-1}$. To describe the cycles we define the maps by identifying the images of Δ_I for each I. We recall (Proposition 10.1.2) that if E is a free submodule of rank r of the free module F with locally free quotient, then the section of $G(r, n)$ associated to E is defined by identifying $\mathrm{Spec}(A)$ with $\mathrm{Proj}(A[T])$ and mapping Δ_I to $\delta_I T$, where δ_I is the corresponding minor of an r-by-r matrix whose columns form a basis for E. This fact is used in the following explicit constructions of the cycles $\alpha(F)$ and $\beta(F)$ for a complex of free modules.

The other fact that will be needed is an algebraic characterization of the closure of a section defined an open set. We use the following simple Lemma.

Lemma 11.3.1 *Let B be a graded integral domain, and let ϕ be a quasi-projective map from $X = \mathrm{Proj}(B)$ to $Y = \mathrm{Proj}(A)$ defined by a map $f : A \to B$ of graded rings. Let x be an element of A of degree zero such that $f(x) \neq 0$, and assume that the map induced by f on localizations from A_x to B_x induces a proper map ϕ_x from $\mathrm{Proj}(B_x)$ to $\mathrm{Proj}(A_x)$. Then the closure of the image of ϕ_x in Y is equal to $V(\mathfrak{p})$, where $\mathfrak{p}$ is the kernel of f.*

Proof Let $\mathfrak{p}$ be the kernel of f; $\mathfrak{p}$ is a prime ideal since B is an integral domain. The hypothesis states that $f(x) \neq 0$, so x is not in $\mathfrak{p}$. After localizing at x, the image of ϕ_x is $\mathrm{Proj}(A_x/\mathfrak{p}_x) = V(\mathfrak{p}_x) \subseteq \mathrm{Proj}(A_x)$. The closure of $V(\mathfrak{p}_x)$ in $\mathrm{Proj}(A)$ must be a closed set that contains $\mathfrak{p}$ and is contained in the set of $\mathfrak{q} \in \mathrm{Proj}(A)$ such that $\mathfrak{q}_x$ contains $\mathfrak{p}_x$; it is clear that this set is $V(\mathfrak{p})$. $\qquad\Box$

Proposition 11.3.2 *Let $F_{\bullet}$ be a bounded complex of free modules over an integral domain A. Choose a basis for each module F_i in $F_{\bullet}$, and let M_i be the matrix defining the map d_i^F with respect to these bases for each i. Let J_i denote the ideal of maximal minors of the matrix $\tilde{M}_i$ with entries in $A[s]$ defined by*

$$\tilde{M}_i = \begin{pmatrix} sI \\ M_i \end{pmatrix}$$

where I is an identity matrix of size equal to the rank of F_i. For each i let R_{G_i} be the graded ring of the Grassmannian $G(f_i, f_i + f_{i-1})$, and let B be the tensor product of the R_{G_i} over A. Let ϕ be the map from B to the multi-Rees ring $R(J_1, \ldots, J_n)$ defined by letting $\phi(\Delta_I)$ be the determinant of the matrix *obtained by taking the rows with indices in I from the matrix $\tilde{M}_i$. Then $\alpha(F)$ is the cycle $[A/\mathfrak{p}]$, where $\mathfrak{p}$ is the kernel of ϕ.*

Proof We apply Lemma 11.3.1 to the map ϕ. After inverting s, the map ϕ defines the product of the sections corresponding to the locally free sheaves $\mathcal{E}_i$ defined in the definition of $\alpha(F)$. We note that the section corresponding to the map to the ith component maps Δ_I to T times the Ith minor of $\tilde{M}_i$ in $A[T]$, and the image of this map is precisely the Rees ring of the ideal generated by the minors of $\tilde{M}_i$. The product is thus defined by the corresponding map to the multi-Rees ring $R(J_1, \ldots, J_n)$. □

A similar computation gives the interpretation of $\beta(F_\bullet)$ in the following Proposition:

Proposition 11.3.3 *Let $F_\bullet$ be a complex of free modules over a domain A as in Proposition 11.3.2. Let Z be the support of $F_\bullet$, and suppose that Z is not all of $X = \mathrm{Spec}(A)$. For each i, let M_i be a matrix that defines d_i, let r_i be the rank of M_{i+1}, and let K_i be an f_i-by-r_i submatrix of M_{i+1} such that at least one r_i-by-r_i minor of K_i is not zero. Let $J(K_i)$ denote the ideal generated by the maximal minors of K_i, and let $R(J(K_i))$ denote the multi-Rees ring of the set of ideals $J(K_i)$. Then the cycle $\beta(\mathcal{F}_\bullet)$ is equal to $[A/\mathfrak{p}]$, where $\mathfrak{p}$ is the kernel of the map from ψ from R_G to $R(J(K_i))$ defined by letting $\psi(\Delta_I)$ be the minor defined in $J(K_i)J(K_{i-1})$ by the Ith rows of the matrix*

$$\begin{pmatrix} K_i & 0 \\ 0 & K_{i-1} \end{pmatrix}. \tag{$*$}$$

Proof The map defining β is the section of the product of Grassmannians defined in the ith component by the direct sum $\mathrm{Ker}(d_i) \oplus \mathrm{Ker}(d_{i-1})$. On $X - Z$, this subsheaf is the same as $\mathrm{Im}(d_{i+1}) \oplus \mathrm{Im}(d_i)$, and since we have chosen the matrices K_i to have the same rank as M_{i+1}, there is a nonempty open set (perhaps smaller than $X - Z$) on which the image is generated by the columns of the matrix $(*)$. The maximal minors of the matrix $(*)$ are products of maximal minors of K_i and K_{i-1}, so the ideal generated by the maximal minors is $J(K_i)J(K_{i-1})$. Hence, using Lemma 11.3.1 and following the same argument as in Proposition 11.3.2, the cycle β is defined by the prime ideal, which is the kernel of the map to the multi-Rees ring $R(J(K_i)J(K_{i-1}))$ defined as in the statement of the Proposition. However, this ring is also contained in the multi-Rees ring $R(J(K_i))$, so we may consider it as a map to $R(J(K_i))$. □

Proposition 11.3.4 *Suppose* $\mathcal{F}_\bullet$ *is exact. Then the cycles* $(s) \cap \alpha(\mathcal{F})$ *and* $(s) \cap$ $\beta(\mathcal{F})$ *are equal.*

Proof We prove this statement by a direct computation, using the previous local descriptions of the cycles $\alpha(\mathcal{F})$ and $\beta(\mathcal{F})$. We can localize and assume that A is a local ring and that $F_\bullet$ is a split exact complex of free modules. If $F_\bullet$ is split exact, the map d_i can be represented by a matrix of the form

$$\begin{pmatrix} 0 & I \\ 0 & 0 \end{pmatrix} \tag{*}$$

where I is an r_{i-1}-by-r_{i-1} identity matrix. In the following we write matrices in 4-by-2 block form, where the rows are in groups of r_i, r_{i-1}, r_{i-1}, and r_{i-2} and the columns are in groups of r_i and r_{i-1}. Then by Proposition 11.3.2 $\alpha(F)$ is the cycle $[A/\mathfrak{p}]$, where $\mathfrak{p}$ is the kernel of the map τ_α that in degree i sends Δ_I to the Ith minor of the matrix

$$\begin{pmatrix} sI & 0 \\ 0 & sI \\ 0 & I \\ 0 & 0 \end{pmatrix}.$$

By Proposition 11.3.3, the cycle $\beta(F)$ is the cycle $[A/\mathfrak{q}]$, where $\mathfrak{q}$ is the kernel of the map τ_β that in degree i sends Δ_I to the Ith minor of the matrix

$$\begin{pmatrix} I & 0 \\ 0 & 0 \\ 0 & I \\ 0 & 0 \end{pmatrix}. \tag{**}$$

To complete the proof we identify the kernels of τ_α and τ_β and show that the resulting cycles intersected with (s) are the same. Let I_0 be the set of nonzero rows of the matrix $(**)$. Then $\tau_\beta(\Delta_{I_0}) = 1$, and if I is any other choice of f_i rows of this matrix, we have $\tau_\beta(\Delta_I) = 0$. Thus, the prime ideal $\mathfrak{q}$ is generated by Δ_I for all $I \neq I_0$.

The map τ_α defining α sends Δ_{I_0} to s^{r_i}. We divide the set of Δ_I for $I \neq I_0$ into two sets. Let S denote the set of Δ_I for I containing a row below the lower identity matrix I in the matrix $(*)$; for $\Delta_I \in S$, $\tau_\alpha(\Delta_I) = 0$. If Δ_I is not in S, we have $\tau_\alpha(\Delta_I) = s^{r_i + k_I}$ for some $k_I > 0$. It follows that the kernel of τ_α is generated by Δ_I for $\Delta_I \in S$ and $\Delta_I - s^{k_I}\Delta_{I_0}$ for I not in S. If we now divide by the ideal generated by s, the latter elements become just Δ_I. Hence the ideals defining $(s) \cap \alpha$ and $(s) \cap \beta$ are both generated by Δ_I for all $I \neq I_0$, and thus the corresponding cycles are equal. $\square$

Denote the intersection of $\alpha(\mathcal{F}) - \beta(\mathcal{F})$ with (s) in the general construction by $\gamma(\mathcal{F})$. It now follows from Proposition 11.3.4 that the support of $\gamma(\mathcal{F})$ is contained in the inverse image in $G_{\mathcal{F}}$ of the support of $\mathcal{F}_\bullet$.

11.4 The Chern Character of a Complex of Locally Free Sheaves

Let Z be a closed subset of X that contains the support of $\mathcal{F}_\bullet$. For each scheme Y, let $A_*(Y)_\mathbb{Q}$ denote the Chow group of Y tensored with $\mathbb{Q}$. The local Chern character of $\mathcal{F}_\bullet$ will be defined as an operator from $A_*(X)_\mathbb{Q}$ to $A_*(Z)_\mathbb{Q}$. If $X = \mathrm{Spec}(A)$ for a local ring A and Z consists only of the maximal ideal, then each matrix M_i defining the map d_i in the complex $\mathcal{F}_\bullet$ (after choosing bases for the free modules) has support $\mathfrak{m}$, and the total Chern character of $\mathcal{F}_\bullet$ will be shown to be $\sum_i (-1)^i \mathrm{ch}(M_i)[A/\mathfrak{m}]$, where $\mathrm{ch}(M_i)$ is defined as in Section 1. As pointed out above, the Chow group of the point $\mathrm{Spec}(A/\mathfrak{m})$ consists only of $A_0(\mathrm{Spec}(A/\mathfrak{m}))_\mathbb{Q} \cong \mathbb{Q}$, and the result can be interpreted as a rational number.

In the previous section we constructed a cycle $\gamma(\mathcal{F})$ on the product of Grassmannians $G_{\mathcal{F}}$ and proved that this cycle has support in Z, so is in fact a cycle in $G_{\mathcal{F}_Z}$, where $\mathcal{F}_Z$ denotes the restriction of $\mathcal{F}$ to Z. We denote $G_{\mathcal{F}_Z}$ by $(G_{\mathcal{F}})_Z$. For each i, let p_i be the projection of $(G_{\mathcal{F}})_Z$ to $(G_i)_Z$, where, as above, $(G_i)_Z$ is the Grassmannian of rank f_i subsheaves of the restriction of $\mathcal{F}_i \oplus \mathcal{F}_{i-1}$ to Z. Let $\mathcal{Q}_i$ denote the pull-back of the canonical quotient sheaf on $(G_i)_Z$ to $(G_{\mathcal{F}})_Z$.

We next apply an alternating sum of the Chern characters $\mathcal{Q}_i$ and push down to $A_*(Z)$. Let p denote the projection from $(G_{\mathcal{F}})_Z$ to Z. We define the local Chern character $\mathrm{ch}(\mathcal{F})([X])$ by the formula

$$\mathrm{ch}(\mathcal{F})([X]) = p_* \left(\sum_{i=0}^{n+1} (-1)^{i+1} \mathrm{ch}(\mathcal{Q}_i)(\gamma(\mathcal{F})) \right).$$

We recall that the Chern character is defined to be a sum of components ch_j in various dimensions. Thus, for each j we have a corresponding component of the local Chern character

$$\mathrm{ch}_j(\mathcal{F})([X]) = p_* \left(\sum_{i=0}^{n+1} (-1)^{i+1} \mathrm{ch}_j(\mathcal{Q}_i)(\gamma(\mathcal{F})) \right).$$

Let $\eta = \sum n_i [X_i]$ be an arbitrary cycle in $A_*(X)_\mathbb{Q}$, where X_i is the cycle corresponding to a prime ideal $\mathfrak{p}_i$ for each i. For each i we let $(\mathcal{F}_\bullet)_{X_i}$ denote the restriction of $\mathcal{F}_\bullet$ to X_i. We now define

$$\mathrm{ch}(\mathcal{F})(\eta) = \sum n_i \mathrm{ch}((\mathcal{F}_\bullet)_{X_i})([X_i])$$

where we identify the classes in $A_*(X_i \cap Z)_\mathbb{Q}$ with their images in $A_*(Z)_\mathbb{Q}$.

An immediate consequence of Proposition 11.3.4 is that the local Chern character defined by an exact complex is zero.

Corollary 11.4.1 *If $\mathcal{F}_\bullet$ is exact, then* $\mathrm{ch}(\mathcal{F}_\bullet) = 0$.

Proof Since $\mathcal{F}_\bullet$ is exact, it follows from Proposition 11.3.4 that the cycle $\gamma(\mathcal{F})$ in the construction of the local Chern character is zero. Hence $\mathrm{ch}(\mathcal{F}_\bullet)(\eta) = 0$ for all η, so $\mathrm{ch}(\mathcal{F}_\bullet) = 0$.   □

While the local Chern character clearly defines a map from cycles on X to rational equivalence classes in $A_*(Z)$, it is not obvious that it is defined on the Chow group of X; that is, that it preserves rational equivalence. This fact will be proven in Section 11.6. We next prove the second property of the trivial cycle, that the above alternating sum on Chern characters vanishes on $\beta(\mathcal{F})$, a fact that is needed in proving that the construction is defined modulo rational equivalence.

Proposition 11.4.2 *Let η be a cycle in $A_*(G_\mathcal{F})_Z$ that is contained in the image of $(H_\mathcal{F})_Z$. Then*

$$\sum_{i=0}^{n+1}(-1)^{i+1}\mathrm{ch}(\mathcal{Q}_i)(\eta) = 0.$$

Proof By restricting to Z, we may assume that $Z = X$ and omit Z from the notation. We recall that the map from $H_\mathcal{F} = \prod G_{r_i}(\mathcal{F}_i)$ to $G_\mathcal{F} = \prod G_{f_i}(\mathcal{F}_i \oplus \mathcal{F}_{i-1})$ was defined to be the map that makes the following diagram commute for all i:

$$
\begin{array}{ccc}
\prod_j G_{r_j}(\mathcal{F}_j) & \longrightarrow & \prod_k G_{f_k}(\mathcal{F}_k \oplus \mathcal{F}_{k-1}) \\
\downarrow & & \downarrow \\
G_{r_i}(\mathcal{F}_i) \oplus G_{r_{i-1}}(\mathcal{F}_{i-1}) & \to & G_{f_i}(\mathcal{F}_i \oplus \mathcal{F}_{i-1})
\end{array}.
$$

Since all of these maps are proper, by the projection formula we can compute the result of the alternating sum of Chern classes of sheaves on the image of $H_\mathcal{F}$ by computing the result of the pull-back of the sheaves on cycles in $H_\mathcal{F}$. Now the pull-back of the canonical quotient sheaf on $G_{f_i}(\mathcal{F}_i \oplus \mathcal{F}_{i-1})$ to $G_{r_i}(\mathcal{F}_i) \oplus G_{r_{i-1}}(\mathcal{F}_{i-1})$ is the direct sum of the canonical quotient sheaves on the factors by Proposition 10.4.4. If we pull these back to $H_\mathcal{F}$ we obtain each summand twice, once with a positive sign and once with a negative sign. Hence in the alternating sum they cancel and the result is zero.   □

In the next proposition we consider the result of replacing α and β by other cycles in the computation of the local Chern character. It will be used to prove that local Chern characters commute with the operations of intersection theory; in many cases the construction of the cycle γ does not quite commute with these operations, but it will commute up to a cycle that will not affect the final result. The next result, which is taken from Fulton [17], shows that there is enough leeway in the construction of the cycles α and β.

Proposition 11.4.3 *Let $\alpha(\mathcal{F})$ and $\beta(\mathcal{F})$ be the cycles constructed from a complex $\mathcal{F}_\bullet$ as in the previous section. We consider the result of intersecting the difference $\alpha(\mathcal{F}) - \beta(\mathcal{F})$ with (s), applying the alternating sum of Chern characters of the canonical sheaves to the intersection, and pushing the resulting cycle down to Z as in the definition of local Chern characters.*

(i) *If $\alpha(\mathcal{F})$ is replaced by $\alpha(\mathcal{F}) + \alpha'$, where α' is supported in the subscheme defined by s in $Z[s]$, then the result is unchanged up to rational equivalence.*
(ii) *If $\beta(\mathcal{F})$ is replaced by $\beta(\mathcal{F}) + \beta'$, where β' is a cycle supported in Z and contained in the image of $(H_\mathcal{F})_Z$, then the result is unchanged up to rational equivalence.*

Proof The first statement follows from the fact that the intersection of any cycle supported in (s) with the divisor defined by s is rationally equivalent to zero. The second assertion follows immediately from Proposition 11.4.2. $\quad\square$

We use Proposition 11.4.3 to prove that local Chern characters commute with proper push-forward and flat pull-back.

Proposition 11.4.4 *Let $\mathcal{F}_\bullet$ be a bounded complex of locally free sheaves on Y with support Z.*

(i) *If $\phi : X \to Y$ is a proper map and $\eta \in Z_*(X)$, then*

$$\phi_*(\mathrm{ch}(\phi^*(\mathcal{F}_\bullet))(\eta)) = \mathrm{ch}(\mathcal{F}_\bullet)(\phi_*(\eta)) \quad \text{in } A_*(Z).$$

(ii) *If $\phi : X \to Y$ is a flat map of relative dimension k, and if $\eta \in Z_*(Y)$, then*

$$\phi^*(\mathrm{ch}(\mathcal{F}_\bullet)(\eta)) = \mathrm{ch}(\phi^*(\mathcal{F}_\bullet))(\phi^*(\eta)) \quad \text{in } A_*(\phi^{-1}(Z)).$$

Proof We prove the first of these statements; the proof of the second is similar. As usual, we may assume that we are computing the image of the class $[X]$.

With notation as in the construction of local Chern characters, we have a commutative diagram

$$\begin{array}{ccc} X[s, s^{-1}] & \to & G_{\phi^*(\mathcal{F})}[s, s^{-1}] \\ \downarrow & & \downarrow \\ Y[s, s^{-1}] & \to & G_{\mathcal{F}}[s, s^{-1}] \end{array}$$

where the horizontal arrows are the sections used in the construction corresponding to $\mathcal{F}_\bullet$ and $\phi^*(\mathcal{F}_\bullet)$, and the vertical arrows are induced by ϕ. Denote the map on the right by ϕ_G. Since the maps from $X[s, s^{-1}]$ and $Y[s, s^{-1}]$ into the corresponding products of Grassmannians are closed embeddings, we can deduce that the cycle defined by the image of $X[s, s^{-1}]$ is mapped by $(\phi_G)_*$ to the cycle defined by the image of $Y[s, s^{-1}]$. Hence, after restricting to the localization at s, we have

$$(\phi_G)_*(\alpha(\phi^*(\mathcal{F}))) = \alpha(\mathcal{F}).$$

Hence the difference $(\phi_G)_*(\alpha(\phi^*(\mathcal{F}))) - \alpha(\mathcal{F})$ in $G_{\mathcal{F}}[s]$ is supported in the divisor defined by s. A similar argument, applied to the diagram

$$\begin{array}{ccccc} X - \phi^{-1}(Z) & \to & \left(H_{\phi^*(\mathcal{F})}\right)_{X - \phi^{-1}(Z)} & \to & \left(G_{\phi^*(\mathcal{F})}\right)_{X - \phi^{-1}(Z)} \\ \downarrow & & \downarrow & & \downarrow \\ Y - Z & \to & (H_{\mathcal{F}})_{Y-Z} & \to & (G_{\mathcal{F}})_{Y-Z} \end{array}$$

shows that $(\phi_H)_*(\beta(\phi^*(\mathcal{F}))) = \beta(\mathcal{F})$ up to a cycle in the image of $H_{\mathcal{F}}$ supported in Z. Thus, by Proposition 11.4.3, we can use the cycles $(\phi_G)_*(\alpha(\phi^*(\mathcal{F})))$ and $(\phi_H)_*(\beta(\phi^*(\mathcal{F})))$ to carry out the construction of $\mathrm{ch}(\mathcal{F}_\bullet)([Y])$. Intersection with the divisor (s) commutes with proper push-forward by Theorem 8.8.2, the pull-backs of the canonical sheaves on $G_{\mathcal{F}}$ are the canonical sheaves on $G_{\phi^*(\mathcal{F})}$ by Proposition 10.4.1, and the Chern characters commute with proper push-forward by Proposition 9.3.1. Combining these facts proves the theorem. $\square$

We next give two examples of local Chern characters. First, we show that if $\mathcal{F}_\bullet$ consists of a single locally free sheaf $\mathcal{F}$ on X, then the local Chern character is just the ordinary Chern character of $\mathcal{F}$ up to a sign that depends on the degree in which $\mathcal{F}$ is situated. We assume that $\mathcal{F}$ is in degree zero; if $\mathcal{F}$ is in degree n, the result is multiplied by $(-1)^n$. In the construction of $\alpha(\mathcal{F})$ we have two nonzero pieces, in $G_r(\mathcal{F} \oplus 0)$ and $G_0(0 \oplus \mathcal{F})$, so that $G_{\mathcal{F}}$ is the product $G_r(\mathcal{F} \oplus 0) \times G_0(0 \oplus \mathcal{F})$. Now $G_r(\mathcal{F} \oplus 0)$ and $G_0(0 \oplus \mathcal{F})$ are the projective schemes defined by the symmetric algebras on $\Lambda^r(\mathcal{F}^*)$ and $\Lambda^0(\mathcal{F}^*) = \mathcal{O}_X$, respectively. Both of these are locally free of rank 1, so the associated projective schemes are isomorphic to X; hence $G_{\mathcal{F}} = X \times_X X = X$. Hence $\alpha(\mathcal{F})$ is the class of $G_{\mathcal{F}} = X$. In the first factor the canonical quotient sheaf is zero, and in the second it is $\mathcal{F}$. Since the support of $\mathcal{F}_\bullet$ is X (except in the trivial case where $\mathcal{F} = 0$), we can let $\beta(\mathcal{F}) = 0$. Thus, the result is simply the Chern character of the canonical sheaf applied to $[G_\mathcal{F}]$, which is equal to the Chern character of $\mathcal{F}$ applied to $[X]$.

As a second example, we work out the local Chern character defined by a map between two rank 1 locally free sheaves. This construction will be used to prove general results using the splitting principle. We first prove a Lemma.

Lemma 11.4.5 *Let $\mathcal{F}_\bullet$ be a complex of locally free sheaves and let $\mathcal{L}$ be a rank 1 locally free sheaf. Then* $\mathrm{ch}(\mathcal{L} \otimes \mathcal{F}_\bullet) = \mathrm{ch}(\mathcal{L})\mathrm{ch}(\mathcal{F}_\bullet)$.

Proof We first show that for every locally free sheaf $\mathcal{F}$, every nonnegative integer r less than or equal to the rank of $\mathcal{F}$, and every rank 1 locally free sheaf $\mathcal{L}$, there is an isomorphism between $G_r(\mathcal{F})$ and $G_r(\mathcal{L} \otimes \mathcal{F})$. Let C be the graded ring defining $G_r(\mathcal{F})$; then C is a quotient of the symmetric algebra on $\Lambda^r(\mathcal{F}^*)$. Similarly, $G_r(\mathcal{L} \otimes \mathcal{F})$ is a quotient of the symmetric algebra on $\Lambda^r((\mathcal{L} \otimes \mathcal{F})^*) = (\mathcal{L}^*)^{\otimes r} \otimes \Lambda^r(\mathcal{F}^*)$. Furthermore, since $\mathcal{L}$ is locally free of rank 1, the Grassmannians are locally isomorphic, so the graded ring defining $G_r(\mathcal{L} \otimes \mathcal{F})$ has $(\mathcal{L}^*)^{\otimes ir} \otimes C_i$ in the ith component.

To construct the isomorphism, we consider the bigraded ring B over A whose component in the ij position is $(\mathcal{L}^*)^{\otimes jr} \otimes C_i$. The ring of the Grassmannian $G_r(\mathcal{L} \otimes \mathcal{F})$ is the subalgebra $\sum_i B_{ii}$. Hence there is a quasi-projective map p from $\mathrm{Proj}(C)$ to $G_r(\mathcal{L} \otimes \mathcal{F})$. There is also a projective map q to $G_r(\mathcal{F})$ corresponding to the inclusion of $\sum B_{0i}$ into B. We claim that both maps are isomorphisms. The map q is an isomorphism since it is defined by the symmetric algebra of a rank 1 locally free sheaf. To check that p is an isomorphism, we use a local computation. Locally, we may assume that $X = \mathrm{Spec}(A)$ for a ring A and that $\mathcal{L} = A$. Then B is the bigraded polynomial ring $C[T]$ for an indeterminate T of degree $(0, 1)$, and p is induced by the map f defined by sending a homogeneous element c of degree i to cT^i. Let x be an element of degree 1 in C. Then x is mapped to xT, and the map induced by f from $C_{(x)}$ to $B_{(xT)}$ maps to c/x^i to $cT^i/(xT)^i$. It is thus clear that f defines an isomorphism onto $B_{(xT)}$, so p defines an isomorphism between the associated projective schemes.

Denote the resulting isomorphism from $G_r(\mathcal{L} \otimes \mathcal{F})$ to $G_r(\mathcal{F})$ by ϕ. Let $\mathcal{H}$ be a locally free subsheaf of $\mathcal{F}$ of rank r with locally free quotient; then $\mathcal{H}$ defines a corresponding subsheaf $\mathcal{L} \otimes \mathcal{H}$ of $\mathcal{L} \otimes \mathcal{F}$. The sections corresponding to $\mathcal{H}$ and $\mathcal{L} \otimes \mathcal{H}$ are induced by the maps from $\Lambda^r(\mathcal{F}^*)$ and $\mathcal{L}^{\otimes r} \otimes \Lambda^r(\mathcal{F}^*)$ to $\Lambda^r(\mathcal{H}^*)$ and $\mathcal{L}^{\otimes r} \otimes \Lambda^r(\mathcal{H}^*)$, and, since both maps are induced by the section on the bigraded ring B that maps $\Lambda^r(\mathcal{F}^*)$ to $\Lambda^r(\mathcal{F}^*)$ and is the identity on $\mathcal{L}^{\otimes r}$, this shows that

$$\phi \sigma_{\mathcal{L} \otimes \mathcal{H}} = \sigma_{\mathcal{H}}.$$

Furthermore, the canonical sheaves are quotients of $\mathcal{L} \otimes \mathcal{F}$ and $\mathcal{F}$, respectively, and by a naturality argument similar to that of Proposition 10.4.1, we have $\mathcal{E}_{\mathcal{L} \otimes \mathcal{F}} = \mathcal{L} \otimes \phi^*(\mathcal{E}_{\mathcal{F}})$ and $\mathcal{Q}_{\mathcal{L} \otimes \mathcal{F}} = \mathcal{L} \otimes \phi^*(\mathcal{Q}_{\mathcal{F}})$.

We now return to the computation of the local Chern character defined by $\mathcal{L} \otimes \mathcal{F}_\bullet$. From the preceding argument, there is an isomorphism, which we again denote ϕ, from $G_{\mathcal{L} \otimes \mathcal{F}}$ to $G_{\mathcal{F}}$. Furthermore, the graphs and kernels of the boundary maps in $\mathcal{L} \otimes \mathcal{F}_\bullet$ are obtained from the corresponding graphs and kernels of boundary maps in $\mathcal{F}_\bullet$ by tensoring with $\mathcal{L}$, so we have $\phi_*(\alpha(\mathcal{L} \otimes \mathcal{F})) = \alpha(\mathcal{F})$ and $\phi_*(\beta(\mathcal{L} \otimes \mathcal{F})) = \beta(\mathcal{F})$, so $\phi_*(\gamma(\mathcal{L} \otimes \mathcal{F})) = \gamma(\mathcal{F})$. Finally, the alternating sum of canonical sheaves on $(G_{\mathcal{L} \otimes \mathcal{F}})_Z$ is the tensor product of the pull-back of $\mathcal{L}$ with the alternating sum of canonical sheaves on $(G_{\mathcal{F}})_Z$ (where Z is the support of both $\mathcal{F}_\bullet$ and of $\mathcal{L} \otimes \mathcal{F}_\bullet$). Combining these results with the multiplicativity formula for Chern characters and the projection formula applied to the Chern character of $\mathcal{L}$ and the projection to Z, we have that

$$\mathrm{ch}(\mathcal{L} \otimes \mathcal{F}_\bullet)([X]) = \mathrm{ch}(\mathcal{L})\mathrm{ch}(\mathcal{F}_\bullet)([X])$$

in $A_*(Z)$, as was to be proven. □

Now let $\mathcal{L}_1 \to \mathcal{L}_0$ be a nonzero map of locally free sheaves of rank 1 on X, where $X = \mathrm{Proj}(A)$ for an integral domain A. We consider $\mathcal{L}_1 \to \mathcal{L}_0$ as a complex that for convenience we take in degrees 1 and 0. We first tensor with $\mathcal{L}_0^{-1}$, resulting in the complex

$$\cdots \to 0 \to \mathcal{L}_0^{-1} \otimes \mathcal{L}_1 \to \mathcal{L}_0^{-1} \otimes \mathcal{L}_0 = \mathcal{O}_X \to 0 \to \cdots.$$

The image of $\mathcal{L}_0^{-1} \otimes \mathcal{L}_1$ in $\mathcal{O}_X$ is a locally principal ideal, so it defines a divisor D. We denote $\mathcal{L}_0^{-1} \otimes \mathcal{L}_1$ by $\mathcal{O}_X(-D)$ and its dual by $\mathcal{O}_X(D)$. Thus, our original complex can be written

$$\mathcal{L}_0 \otimes (\cdots 0 \to \mathcal{O}_X(-D) \to \mathcal{O}_X \to 0 \cdots).$$

By Lemma 11.4.5, it suffices to compute the local Chern character of the complex $\mathcal{O}_X(-D) \to \mathcal{O}_X$. We denote this complex $\mathcal{F}_\bullet$.

We first note that $G_{\mathcal{F}}$ is the product

$$G_0(\mathcal{O}_X(-D)) \times G_1(\mathcal{O}_X(-D) \oplus \mathcal{O}_X) \times G_1(\mathcal{O}).$$

The first and third of these factors are isomorphic to X, so we have that $G_{\mathcal{F}} = G_1(\mathcal{O}_X(-D) \oplus \mathcal{O}_X)$. By the local computation at the end of the first section of this chapter, the cycle $\gamma(\mathcal{F})$ is simply the class of $G_{\mathcal{F}}$ restricted to

D, which we denote $[(G_{\mathcal{F}})_D]$. We now have to compute the result of applying the Chern characters of the canonical sheaves to this cycle.

As in the above computation of the Chern character of a single sheaf, the pull-backs of canonical sheaves on the first and third factors are pull-backs of sheaves on X. Since $\gamma(\mathcal{F})$ is supported in D and has dimension 1 more than that of D, we have $p_*(\gamma(\mathcal{F})) = 0$. Thus, by the projection formula, the contributions of these factors to the local Chern character are zero.

To compute the canonical quotient sheaf of $G_1(\mathcal{O}_X(-D) \oplus \mathcal{O}_X)$, we again consider the local situation. Suppose that the ideal $\mathcal{O}_X(-D)$ defining D is generated locally by x, so $\mathcal{O}_X(D)$ is generated locally by $1/x$. The Grassmannian $G_1(\mathcal{O}_X(-D) \oplus \mathcal{O}_X)$ is the projective scheme defined by the symmetric algebra on $(\mathcal{O}_X(-D) \oplus \mathcal{O}_X)^* = \mathcal{O}_X(D) \oplus \mathcal{O}_X$. Let p denote the projection of $G_{\mathcal{F}}$ onto X. By the definition in Chapter 10, the canonical quotient sheaf is the quotient of $\mathrm{Sym}((\mathcal{O}_X(-D) \oplus \mathcal{O}_X)^*)$ modulo $e_1^* \otimes e_1 + e_2^* \otimes e_2$, where e_1 and e_2 are local generators for $\mathcal{O}_X(-D)$ and $\mathcal{O}_X$ and e_1^* and e_2^* are their duals. We take $e_1 = x$ and $e_2 = 1$. Let S and T correspond to e_1^* and e_2^*, respectively, so that the symmetric algebra on $(\mathcal{O}_X(D) \oplus \mathcal{O}_X)$ is a polynomial ring $A[S, T]$. Under this identification, $e_1^* \otimes e_1 + e_2^* \otimes e_2$ corresponds to (xS, T) in $p^*(\mathcal{O}_X(-D) \oplus \mathcal{O}_X)$. Locally on X, (xS, T) generates the kernel of the map f from $A[S, T] \oplus A[S, T]$ to $A[S, T][1]$ that sends the first generator to $-T$ and the second to xS. We now interpret this map as a map from the component of degree 0 of $p^*(\mathcal{O}_X(-D) \oplus \mathcal{O}_X)$, which is just $\mathcal{O}_X(-D) \oplus \mathcal{O}_X$, to the component of degree 1 of $\mathrm{Sym}(\mathcal{O}_X(D) \oplus \mathcal{O}_X)$, which is $\mathcal{O}_X(D) \oplus \mathcal{O}_X$. The first generator $(x, 0)$ is sent to $-Tx$, which is $-x$ in the second component. The second generator $(0, 1)$ is sent to xS, which is x times the generator of $\mathcal{O}_x(D)$. Hence this map is a surjection onto $\mathcal{O}_X(-D) \otimes (\mathcal{O}_X(D) \oplus \mathcal{O}_X)$. Thus, the result of this computation is that the canonical quotient sheaf is $p^*(\mathcal{O}_X(-D)) \otimes \mathcal{O}_{G_{\mathcal{F}}}(1)$. Let h be the hyperplane on $G_{\mathcal{F}}$. Since h is a section of $\mathcal{O}_{G_{\mathcal{F}}}(1)$, we have that the Chern character of the canonical quotient sheaf is the product of the Chern character of the pull-back of $\mathcal{O}_X(-D)$ and the Chern character of the hyperplane.

We next compute the Chern classes of powers of the hyperplane. Since a generator of $\mathcal{O}_X$ has degree 0, we may divide by the ideal it generates in degree 1 in $\mathrm{Sym}(\mathcal{O}_X(D) \oplus \mathcal{O}_X)$ to compute $h([(G_{\mathcal{F}})_D])$. The quotient is the symmetric algebra on $\mathcal{O}_X(D)$, and a section of $\mathcal{O}_X(D)$ defines intersection with D. Hence we have

$$p_*(h^r([(G_{\mathcal{F}})_D])) = D^{r-1}([D])$$

for $r \geq 1$, where D^{r-1} denotes the result of intersection $r - 1$ times with the divisor D. In particular, $p_*(h([(G_{\mathcal{F}})_D])) = [D]$; the higher powers are rational equivalence classes of cycles supported in D.

We now combine these results to produce the formula for the local Chern character of $\mathcal{F}_\bullet = (\mathcal{O}_X(-D) \to \mathcal{O}_X)$. We have

$$\mathrm{ch}(\mathcal{F}_\bullet)([X]) = p_*(\mathrm{ch}(p^*(\mathcal{O}_X(-D)))e^h([(G_\mathcal{F})_D]))$$
$$= \mathrm{ch}(\mathcal{O}_X(-D))p_*((1 + h + h^2/2! + h^3/3! + \cdots)([(G_\mathcal{F})_D]))$$
$$= e^{-D}(1 + D/2! + D^2/3! + \cdots)([D])$$
$$= e^{-D}\left(\frac{e^D - 1}{D}\right)([D]) = 1 - e^{-D}.$$

The result is a power series in D with zero constant term, and the power D^n is defined to be the result of intersection with D^{n-1} on the class of D as above; the result is an element of $A_*(D)$.

We now combine this result with Lemma 11.4.5 and state it as a theorem.

Theorem 11.4.6 *Let* $\mathcal{F}_\bullet = 0 \to \mathcal{L}_1 \to \mathcal{L}_0 \to 0$ *be a complex of locally free sheaves of rank 1. Then if D is the divisor such that*

$$0 \to \mathcal{L}_1 \to \mathcal{L}_0 \to 0 = \mathcal{L}_0 \otimes (0 \to O(-D) \to 0 \to 0$$

we have

$$\mathrm{ch}(\mathcal{F}_\bullet)(\eta) = \mathrm{ch}(\mathcal{L}_0)(1 - e^{-D})(\eta)$$

for all $\eta \in A_*(X)_\mathbb{Q}$.

Note that in the case of local rings, D^2 and higher powers vanish and of course $\mathcal{L}_0$ is trivial, so the local Chern character is simply intersection with D.

11.5 Complexes of Free Modules over a Local Ring

In this section we prove some special properties of local Chern characters in the case of a complex of free modules over a local integral domain. We first show that the definition given in the first section agrees with the more general definition. Recall that in that section we defined the local Chern character of a matrix with support in the maximal ideal. Let A be a local domain of dimension d, and let $F_\bullet$ be a bounded complex of free A-modules with support equal to $\mathrm{Spec}(A/\mathfrak{m})$. We assume that the dimension d is at least one, so that the support of $F_\bullet$ is not all of $\mathrm{Spec}(A)$. For each i choose a basis of F_i, and let M_i be a matrix defining the map d_i^F with respect to the chosen bases. Then each matrix M_i has support in $\mathfrak{m}$ in the sense used in Section 1, so we have a local Chern character $\mathrm{ch}(M_i)$, which is a rational number, defined for each i. On the other hand, we defined a local Chern character $\mathrm{ch}_d(F_\bullet)$, and, applying this operator to the class $[A]$, this defines an element of $A_0(A/\mathfrak{m})_\mathbb{Q}$. The first result we show is that the coefficient of $[A/\mathfrak{m}]$ in $\mathrm{ch}_d([A])$ is the alternating sum of the Chern characters $\mathrm{ch}(M_i)$.

Proposition 11.5.1 *With the above notation, we have*

$$\mathrm{ch}_d(F_\bullet)([A]) = \left(\sum (-1)^i \mathrm{ch}(M_i)\right) [A/\mathfrak{m}].$$

Proof The first step in this computation is to divide the Chern character in the general construction into its components. In that construction, we defined cycles $\alpha(F)$ and $\beta(F)$ in the product of the G_i. For each i, let $\alpha_i(F)$ and $\beta_i(F)$ denote the projection onto the ith factor. Since both $\alpha(F)$ and $\beta(F)$ are defined by taking the closure of a section that is the product of sections on the factors, $\alpha_i(F)$ and $\beta_i(F)$ are the closures of the projections onto the ith factor. Furthermore, the Chern character used in the definition is the alternating sum of the pull-backs of the Chern characters of the canonical sheaves on the factors. Hence by the projection formula, the ith term in this sum is equal to

$$\mathrm{ch}_d(\mathcal{Q}_i)(\alpha_i(F) - \beta_i(F))$$

where $\mathcal{Q}_i$ denotes the canonical quotient sheaf on G_i. Thus, it suffices to show that $\mathrm{ch}_d(\mathcal{Q}_i)(\alpha_i(F) - \beta_i(F)) = \mathrm{ch}(M_i)$ for each i.

The proof of the latter statement follows from the explicit construction of the cycles $\alpha_i(F)$ and $\beta_i(F)$ in the local case computed in Propositions 11.3.2 and 11.3.3. It is shown in Proposition 11.3.2 that the cycle $\alpha_i(F)$ is defined by the map from the ring defining the Grassmannian G_i onto the Rees ring of the ideal of maximal minors of the matrix $\tilde{M}_i = \left(\begin{smallmatrix} sI \\ M_i \end{smallmatrix}\right)$. The cycle β_i was shown in Proposition 11.3.3 to be defined by the map to the Rees ring of the matrix of maximal minors of the matrix $\left(\begin{smallmatrix} K_i & 0 \\ 0 & K_{i-1} \end{smallmatrix}\right)$, where K_i and K_{i-1} are submatrices consisting of columns of M_{i+1} and M_i such that the ranks of K_i and K_{i-1} are equal to the ranks of M_i and M_{i-1}. Since $\beta_i(F)$ is defined as the closure of the corresponding section on the open set where $F_\bullet$ is exact, it suffices to show that this defines the correct cycle when $F_\bullet$ is exact. In this case, the computation of Proposition 11.3.4 shows that the corresponding ideal is generated by s^k for k greater than the rank of M_i. Thus, this is precisely the cycle discarded in the construction of Section 1. Since the remainder of the construction consisted of applying the Chern character of the canonical sheaf and pushing down to $\mathrm{Spec}(A/\mathfrak{m})$ just as in the general construction, this completes the proof. $\square$

The next proposition is useful in computations.

250   *Chern Classes*

Proposition 11.5.2 *Let J be the ideal of minors of $\tilde{M}$ as in the construction of Section 1. Let K be an ideal of $A[s]$ that is integral over J. Then we can compute the local Chern character of M using the ring $R(K)$ in place of $R(J)$.*

Proof Since K is integral over J, the Rees ring $R(K)$ is a finitely generated module over the Rees ring $R(J)$. Hence it also defines a finitely generated module over the ring R_G of the Grassmannian, so it defines cycles $[R(K)]_{d+1}$ and $[R(K)/sR(K)]_d$.

We must show that the two cycles $[R(J)/sR(J)]_d$ and $[R(K)/sR(K)]_d$ are equal. We show that for all minimal primes $\mathfrak{p}$ of $R(K)/sR(K)$, the length of $(R(K)/sR(K))_\mathfrak{p}$ equals the length of $(R(J)/sR(J))_\mathfrak{p}$. Consider the short exact sequence

$$0 \to R(J) \to R(K) \to N \to 0.$$

Since $s^n \in J$, for each element x of $R(K)$ there is some integer k for which $s^k x$ is in $R(J)$. Since $R(K)$ is a finitely generated module, there is a fixed k that works for all $x \in R(K)$, so that $s^k N = 0$. If we tensor this sequence with $A[s]/sA[s]$ we get the exact sequence

$$0 \to \text{Tor}_1(N, A[s]/sA[s]) \to R(J)/sR(J)$$
$$\to R(K)/sR(K) \to N \otimes A[s]/sA[s] \to 0.$$

To conclude the proof we must show that the length of the module $\text{Tor}_1(N, A[s]/sA[s])_\mathfrak{p}$ is equal to the length of $N \otimes A[s]/sA[s]_\mathfrak{p}$. If we have $sN = 0$ this is clear, since the two modules are both isomorphic to $N_\mathfrak{p}$. Thus, since the difference of these two lengths is an additive function on short exact sequences in N, the result follows by induction on the integer k for which $s^k N = 0$. $\square$

The above result can often be used in examples to replace a graded ring by one that is Cohen-Macaulay.

11.6 Compatibility with Intersection with Divisors

One of the most important properties of local Chern characters is that they commute with intersection with divisors. Let X be a projective scheme, let η be a cycle in $Z_k(X)$, let $\mathcal{F}_\bullet$ be a bounded complex of locally free sheaves with support Z, and let D be a divisor on X.

Theorem 11.6.1 *With the above notation, we have*

$$D \cap \text{ch}_i(\mathcal{F}_\bullet)(\eta) = \text{ch}_i((\mathcal{F}_\bullet)_D)(D \cap (\eta))$$

in $A_{k-i-1}(Z \cap D)$.

The proof of this theorem will be a step-by-step comparison of the cycles at each stage in the construction. Applying this theorem to the case of principal divisors, it will follow that local Chern characters are defined modulo rational equivalence and thus define operations on the Chow group.

For the most part, this theorem follows from the commutativity of intersection with divisors with the various operations that we have discussed in previous chapters. However, in the first step we need a stronger version of the commutativity of intersection with divisors than the original version proven in Proposition 8.9.1. The problem is that, when we intersect the difference $\alpha(\mathcal{F}) - \beta(\mathcal{F})$ with the divisor defined by s, we need to consider the result in the inverse image of the support of $\mathcal{F}_\bullet$. The theorem on commutativity of intersection with divisors states that intersection with divisors D and D' commutes up to rational equivalence in the intersection $D \cap D'$ of the supports of the divisors. In our case we need the intersection to commute up to rational equivalence in the intersection of $D \cap D'$ with the support of the complex $\mathcal{F}_\bullet$. Although the cycle $(s) \cap \gamma(\mathcal{F})$ used in the definition is supported in this set, the individual components of this cycle are not, so that the theorem on commutativity of intersection does not imply the stronger result as it stands. However, the behavior of the intersection of $\alpha(\mathcal{F}) - \beta(\mathcal{F})$ with (s) outside the support of $\mathcal{F}_\bullet$ has been described in detail, and we will show that the stronger version of commutativity holds in this situation.

For the purposes of the next proof we introduce a new definition. We say first that a divisor D intersects a cycle of the form $[Z] = [A/\mathfrak{q}]$ (so $Z = \mathrm{Proj}(A/\mathfrak{q})$) transversally at $\mathfrak{p}$ if Z is not contained in D, if $\mathfrak{p}$ is minimal over a local generator x of the divisor D in Z, and x generates $\mathfrak{p}$ locally in Z (so in particular the local ring $(A/\mathfrak{q})_{(\mathfrak{p})}$ is a discrete valuation ring). If W is a closed subset of X, we say that $D \cap \eta$ is *strongly equivalent to zero outside of* W if D meets every component of η in codimension at least 1 and if for every $\mathfrak{p}$ in the support of $D \cap \eta$ of dimension $i - 1$ such that $\mathfrak{p} \notin W$, η can be written as a sum of cycles $[A/\mathfrak{q}_i] - [A/\mathfrak{q}'_i]$ where D intersects $[A/\mathfrak{q}_i]$ and $[A/\mathfrak{q}'_i]$ transversally at $\mathfrak{p}$ together with a sum of cycles $[A/\mathfrak{q}]$ where $\mathfrak{q} \not\subseteq \mathfrak{p}$. It follows that $D \cap \eta$ is a well-defined cycle and that its support is contained in W. Our computations show that this condition holds for $(s) \cap (\alpha(\mathcal{F}) - \beta(\mathcal{F}))$ outside the support of $\mathcal{F}_\bullet$, since s is itself a prime ideal when restricted to either $\alpha(\mathcal{F})$ or $\beta(\mathcal{F})$ outside the support of $\mathcal{F}_\bullet$ and intersects both components in the same prime ideal of codimension 1.

Lemma 11.6.2 *Let X be an integral scheme, and let η be a cycle in $Z_k(X)$. Let D and D' be divisors on X, and let W and W' be closed subsets of X. Assume that $D \cap \eta$ is strongly equivalent to zero outside of W and $D' \cap \eta$ is*

strongly equivalent to zero outside of W'. *Then the intersection* $D \cap (D' \cap \eta)$
is rationally equivalent to $D' \cap (D \cap \alpha)$ *in* $Z_{k-2}(D \cap D' \cap W \cap W')$.

Proof As stated above, the difference between this Lemma and Proposition 8.9.1
is that the rational equivalence is in $D \cap D' \cap W \cap W'$ and not merely in $D \cap D'$.
We show, in fact, that the same proof goes through; we briefly recall the proof of
Proposition 8.9.1. If every component of η meets $D \cap D'$ in codimension 2, then
both sides of the equation agree as cycles, so the result is clear. For the general
case, we assumed in Proposition 8.9.1 that η was the class of X. We then blew
up the intersection $D \cap D'$ to reduce the value of the sum of products of orders
of D and D' at prime ideals of codimension 1 in X. Let Y denote this blow-up,
and let p be the projective map from Y to X. If D and D' are generated locally
by x and y, respectively, then Y is locally Proj(R), where R is the Rees ring
of the ideal (x, y). Let $\tilde{D}$ and $\tilde{D}'$ be the divisors generated locally by x and y,
respectively, in degree 1, and let E be the divisor generated locally by (x, y) in
degree 0. The proof was then completed by showing that $p^*(D) = \tilde{D} + E$ and
$p^*(D') = \tilde{D}' + E$, using the induction hypothesis on each of the four terms of
the intersection $(\tilde{D} + E) \cap (\tilde{D}' + E)$ and pushing the result down to X.

We now discuss how the proof goes over to our new situation. First, we
cannot assume that η is the class of X, but it follows from the hypotheses that
both D and D' meet every component of η in codimension at least 1. Let Z be
an irreducible closed subset of X such that $[Z]$ has nonzero coefficient in η. If
we let $\tilde{Z}$ be the proper transform of Z, then $\tilde{Z}$ is the blow-up of Z along the
intersection $D \cap D' \cap Z$. Thus, if $\eta = \sum n_i [Z_i]$, where each Z_i is a component,
and if we let $\tilde{\eta} = \sum n_i [\tilde{Z}_i]$, where $\tilde{Z}_i$ is the proper transform of Z_i, we may
apply the method of Proposition 8.9.1 to each component and the induction part
of the argument goes through as before.

To complete the proof, we must show that we have commutativity in each term
of the intersection $(\tilde{D} + E) \cap (\tilde{D}' + E)$ in $A_*(p^{-1}(D) \cap p^{-1}(D') \cap p^{-1}(W) \cap$
$p^{-1}(W'))$. It suffices to verify that $\tilde{D} \cap \tilde{\eta}$, $\tilde{D}' \cap \tilde{\eta}$, and $E \cap \tilde{\eta}$ are strongly equivalent to zero outside $p^{-1}(W)$, $p^{-1}(W')$, and $p^{-1}(W) \cap p^{-1}(W')$, respectively.
If this is true, then we can use induction (and the fact that E commutes with
itself) to conclude the result. Let q be a prime ideal of Y of dimension 1 less
than that of $\tilde{\eta}$ in $\tilde{D}$ or E such that q $\not\subseteq p^{-1}(W)$. Then q contracts to a prime
ideal p of X that is in D but not in W. We now use the hypothesis that $D \cap \eta$ is
strongly equivalent to zero outside W. We can write η as a sum of cycles $\eta_i - \eta'_i$
where η_i and η'_i are irreducible and D meets them transversally together with
components that do not contain p. The proper transforms of the components
that do not contain p will not contain q, so they do not have to be considered.
Let $\eta_i = [Z_i]$ be a component such that D meets η_i transversally. Then the

localization of $\mathfrak{p}$ in Z_i is a discrete valuation ring with maximal ideal generated by a local equation for D, which we call x. Locally at $\mathfrak{p}$, the map from the proper transform $\tilde{Z}_i$ to Z_i is an isomorphism, and under this isomorphism D and E correspond to the same divisor, and $\tilde{D}$ is trivial. Thus, q is in fact not in the support of $\tilde{D}$, while E meets $\tilde{\eta}_i$ transversally. Hence $\tilde{D}$ does not meet the support of $\tilde{\eta}$ outside $p^{-1}(W)$, while an expression for η as a sum of $\eta_i - \eta'_i$ where D meets η_i and η'_i gives a corresponding expression for $\tilde{\eta}$ as a sum of $\tilde{\eta}_i - \tilde{\eta}'_i$ where E intersects $\tilde{\eta}_i$ and $\tilde{\eta}'_i$ transversally. Thus, the conditions still hold, and they hold similarly for D'. Hence we have $p^*(D) \cap p^*(D') \cap \tilde{\eta} = p^*(D') \cap p^*(D) \cap \tilde{\eta}$ in $A_*(p^{-1}(D) \cap p^{-1}(D') \cap p^{-1}(W) \cap p^{-1}(W'))$ by induction, and thus by pushing down to X we have that $D \cap D' \cap \eta = D' \cap D \cap \eta$ in $A_*(D \cap D' \cap W \cap W')$. $\qquad\square$

We now continue with the proof that local Chern characters commute with intersection with divisors. As usual, we may assume that $\eta = [X]$. Consider the cycle $\alpha(\mathcal{F}) - \beta(\mathcal{F})$ as defined in the general construction. As discussed above, we have that the intersection $(s) \cap (\alpha(\mathcal{F}) - \beta(\mathcal{F}))$ is strongly equivalent to zero outside the support Z of $\mathcal{F}_\bullet$. Hence, by Lemma 11.6.2, the cycles $D \cap (s) \cap (\alpha(\mathcal{F}) - \beta(\mathcal{F}))$ and $(s) \cap D \cap (\alpha(\mathcal{F}) - \beta(\mathcal{F}))$ are rationally equivalent in $(G_{\mathcal{F}})_Z$. Furthermore, if we localize by inverting s, then $D \cap (\alpha(\mathcal{F}))$ is equal to the cycle $\alpha(\mathcal{F}_D)$, and if we restrict to $X - Z$, then $D \cap (\beta(\mathcal{F}))$ is equal to $\beta(\mathcal{F}_D)$. Hence by Proposition 11.4.3 we can use $D \cap (\alpha(\mathcal{F}) - \beta(\mathcal{F}))$ in the computation of the local Chern character of $\mathcal{F}_D$. Since the other parts of the construction have already been shown to commute with intersection with divisors, this completes the proof.

As mentioned above, we have the following corollary.

Corollary 11.6.3 *Local Chern characters are well-defined modulo rational equivalence.*

Proof This corollary follows from Theorem 11.6.1 applied to a divisor of the form $\mathrm{div}(\mathfrak{p}, x) - \mathrm{div}(\mathfrak{p}, y)$, where x and y are homogeneous elements not in $\mathfrak{p}$ of the same degree. $\qquad\square$

11.7 Composition of Local Chern Characters

The most important use of local Chern characters in commutative algebra is to compute and prove results on intersection multiplicities. For this purpose it is necessary to discuss the composition of local Chern characters defined by two different complexes. In this section we present this construction in detail. In fact, we present an alternative construction that may be considered as a kind of generalization from the case of one complex to a definition of local Chern characters for two complexes. It is analogous to the construction of a product of Chern classes of locally free sheaves described in Section 9.4.

Let $\mathcal{E}_\bullet$ and $\mathcal{F}_\bullet$ be bounded complexes of locally free sheaves with supports V and W, respectively. Let i and j be positive integers. We wish to discuss different constructions of $\mathrm{ch}_i(\mathcal{E}_\bullet)\mathrm{ch}_j(\mathcal{F}_\bullet)(\eta)$, where η is a cycle. We may assume as usual that $\eta = [X]$.

Let $G_\mathcal{E}$ and $G_\mathcal{F}$ be the products of Grassmannians in the constructions of $\mathrm{ch}_i(\mathcal{E}_\bullet)$ and $\mathrm{ch}_j(\mathcal{F}_\bullet)$. In the previous construction we introduced an auxiliary variable s; here we introduce two indeterminates, s and t. We recall that the first step in the construction (for $\mathcal{E}_\bullet$) is to take the difference of the closures of two sections of $G_\mathcal{E}$ and then intersect with the divisor (s). In this case we first take a combination of four sections of the product $G_\mathcal{E} \times G_\mathcal{F}$.

Let $\sigma_\mathcal{E}$ be the section of $G_\mathcal{E}$ on $X[s, s^{-1}]$ used in the construction of $\alpha(\mathcal{E})$ and let $\overline{\sigma}_\mathcal{E}$ denote the section of $G_\mathcal{E}$ on $(X - V)[s]$ used in the construction of $\beta(\mathcal{E})$. Define the sections $\sigma_\mathcal{F}$ and $\overline{\sigma}_\mathcal{F}$ of $G_\mathcal{F}$ similarly on open subsets of $X[t]$. We then have sections $\sigma_\mathcal{E} \times \sigma_\mathcal{F}$, $\sigma_\mathcal{E} \times \overline{\sigma}_\mathcal{F}$, $\overline{\sigma}_\mathcal{E} \times \sigma_\mathcal{E}$, and $\overline{\sigma}_\mathcal{E} \times \overline{\sigma}_\mathcal{F}$ of the projection from $G_\mathcal{E} \times G_\mathcal{F}[s, t]$ to $X[s, t]$ defined over certain open subsets of $X[s, t]$. We denote the closures of these sections $\alpha(\mathcal{E})\alpha(\mathcal{F})$, $\alpha(\mathcal{E})\beta(\mathcal{F})$, $\beta(\mathcal{E})\alpha(\mathcal{F})$, and $\beta(\mathcal{E})\beta(\mathcal{F})$. Let

$$\tilde{\gamma}(\mathcal{E}, \mathcal{F}) = \alpha(\mathcal{E})\alpha(\mathcal{F}) - \alpha(\mathcal{E})\beta(\mathcal{F}) - \beta(\mathcal{E})\alpha(\mathcal{F}) + \beta(\mathcal{E})\beta(\mathcal{F}).$$

Then $\tilde{\gamma}(\mathcal{E}, \mathcal{F})$ is a cycle in $(G_\mathcal{E} \times G_\mathcal{F})[s, t]$.

We next intersect with the divisors defined by s and t. We use Lemma 11.6.2, where (s) and (t) are the divisors D and D' and the closed sets W and W' of that theorem are the supports V and W of $\mathcal{E}_\bullet$ and $\mathcal{F}_\bullet$, respectively. We write $\tilde{\gamma}(\mathcal{E}, \mathcal{F})$ in the form

$$(\alpha(\mathcal{E})\alpha(\mathcal{F}) - \beta(\mathcal{E})\alpha(\mathcal{F})) + (-\alpha(\mathcal{E})\beta(\mathcal{F}) + \beta(\mathcal{E})\beta(\mathcal{F})).$$

At points where $\mathcal{E}_\bullet$ is exact, s generates the prime ideal that defines $\alpha(\mathcal{F})$ in each of the first two components of this cycle, and it defines the same prime ideal defining $\beta(\mathcal{F})$ in each of the last two components. Hence, in the terminology of Lemma 11.6.2, intersection with (s) is strongly equivalent to zero outside V; similarly, intersection with (t) is strongly equivalent to zero outside W. Thus, by Lemma 11.6.2 the intersections are rationally equivalent in $(G_\mathcal{E} \times G_\mathcal{F})_{V \cap W}$.

Let p_m and q_n denote the projections from $G(\mathcal{E}) \times G(\mathcal{F})_{V \cap W}$ onto the mth factor of $G_\mathcal{E}$ and the nth factor of $G_\mathcal{F}$, respectively. We denote the canonical quotient sheaves on the factors by $\mathcal{Q}_m(\mathcal{E})$ and $\mathcal{Q}_n(\mathcal{F})$. We denote the projection to $V \cap W$ by p. Then we define the i, j component of the local Chern character of $\mathcal{E}$ and $\mathcal{F}$, which we denote $\mathrm{ch}_{i,j}(\mathcal{E}_\bullet, \mathcal{F}_\bullet)([X])$, to be

$$p_* \left(\sum_{m,n} (-1)^{m+n} \mathrm{ch}_i \left(p_m^*(\mathcal{Q}_m(\mathcal{E})) \right) \mathrm{ch}_j \left(q_n^*(\mathcal{Q}_n(\mathcal{F})) \right) ((s) \cap (t) \cap \bar{\gamma}(\mathcal{E}, \mathcal{F})) \right).$$

Proposition 11.7.1 *We have*

$$\mathrm{ch}_{i,j}(\mathcal{E}_\bullet, \mathcal{F}_\bullet) = \mathrm{ch}_i(\mathcal{E}_\bullet)\mathrm{ch}_j(\mathcal{F}_\bullet) = \mathrm{ch}_j(\mathcal{F}_\bullet)\mathrm{ch}_i(\mathcal{E}_\bullet)$$

as operators from $A_*(X)$ to $A_*(V \cap W)$.

Proof We prove the first equality; the second is proven the same way. We first consider the following diagram:

$$
\begin{array}{ccc}
Z_*(G_\mathcal{E} \times G_\mathcal{F}[s,t]) & \overset{\cap s}{\to} & Z_*(G_\mathcal{E} \times G_\mathcal{F}[t]_V) \to \\
\downarrow & & \downarrow \\
Z_*(G_\mathcal{E} \times G_\mathcal{F}[s]_W) & \overset{\cap s}{\to} & Z_*((G_\mathcal{E} \times G_\mathcal{F})_{V \cap W}) \to \\
\downarrow & & \downarrow \\
A_* G_\mathcal{E} \times G_\mathcal{F}[s]_W) & \overset{\cap s}{\to} & A_*((G_\mathcal{E} \times G_\mathcal{F})_{V \cap W}) \to \\
\downarrow & & \downarrow \\
A_*(G_\mathcal{E}[s]_W) & \overset{\cap s}{\to} & A_*((G_\mathcal{E})_{V \cap W}) \to
\end{array}
$$

$$
\begin{array}{ccc}
\to & A_*(G_\mathcal{E} \times G_\mathcal{F}[t]_V) & \to & A_*(G_\mathcal{F}[t]_V) \\
& \downarrow & & \downarrow \\
\to & A_*((G_\mathcal{E} \times G_\mathcal{F})_{V \cap W}) & \to & A_*((G_\mathcal{F})_{V \cap W}) \\
& \downarrow & & \downarrow \\
\to & A_*((G_\mathcal{E} \times G_\mathcal{F})_{V \cap W}) & \to & A_*((G_\mathcal{F})_{V \cap W}) \\
& \downarrow & & \downarrow \\
\to & A_*((G_\mathcal{E})_{V \cap W}) & \to & A_*(V \cap W)
\end{array}
$$

The horizontal arrows are intersection with (s), application of the Chern characters of the pull-backs of the canonical bundles on $G_\mathcal{E}$, and push-down by the projection from $G_\mathcal{E}$ to V (or maps induced by this projection). The vertical maps are defined similarly using (t) and $G_\mathcal{F}$. We have just shown that the upper left square commutes up to rational equivalence, and the remaining squares commute by various theorems that establish the compatibility of the operations involved.

The bottom row of this diagram is the sequence of operations in the definition of the local Chern character of $\mathcal{E}_\bullet$. Hence to prove the result it will suffice to show that the image of $\alpha(\mathcal{E})\alpha(\mathcal{F}) - \alpha(\mathcal{E})\beta(\mathcal{F})$ under the composition of maps in the column on the left can be used for $\alpha(\mathcal{E})$ in the construction of the local Chern character of $\mathcal{E}_\bullet$ applied to a cycle representing $\mathrm{ch}_n(\mathcal{F}_\bullet)([X])$

and similarly that the image $\beta(\mathcal{E})\alpha(\mathcal{F}) - \beta(\mathcal{E})\beta(\mathcal{F})$ can be used for $\beta(\mathcal{E})$. We prove the first statement. By Proposition 11.4.3 we may localize at s and show that the image of $\alpha(\mathcal{E})\alpha(\mathcal{F}) - \alpha(\mathcal{E})\beta(\mathcal{F})$ restricts to the correct cycle in the localization. After localization, we have a section from $X[s, s^{-1}]$ to $G_{\mathcal{E}}[s, s^{-1}]$. Taking the product with $G_{\mathcal{F}}[t]$, this section induces a section from $G_{\mathcal{F}}[s, s^{-1}, t]$ to $G_{\mathcal{E}} \times G_{\mathcal{F}}[s, s^{-1}, t]$. Furthermore, by construction this section takes $\alpha(\mathcal{F}) - \beta(\mathcal{F})$ to $\alpha(\mathcal{E})\alpha(\mathcal{F}) - \alpha(\mathcal{E})\beta(\mathcal{F})$. We now have a commutative diagram:

$$Z_*(G_{\mathcal{F}}[s, s^{-1}, t]) \to Z_*(G_{\mathcal{E}} \times G_{\mathcal{F}}[s, s^{-1}, t])$$
$$\downarrow \qquad\qquad\qquad \downarrow$$
$$Z_*(G_{\mathcal{F}}[s, s^{-1}]_w) \to Z_*(G_{\mathcal{E}} \times G_{\mathcal{F}}[s, s^{-1}]_w)$$
$$\downarrow \qquad\qquad\qquad \downarrow$$
$$A_*(G_{\mathcal{F}}[s, s^{-1}]_w) \to A_*(G_{\mathcal{E}} \times G_{\mathcal{F}}[s, s^{-1}]_w)$$
$$\downarrow \qquad\qquad\qquad \downarrow$$
$$A_*(W[s, s^{-1}]) \to A_*(G_{\mathcal{E}}[s, s^{-1}]_w)$$

The maps in the left-hand column are those in the construction of $\text{ch}_j(\mathcal{F})([X])$ pulled back to $X[s, s^{-1}]$, while those on the right-hand side are the maps in the previous diagram. Hence the result of applying the vertical maps in that diagram gives a class that represents $\alpha(\mathcal{E})$ for the cycle $\text{ch}_j(\mathcal{F}_\bullet)([X])$. A similar argument, restricting to $X - V$, shows that the image of $\beta(\mathcal{E})\alpha(\mathcal{F}) - \beta(\mathcal{E})\beta(\mathcal{F})$ can be used in the construction of $\beta(\mathcal{E})$ for $\text{ch}_j(\mathcal{F}_\bullet)([X])$. Since the maps in the bottom row are those used in the construction of the local Chern character of $\mathcal{E}_\bullet$, this proves that $\text{ch}_{i,j}(\mathcal{E}_\bullet, \mathcal{F}_\bullet) = \text{ch}_i(\mathcal{E}_\bullet)\text{ch}_j(\mathcal{F}_\bullet)$. A similar argument, reversing the roles of $\mathcal{E}$ and $\mathcal{F}$, shows that $\text{ch}_{i,j}(\mathcal{E}_\bullet, \mathcal{F}_\bullet) = \text{ch}_j(\mathcal{F}_\bullet)\text{ch}_i(\mathcal{E}_\bullet)$. $\quad\square$

It follows from Proposition 11.7.1 that Chern classes commute with each other. This is an important property, and we state it as a separate theorem.

Theorem 11.7.2 *Let $\mathcal{E}_\bullet$ and $\mathcal{F}_\bullet$ be complexes with supports V and W, respectively. Let i and j be nonnegative integers, and Let η be an element of $A_*(X)$. Then*

$$\text{ch}_i(\mathcal{E}_\bullet)\text{ch}_j(\mathcal{F}_\bullet)(\eta) = \text{ch}_j(\mathcal{F}_\bullet)\text{ch}_i(\mathcal{E}_\bullet)(\eta)$$

in $A_*(V \cap W)$.

Exercises

11.1 Verify that the ideal $\mathfrak{a}$ in Section 1 can be taken to be the ideal generated by I_r.

11.2 Let $A = k[[X, Y]]$, and let M be the matrix

$$M = \begin{pmatrix} X^2 & Y^2 & 0 \\ 0 & -X & Y \end{pmatrix}.$$

Show that M has support in the maximal ideal of A as defined in Section 11.1 and compute the local Chern character $ch_2(M)$.

11.3 Let N be the cokernel of the matrix in the previous problem. Complete M to a minimal free resolution of N, and let M' be the second matrix in the resolution. Compute $ch_2(M')$ and show that $ch_2(M) - ch_2(M')$ is equal to the length of the module N.

11.4 Carry out the construction of the local Chern character in the case of a Koszul complex on two elements.

11.5 Follow the procedure preceding Theorem 11.4.6 to compute the local Chern character of the complex $\mathcal{O}_X \to \mathcal{O}_X(D)$ and verify that

$$ch(\mathcal{O}_X \to \mathcal{O}_X(D)) = ch(\mathcal{O}_X(D))ch(\mathcal{O}_X(-D) \to \mathcal{O}_X).$$

11.6 Let $A = k[[X, Y]]$. Carry out the computation of $ch_{1,1}(\mathcal{E}_\bullet, \mathcal{F}_\bullet)$ as in Proposition 11.7.1 for

$$\mathcal{E}_\bullet = 0 \to A \xrightarrow{X} A \to 0$$

and

$$\mathcal{F}_\bullet = 0 \to A \xrightarrow{Y} A \to 0.$$

12

Properties of Local Chern Characters

In this chapter we prove the properties of local Chern characters that are needed for our applications. These include additivity and multiplicativity formulas as well as the local Riemann-Roch formula. To prove the multiplicativity formula and the local Riemann-Roch formula we use a version of the splitting principle for complexes. We also show that in positive characteristic local Chern characters can be computed using Dutta multiplicities.

12.1 The Additivity Formula

In this section we show that local Chern characters are additive on short exact sequences of complexes.

We first recall the additivity formula for Grassmannians. This theorem, which is proven in Section 10.4, states that given two locally free sheaves $\mathcal{E}$ and $\mathcal{F}$ we have a map $\psi_{\oplus}$ from the product $G_k(\mathcal{E}) \times G_r(\mathcal{F})$ to $G_{k+r}(\mathcal{E} \oplus \mathcal{F})$ induced by taking direct sums. The pull-back by $\psi_{\oplus}$ of the canonical sheaf on $G_{k+r}(\mathcal{E} \oplus \mathcal{F})$ is the direct sum of the pull-backs of the canonical sheaves on the factors of the product, and the section defined by a direct sum of subsheaves is the product of the corresponding sections on the factors.

We first prove additivity in the case where the short exact sequence is split.

Proposition 12.1.1 *Let* $\mathcal{F}_{\bullet} = \mathcal{F}'_{\bullet} \oplus \mathcal{F}''_{\bullet}$. *Then* $\mathrm{ch}(\mathcal{F}_{\bullet}) = \mathrm{ch}(\mathcal{F}'_{\bullet}) + \mathrm{ch}(\mathcal{F}''_{\bullet})$.

Proof We assume that $X = \mathrm{Proj}(A)$ where A is an integral domain and that we are computing the Chern characters on the class $[X]$. Let Z be a closed subset of X that contains the support of $\mathcal{F}'_{\bullet} \oplus \mathcal{F}''_{\bullet}$. We note that the support of $\mathcal{F}_{\bullet}$ is the union of the supports of $\mathcal{F}'_{\bullet}$ and $\mathcal{F}''_{\bullet}$, so Z will also contain the supports of $\mathcal{F}'_{\bullet}$ and $\mathcal{F}''_{\bullet}$.

The main point of the proof is that for each i, the graph of $d_i^{\mathcal{F}}$ is the direct sum of the graphs of $d_i^{\mathcal{F}'}$ and $d_i^{\mathcal{F}''}$. Let $G_{\mathcal{F}'}$, $G_{\mathcal{F}''}$, and $G_{\mathcal{F}'\oplus\mathcal{F}''}$ denote the products of Grassmannians used in the constructions of the cycles α and β for $\mathcal{F}'$, $\mathcal{F}''$, and $\mathcal{F}' \oplus \mathcal{F}''$. Let $\psi_{\oplus}$ denote the map from $G_{\mathcal{F}'} \times G_{\mathcal{F}''}$ to $G_{\mathcal{F}'\oplus\mathcal{F}''}$ induced by the direct sum maps

$$G_{f_i'}\left(\mathcal{F}_i' \oplus \mathcal{F}_{i-1}'\right) \times G_{f_i''}\left(\mathcal{F}_i'' \oplus \mathcal{F}_{i-1}''\right) \to G_{f_i'+f_i''}\left(\left(\mathcal{F}_i' \oplus \mathcal{F}_i''\right) \oplus \left(\mathcal{F}_{i-1}' \oplus \mathcal{F}_{i-1}''\right)\right)$$

in the ith factor. Then if $\sigma_{\mathcal{F}'} \times \sigma_{\mathcal{F}''}$ denotes the product of the sections induced by the graphs on $X[s, s^{-1}]$ in the construction of α, we have

$$\psi_{\oplus}(\sigma_{\mathcal{F}'} \times \sigma_{\mathcal{F}''}) = \sigma_{\mathcal{F}'\oplus\mathcal{F}''}.$$

A similar argument applies to the cycle β, using the fact that the kernel of $d_i^{\mathcal{F}'\oplus\mathcal{F}''}$ is the direct sum of the kernels of $d_i^{\mathcal{F}'}$ and $d_i^{\mathcal{F}''}$. Let $\alpha(\mathcal{F}', \mathcal{F}'')$ be the closure of the image of $\sigma_{\mathcal{F}'} \times \sigma_{\mathcal{F}''}$ in $G_{\mathcal{F}'} \times G_{\mathcal{F}''}$, and define $\beta(\mathcal{F}', \mathcal{F}'')$ similarly using the sections defined by the kernels of the boundary maps (if $Z = X$, we let $\beta(\mathcal{F}', \mathcal{F}'') = 0$). Let $\gamma(\mathcal{F}', \mathcal{F}'') = (s) \cap (\alpha(\mathcal{F}', \mathcal{F}'') - \beta(\mathcal{F}', \mathcal{F}''))$. Let p' and p'' denote the projections from $G_{\mathcal{F}'} \times G_{\mathcal{F}''}$ to $G_{\mathcal{F}'}$ and $G_{\mathcal{F}''}$, respectively. Then, using Proposition 11.4.3, we have, up to cycles that do not affect the result of computing local Chern characters

$$p_*'(\gamma(\mathcal{F}', \mathcal{F}'')) = \gamma(\mathcal{F}') \quad \text{and} \quad p_*''(\gamma(\mathcal{F}', \mathcal{F}'')) = \gamma(\mathcal{F}'');$$

$$\psi_{\oplus}(\gamma(\mathcal{F}', \mathcal{F}'')) = \gamma(\mathcal{F}' \oplus \mathcal{F}'').$$

These equalities hold in the restrictions of the respective Grassmannians to Z.

It remains to consider the alternating sums of canonical sheaves. Let $\mathcal{Q}(\mathcal{F}')$ and $\mathcal{Q}(\mathcal{F}'')$ denote the alternating sums of canonical sheaves on $(G_{\mathcal{F}'})_Z$ and $(G_{\mathcal{F}''})_Z$, respectively, used in the definition of the local Chern characters defined by $\mathcal{F}_\bullet'$ and $\mathcal{F}_\bullet''$. Let $\mathcal{Q}(\mathcal{F}' \oplus \mathcal{F}'')$ denote the corresponding alternating sum of pull-backs of the canonical sheaves on $(G_{\mathcal{F}'\oplus\mathcal{F}''})_Z$. By Proposition 10.4.4, we thus have

$$\psi_{\oplus}^*(\mathcal{Q}(\mathcal{F}' \oplus \mathcal{F}'')) = p^*(\mathcal{Q}(\mathcal{F}')) \oplus (p')^*(\mathcal{Q}(\mathcal{F}'')).$$

Let q denote any of the projections from the Grassmannians under consideration to Z. Then, combining these formulas and using the functorial properties of local Chern characters, we have

$$\mathrm{ch}\big(\mathcal{F}_\bullet'\big)([X]) + \mathrm{ch}\big(\mathcal{F}_\bullet''\big)([X])$$

$$= q_*(\mathrm{ch}(\mathcal{Q}(\mathcal{F}'))(\gamma(\mathcal{F}'))) + q_*(\mathrm{ch}(\mathcal{Q}(\mathcal{F}''))(\gamma(\mathcal{F}'')))$$

$$= q_*(\mathrm{ch}(\mathcal{Q}(\mathcal{F}'))(p_*(\gamma(\mathcal{F}', \mathcal{F}'')))) + q_*\big(\mathrm{ch}(\mathcal{Q}(\mathcal{F}''))\big(p_*'(\gamma(\mathcal{F}', \mathcal{F}''))\big)\big)$$

$$= q_*(\mathrm{ch}(p^*(\mathcal{Q}(\mathcal{F}')))(\gamma(\mathcal{F}', \mathcal{F}''))) + q_*(\mathrm{ch}((p')^*(\mathcal{Q}(\mathcal{F}'')))(\gamma(\mathcal{F}', \mathcal{F}'')))$$

$$= q_*(\mathrm{ch}(p^*(\mathcal{Q}(\mathcal{F}'))) \oplus (p')^*(\mathcal{Q}(\mathcal{F}'')))(\gamma(\mathcal{F}', \mathcal{F}'')))$$

$$= q_*\big(\mathrm{ch}\big(\psi_\oplus^*(\mathcal{Q}(\mathcal{F}' \oplus \mathcal{F}''))\big)\big)(\gamma(\mathcal{F}', \mathcal{F}'')))$$

$$= q_*(\mathrm{ch}(\mathcal{Q}(\mathcal{F}' \oplus \mathcal{F}''))((\psi_\oplus)_*(\gamma(\mathcal{F}', \mathcal{F}''))))$$

$$= q_*(\mathrm{ch}(\mathcal{Q}(\mathcal{F}' \oplus \mathcal{F}''))(\gamma(\mathcal{F}' \oplus \mathcal{F}''))) = \mathrm{ch}\big(\mathcal{F}'_\bullet \oplus \mathcal{F}''_\bullet\big)([X]).$$

$$\square$$

From this result we may deduce the general additivity formula by deforming to the split case.

Theorem 12.1.2 *Let* $X = \mathrm{Proj}(A)$, *and let*

$$0 \to \mathcal{F}'_\bullet \to \mathcal{F}_\bullet \to \mathcal{F}''_\bullet \to 0$$

be a short exact sequence of bounded free complexes on X. *Let* Z *be a closed subset of* X *containing the support of* $\mathcal{F}_\bullet$. *Then*

$$\mathrm{ch}(\mathcal{F}_\bullet) = \mathrm{ch}\big(\mathcal{F}'_\bullet\big) + \mathrm{ch}\big(\mathcal{F}''_\bullet\big)$$

in $A_*(Z)$.

Proof We assume that A is an integral domain and that we are computing the result on the class $[X]$. Let T be an indeterminate of degree 0, and let ϕ be the map from $Y = \mathrm{Proj}(A[T])$ to $X = \mathrm{Proj}(A)$ induced by the inclusion of A in $A[T]$ (ϕ is a quasi-projective map but is not proper). Let h be the surjection of complexes of locally free sheaves $h : \phi^*(\mathcal{F}_\bullet) \oplus \phi^*(\mathcal{F}''_\bullet) \to \phi^*(\mathcal{F}''_\bullet)$ defined in degree i by

$$h\big(f_i, f''_i\big) = p(f_i) - T f''_i$$

where p is the original map from $\mathcal{F}_i$ to $\mathcal{F}''_i$. Let $\mathcal{E}_\bullet$ be the kernel of h, so that we have a short exact sequence

$$0 \to \mathcal{E}_\bullet \to \phi^*(\mathcal{F}_\bullet) \oplus \phi^*\big(\mathcal{F}''_\bullet\big) \to \phi^*\big(\mathcal{F}''_\bullet\big) \to 0.$$

Let h_0 be the result of tensoring the map h with $A[T]/T$. Then h_0 is the direct sum of the original map p and the zero map, so the kernel of h_0 is $\mathcal{F}'_\bullet \oplus \mathcal{F}''_\bullet$. Let h_1 be the result of tensoring h with $A[T]/(T-1)$. Then $h_1(f, f'') = p(f) - f''$, which is split by the map k defined by $k(f'') = (0, f'')$, so the kernel of h_1 is $\mathcal{F}_\bullet$.

To complete the proof we must interpret these results as intersections with the divisors (T) and $(T - 1)$ and show that they are rationally equivalent. To do this we replace Y with $\tilde{Y} = \mathrm{Proj}(A[S, V])$, where S and V have degree 1 over A (more precisely, S and V have degree $(0, \ldots, 0, 1) \in \mathbb{Z}^{n+1}$, where A is a $\mathbb{Z}^n$-graded ring), so that the map $\tilde{\phi}$ from $\tilde{Y}$ to X is projective. We consider the map $\tilde{h}$ from $\tilde{\phi}^*(\mathcal{F}_\bullet) \oplus \tilde{\phi}^*(\mathcal{F}''_\bullet)$ to $\tilde{\phi}^*(\mathcal{F}''_\bullet)[1]$ defined in each degree by letting

$$\tilde{h}(f, f'') = Sp(f) - Vf''.$$

The supports of the divisors (V) and $(S - V)$ are contained in the basic open set $U_{(S)}$, which is isomorphic to $\mathrm{Proj}(A[V/S]) \cong \mathrm{Proj}(A[T])$. Under this isomorphism the map $\tilde{h}$ corresponds to the map h defined above, and the divisors (V) and $(S - V)$ correspond to (T) and $(T - 1)$. Let $\tilde{\mathcal{E}}_\bullet$ be the kernel of $\tilde{h}$. Then the above computations show that the restriction of $\tilde{\mathcal{E}}_\bullet$ to (V) is isomorphic to $\mathcal{F}'_\bullet \oplus \mathcal{F}''_\bullet$ and the restriction of $\tilde{\mathcal{E}}_\bullet$ to $(S - V)$ is isomorphic to $\mathcal{F}_\bullet$. We note that $\tilde{\phi}$ restricted to either of the divisors (V) or $(S - V)$ is an isomorphism onto X, and we can identify the schemes and the corresponding sheaves. Making these identifications, and using the above facts about the restrictions of $\tilde{\mathcal{E}}_\bullet$, we have

$$\mathrm{ch}(\mathcal{F}_\bullet)([X]) = \tilde{\phi}_*\big(\mathrm{ch}\big((\tilde{\mathcal{E}}_\bullet)_{(S-V)}\big)((S - V) \cap [\tilde{Y}])\big)$$

and

$$\mathrm{ch}\big(\mathcal{F}'_\bullet \oplus \mathcal{F}''_\bullet\big)([X]) = \tilde{\phi}_*\big(\mathrm{ch}\big((\tilde{\mathcal{E}}_\bullet)_{(V)}\big)((V) \cap [\tilde{Y}])\big).$$

On the other hand, using the commutativity of local Chern characters with intersection with divisors, we have that $\mathrm{ch}((\tilde{\mathcal{E}}_\bullet)_{(S-V)})((S - V) \cap [\tilde{Y}]) = (S - V) \cap (\mathrm{ch}(\tilde{\mathcal{E}}_\bullet)[\tilde{Y}])$, and a similar formula for the intersection with (V). Since V and $S - V$ have the same degree, $(S - V) \cap (\mathrm{ch}(\tilde{\mathcal{E}}_\bullet)[\tilde{Y}])$ and $(V) \cap (\mathrm{ch}(\tilde{\mathcal{E}}_\bullet)[\tilde{Y}])$ are rationally equivalent. Hence, combining these formulas and using the fact that $\tilde{\phi}_*$ preserves rational equivalence, we have that $\mathrm{ch}(\mathcal{F}_\bullet)([X])$ and $\mathrm{ch}(\mathcal{F}'_\bullet \oplus \mathcal{F}''_\bullet)([X])$ are rationally equivalent. Thus, from the split case, we deduce that

$$\mathrm{ch}(\mathcal{F}_\bullet)[X] = \mathrm{ch}\big(\mathcal{F}'_\bullet\big)[X] + \mathrm{ch}\big(\mathcal{F}''_\bullet\big)[X].$$

$\square$

One consequence of the additivity formula is that quasi-isomorphic complexes define the same local Chern character.

Proposition 12.1.3 *Let* $f_\bullet : \mathcal{E}_\bullet \to \mathcal{F}_\bullet$ *be a quasi-isomorphism. Then* $\mathrm{ch}(\mathcal{E}_\bullet) = \mathrm{ch}(\mathcal{F}_\bullet)$.

Proof Let $C_\bullet$ be the mapping cone of $f_\bullet$. Since we are assuming that $f_\bullet$ is a quasi-isomorphism, $C_\bullet$ is exact, and we have a short exact sequence of complexes

$$0 \to \mathcal{F}_\bullet \to C_\bullet \to \mathcal{E}_\bullet[-1] \to 0.$$

By Corollary 11.4.1, $\text{ch}(C_\bullet) = 0$. Hence the additivity formula implies that

$$0 = \text{ch}(C_\bullet) = \text{ch}(\mathcal{F}_\bullet) + \text{ch}(\mathcal{E}_\bullet[-1]) = \text{ch}(\mathcal{F}_\bullet) - \text{ch}(\mathcal{E}_\bullet).$$

Hence $\text{ch}(\mathcal{E}_\bullet) = \text{ch}(\mathcal{F}_\bullet)$. $\qquad\qquad\square$

For Chern characters defined by an individual matrix over a local ring as in Section 1 of Chapter 11, the proof of Lemma 11.1.1 goes through to show that we have additivity on direct sums, but the Chern characters defined by the individual maps are not additive on short exact sequences. In fact, we can see this with an example of multiplicities; if $A = k[[X, Y]]$, we have a short exact sequence

$$0 \to A/(X, Y) \xrightarrow{XY} A/(X^2, Y^2) \to A/(X^2, XY, Y^2) \to 0.$$

One can resolve the middle module by combining resolutions of the outer ones, and, if local Chern characters defined by matrices were additive in this sense, we would conclude that the multiplicity of (X^2, Y^2) was the sum of the multiplicities of (X, Y) and (X^2, XY, Y^2). However, the multiplicities of (X^2, Y^2) and (X^2, XY, Y^2) are equal, while the multiplicity of (X, Y) is 1. This is analogous to the fact that the length of homology of a complex is additive on direct sums, but for additivity on general short exact sequences it is necessary to take the Euler characteristic.

12.2 The Splitting Principle for Complexes

In Chapter 9 we used a splitting principle for locally free sheaves, which enabled us to reduce computations involving Chern classes by combining simple computations with symmetric functions with results for rank 1 locally free sheaves. The main theorem that makes this procedure possible, Theorem 9.6.1, states that we can find a proper map that is surjective on Chow groups and such that the pull-back of the sheaf has a filtration with quotients of rank 1. A similar process works for complexes. We define an *elementary complex* to be a complex of either of the following two types.

(i) A complex

$$\mathcal{L}_\bullet = \cdots 0 \to \mathcal{L}_i \to \mathcal{L}_{i-1} \to 0 \to \cdots$$

where $\mathcal{L}_i$ and $\mathcal{L}_{i-1}$ are locally free sheaves of rank 1.

(ii) A complex

$$\mathcal{L}_\bullet = \cdots 0 \to \mathcal{L}_i \to 0 \to \cdots$$

where $\mathcal{L}_i$ is a locally free sheaf of rank 1.

The main result of this section states that for any bounded complex $\mathcal{F}_\bullet$ of locally free sheaves on $X = \text{Proj}(A)$, we can find a projective scheme Y and a projective map from Y to X such that the pull-back of $\mathcal{F}_\bullet$ to Y has a filtration whose quotients are elementary complexes.

Theorem 12.2.1 *Let $\mathcal{F}_\bullet$ be a bounded complex of locally free sheaves over a projective scheme $\text{Proj}(A)$, where A is an integral domain. Let Z be the support of $\mathcal{F}_\bullet$. Then there exists a projective map $\phi : Y \to X$ such that*

(i) *The map ϕ induces a surjective map on groups of cycles over $\mathbb{Q}$ (and hence on the rational Chow groups).*
(ii) *The pull-back $\phi^*(\mathcal{F}_\bullet)$ has a filtration by complexes $\mathcal{F}_\bullet^i$ such that the quotients in the filtration are elementary complexes with support in $\phi^{-1}Z$.*

Proof By induction on the sum of the ranks of $\mathcal{F}_i$, it suffices to show that we can find a Y satisfying the required properties such that the pull-back of $\mathcal{F}_\bullet$ to Y has a quotient that is an elementary complex. For convenience of notation, assume that 0 is the smallest integer i such that $\mathcal{F}_i$ is not equal to zero. We first construct a projective map such that the pull-back of $\mathcal{F}_0$ has a rank 1 quotient $\mathcal{L}_0$; the existence of such a map, together with the fact that it satisfies condition (i), follows from the splitting principle of Theorem 9.6.1. Replacing X with the new scheme we may assume that $\mathcal{F}_0$ has a locally free rank 1 quotient, which we denote $\mathcal{L}_0$. If the composition $\mathcal{F}_1 \to \mathcal{F}_0 \to \mathcal{L}_0$ is zero, the elementary complex $\cdots 0 \to \mathcal{L}_0 \to 0 \cdots$ is a quotient of $\mathcal{F}_\bullet$ and we are done. The condition that the support of this elementary complex is contained in $\phi^{-1}(Z)$ holds in this case because the fact that the composition $\mathcal{F}_1 \to \mathcal{F}_0 \to \mathcal{L}_0$ is zero implies that the support of $\mathcal{F}_\bullet$ is all of X. We thus consider the case where this composition is not zero.

Let $P(\mathcal{F}_1)$ denote $\text{Proj}(\text{Sym}(\mathcal{F}_1))$, where $\text{Sym}(\mathcal{F}_1)$ is the symmetric algebra on $\mathcal{F}_1$. Let ψ denote the projection from $P(\mathcal{F}_1)$ onto X. Let $\mathcal{L}_1$ denote the canonical rank 1 quotient of $\psi^*(\mathcal{F}_1)$, and let $\mathcal{G}$ denote the kernel of the map from $\psi^*(\mathcal{F}_1)$ to $\mathcal{L}_1$. To simplify notation we use the same notation for $\mathcal{F}_i$ and $\mathcal{L}_0$ and for their pull-backs to $P(\mathcal{F}_1)$. We then have a diagram

$$
\begin{array}{ccc}
& \mathcal{G} & \\
& \downarrow & \\
\mathcal{F}_2 \to & \mathcal{F}_1 & \to \mathcal{F}_0 \\
& \downarrow & \downarrow \\
& \mathcal{L}_1 & \mathcal{L}_0
\end{array}
$$

in which the middle column is exact at $\mathcal{F}_1$. We now wish to show that there is a map from $\mathcal{L}_1$ to $\mathcal{L}_0$ induced by the map from $\mathcal{F}_1$ to $\mathcal{F}_0$ such that the result is a quotient complex. For this to be true it is necessary for two conditions to be satisfied. First, the image of $\mathcal{G}$ in $\mathcal{F}_1$ must be mapped to zero in $\mathcal{L}_0$ so that there is a map induced from the quotient $\mathcal{L}_1$ to $\mathcal{L}_0$. Second, the image of $\mathcal{F}_2$ in $\mathcal{F}_1$ must be mapped to zero in $\mathcal{L}_1$ so that the result is a quotient complex. There is no reason why either of these conditions should be satisfied, but we show next that there is a closed subscheme of $P(\mathcal{F}_1)$ on which they do hold.

Let $\mathcal{M}$ denote the image of the composition $\mathcal{G} \to \mathcal{F}_1 \to \mathcal{F}_0 \to \mathcal{L}_0$ in $\mathcal{L}_0$. By tensoring the inclusion of $\mathcal{M}$ in $\mathcal{L}_0$ with $\mathcal{L}_0^{-1}$ we obtain a map

$$\mathcal{M} \otimes \mathcal{L}_0^{-1} \to \mathcal{L}_0 \otimes \mathcal{L}_0^{-1} \cong \mathcal{O}_{P(\mathcal{F}_1)}.$$

Let $\mathfrak{a}$ be the ideal of $\mathrm{Sym}(\mathcal{F}_1)$ that defines the image of $\mathcal{M} \otimes \mathcal{L}_0^{-1}$ in $\mathcal{O}_{P(\mathcal{F}_1)}$. Let $\mathfrak{b}$ be the ideal defined similarly by the image of $\mathcal{F}_2 \to \mathcal{F}_1 \to \mathcal{L}_1$ in $\mathcal{L}_1$. Let $Y = \mathrm{Proj}(\mathrm{Sym}(\mathcal{F}_1)/(\mathfrak{a} + \mathfrak{b}))$. When restricted to Y, both of the maps we are considering become zero. Hence if we let ϕ denote the composition of ψ with the inclusion of Y in $P(\mathcal{F}_1)$, $\phi^*(\mathcal{F}_\bullet)$ has an elementary quotient. It remains to show that the elementary quotient has support in $\phi^{-1}(Z)$ and that ϕ satisfies the other required conditions.

We first show that the support of $0 \to \mathcal{L}_1 \to \mathcal{L}_0 \to 0$ is contained in $f^{-1}(Z)$. To do this, it suffices to show that if $\mathcal{F}_\bullet$ is exact, then the map from $\mathcal{L}_1$ to $\mathcal{L}_0$ is an isomorphism. Looking at the construction locally, we see that the map from $\mathcal{L}_1$ to $\mathcal{L}_0$ is induced by a surjection on quotient sheaves and is therefore surjective; since it is locally a map from a ring to itself it follows that it is also injective. Thus, this map is an isomorphism.

It remains to show that the map ϕ_* is surjective on cycles over $\mathbb{Q}$ (so far it is not even obvious that Y is nonempty), and to do this it suffices to show that for every $\mathfrak{p}$ in $\mathrm{Proj}(A)$ there is a $\mathfrak{q}$ in $\mathrm{Proj}(\mathrm{Sym}(\mathcal{F}_1)/(\mathfrak{a} + \mathfrak{b}))$ such that $\phi(\mathfrak{q}) = \mathfrak{p}$ and $k(\mathfrak{q})$ is a finite extension of $k(\mathfrak{p})$. Localizing at $\mathfrak{p}$, we may assume that $X = \mathrm{Spec}(A_{(\mathfrak{p})})$. To simplify notation, we denote $A_{(\mathfrak{p})}$ by A and assume that A is a local ring with maximal ideal $\mathfrak{m}$. We then have to show that $\mathfrak{m}$ is in the image of ϕ.

Since $X = \mathrm{Spec}(A)$ for a local ring A, we may assume that the complex $\cdots \to \mathcal{F}_2 \to \mathcal{F}_1 \to \mathcal{F}_0 \to 0 \to \cdots$ is of the form $\cdots \to A^m \to A^n \to A^p \to 0 \to \cdots$, and that $\mathcal{L}_0$ is simply a rank 1 free quotient of A^p. The scheme $P(\mathcal{F}_1)$ is then $\mathrm{Proj}(A[X_1, \ldots, X_n])$, which we denote $\mathrm{Proj}(A[X_i])$, and the diagram of sheaves we obtain is

$$
\begin{array}{ccccc}
& & \mathcal{G} & & \\
& & \downarrow & & \\
A[X_i]^m & \to & A[X_i]^n & \to & A[X_i]^p \\
& & \downarrow & \searrow & \downarrow \\
& & A[X_i][1] & & A[X_i]
\end{array}
$$

where the vertical map from $A[X_i]^n$ to $A[X_i][1]$ is defined by the matrix $(X_1, \ldots, X_n)$, the map from $\mathcal{G}$ to $A[X_i]^n$ is the kernel of this map, and the other maps are pull-backs of maps defined over A and so are defined by matrices with coefficients in A. Let f denote the map represented by the diagonal arrow from $A[X_i]^n$ to $A[X_i]$, let $e_1, \ldots, e_n$ denote the basis of $A[X_i]^n$, and let $a_i = f(e_i)$. Note that a_i is an element of A for each i. In the general construction we divided by the ideal generated by the image of $\mathcal{G}$ in $\mathcal{L}_0$ and by the image of $\mathcal{F}_2$ in $\mathcal{L}_1$. Now $\mathcal{G}$ is generated by elements of the form $X_i e_j - X_j e_i$, and $X_i e_j - X_j e_i$ is mapped to $X_i a_j - X_j a_i$ in $\mathcal{L}_0$. Using the above notation, $\mathfrak{a}$ is the ideal of $A[X_i]$ generated by $X_i a_j - X_j a_i$ for all i and j. The image of $\mathcal{F}_2$ in $\mathcal{L}_1$ is generated by elements of the form $b_1 X_1 + \cdots + b_n X_n$ where $b_1 e_1 + \cdots + b_n e_n$ is in the image of d_2, and $\mathfrak{b}$ is the ideal generated by these elements. We claim that $\mathfrak{a} + \mathfrak{b}$ is contained in a homogeneous prime ideal $\mathfrak{q}$ that maps to $\mathfrak{m}$ in $\mathrm{Spec}(A)$ and such that $k(\mathfrak{q})$ is a finite extension of $k(\mathfrak{m})$.

To see this, we define a map g from $A[X_i]$ to $A[T]$ letting $g(X_i) = a_i$. We will show that the kernel of g contains both $\mathfrak{a}$ and $\mathfrak{b}$. The kernel of g contains $\mathfrak{a}$ since

$$g(X_i a_j - X_j a_i) = a_i a_j T - a_j a_i T = 0$$

for all i and j. It contains $\mathfrak{b}$ since $a_1 b_1 + \cdots + a_n b_n = 0$ whenever $b_1 e_1 + \cdots + b_n e_n$ is in the image of $\mathcal{F}_2$, since $F_\bullet$ is a complex. We are assuming that not all of the a_i are zero, since we have already dealt with the case where the composition $\mathcal{F}_1 \to \mathcal{F}_0 \to \mathcal{L}_0$ is zero. Thus, the image of $A[X_i]$ in $A[T]$ is a nonzero graded ring B, and since A is an integral domain, we can conclude that $B/\mathfrak{m}B$ is a graded ring over $A/\mathfrak{m}$ such that $\mathrm{Proj}(B/\mathfrak{m}B)$ is not empty. Hence, since $B/\mathfrak{m}B$ is a finitely generated algebra over the field $A/\mathfrak{m}$, there is a prime ideal $\mathfrak{q}$ of $B/\mathfrak{m}B$ such that $k(\mathfrak{q})$ is a finite extension of $A/\mathfrak{m} = k(\mathfrak{m})$. The inverse image of $\mathfrak{q}$ in $A[X_i]$ then satisfies the required conditions. $\quad\square$

A map constructed as in Theorem 12.2.1 will be called a *splitting map* for $\mathcal{F}_\bullet$.

We work out one example of this construction, that of a Koszul complex in two elements x and y over a nongraded ring A. In this case $\mathcal{F}_0$ already has rank 1. To get $\mathcal{L}_1$ as in the theorem, we take projective space on two elements S, T. The map from $\mathcal{F}_1$ to $\mathcal{L}_1 = \mathcal{O}(1)$ maps the generators of the free module of rank 2 over $A[S, T]$ to S and T, and the kernel $\mathcal{G}$ of this map is generated by $(-T, S)$, so the kernel of $\mathcal{G}$ is the locally free rank 1 sheaf $\mathcal{O}(-1)$. The map from $\mathcal{G}$ to $\mathcal{L}_0 = \mathcal{E}_0$ takes $(-T, S)$ to $-xT + yS$. Thus, the ideal $\mathfrak{a}$ is generated by $-xT + yS$. The map from $\mathcal{F}_2$ to $\mathcal{L}_1$ takes a generator of $\mathcal{F}_2 = A[S, T]$ first to $(-y, x) \in \mathcal{F}_1$ and then to $-yS + xT$ in $\mathcal{L}_1$. Hence $\mathfrak{b} = \mathfrak{a}$, and Y is equal to

$\text{Proj}(A[S, T]/(xT - yS))$. Thus, since $\mathcal{G} = \mathcal{O}(-1)$ and $\mathcal{L}_1$ is $\mathcal{O}(1)$, the two quotient complexes in the filtration are $\mathcal{O} \rightarrow \mathcal{O}(-1)$ in degrees 2 and 1 and $\mathcal{O}(1) \rightarrow \mathcal{O}$ in degrees 1 and 0.

In this example the ring $(A[S, T]/(xT - yS))$ maps onto the Rees ring $R((x, y))$ of the ideal generated by x and y, so we can also pull-back further to the projective scheme defined by $R((x, y))$, which is the blow-up of the ideal (x, y). We denote $\text{Proj}(R((x, y)))$ by Y'. On Y', the two complexes in the filtration are still $\mathcal{O} \rightarrow \mathcal{O}(-1)$ in degrees 2 and 1 and $\mathcal{O}(1) \rightarrow \mathcal{O}$ in degrees 1 and 0.

We give one application of the splitting principle immediately.

Proposition 12.2.2 *Let $\mathcal{F}_\bullet$ be a bounded complex of locally free sheaves over X. Let $\mathcal{F}_\bullet^*$ denote the complex $\text{Hom}(\mathcal{F}_\bullet, \mathcal{O}_X)$. Then for each integer k we have*

$$\text{ch}_k\left(\mathcal{F}_\bullet^*\right) = (-1)^k \text{ch}_k(\mathcal{F}_\bullet).$$

Proof Let η be a cycle in $A_*(X)$. By the splitting principle we can find a projective map $\phi : Y \rightarrow X$ over which the pull-back of $\mathcal{F}_\bullet$ has a filtration with quotients elementary complexes and a cycle η' in $A_*(Y)$ with $\phi_*(\eta') = \eta$. Since the pull-back of locally free sheaves preserves duals, and taking the dual preserves short exact sequences for locally free sheaves, the projection formula allows us to reduce to the case where $\mathcal{F}_\bullet$ is an elementary complex.

If there is only one nonzero sheaf in $\mathcal{F}_\bullet$, then the local Chern character is the Chern character of this sheaf, and the result follows from Theorem 9.7.2. If $\mathcal{F}_\bullet$ is of the form $\mathcal{L} \otimes (\mathcal{O}(-D) \rightarrow \mathcal{O})$, then $\mathcal{F}_\bullet^*$ is $\mathcal{L}^* \otimes (\mathcal{O} \rightarrow \mathcal{O}(D)) = \mathcal{L}^* \otimes \mathcal{O}(D) \otimes (\mathcal{O}(-D) \rightarrow \mathcal{O})$ with the degrees shifted by one (from 1 and 0 to 0 and -1). Let α denote the Chern class of $\mathcal{L}$. By Theorem 11.4.6, the local Chern character of $\mathcal{L} \otimes (\mathcal{O}(-D) \rightarrow \mathcal{O})$ is $e^\alpha(1 - e^{-D}) = e^\alpha - e^{\alpha - D}$. Similarly, the local Chern character of $\mathcal{L}^* \otimes \mathcal{O}(D) \otimes (\mathcal{O}(-D) \rightarrow \mathcal{O})$ with the degrees shifted by one is $-(e^{-\alpha}e^D(1 - e^{-D})) = e^{-\alpha}e^D(e^{-D} - 1) = e^{-\alpha}(1 - e^D) = e^{-\alpha} - e^{D-\alpha}$. Since the ith term in the power series expansion of e^{-x} is obtained from the corresponding term in the power series for e^x by multiplying by $(-1)^i$, this completes the proof. $\square$

12.3 The Multiplicativity Formula

We next use the splitting principle to prove the multiplicativity formula.

Theorem 12.3.1 *Let $\mathcal{E}_\bullet$ and $\mathcal{F}_\bullet$ be bounded complexes of locally free sheaves of X with supports Z and W, respectively. Then for any element η of $A_*(X)$*

we have

$$\text{ch}(\mathcal{E}_{\bullet} \otimes \mathcal{F}_{\bullet})(\eta) = \text{ch}(\mathcal{E}_{\bullet})\text{ch}(\mathcal{F}_{\bullet})(\eta)$$

in $A_*(Z \cap W)$.

Proof Using the splitting principle, we can assume that both $\mathcal{E}_{\bullet}$ and $\mathcal{F}_{\bullet}$ have filtrations with quotients elementary complexes. Using the additivity, along with the fact that a short exact sequence in either variable gives a short exact sequence on the tensor products, we see that it suffices to prove the result in the case $\mathcal{E}_{\bullet}$ and $\mathcal{F}_{\bullet}$ are themselves elementary complexes. If either complex consists of one locally free rank 1 sheaf, the result follows from Lemma 11.4.5. Hence we may assume that each complex consists of a map between two locally free sheaves of rank 1. In this case they are of the form $\mathcal{L} \otimes (0 \to \mathcal{O}_X(-D) \to \mathcal{O}_X \to 0)$, and, again using Lemma 11.4.5, we have that

$$\text{ch}(\mathcal{L} \otimes (0 \to \mathcal{O}_X(-D) \to \mathcal{O}_X \to 0))$$
$$= \text{ch}(\mathcal{L})\text{ch}(0 \to \mathcal{O}_X(-D) \to \mathcal{O}_X \to 0)$$

and similarly for the other complex, so we may assume that we have

$$\mathcal{E}_{\bullet} = 0 \to \mathcal{O}_X(-D) \to \mathcal{O}_X \to 0$$

and

$$\mathcal{F}_{\bullet} = 0 \to \mathcal{O}_X(-D') \to \mathcal{O}_X \to 0.$$

The remainder of a proof is a direct computation in this case. By Theorem 11.4.6 we have that the local Chern character of

$$0 \to \mathcal{O}_X(-D) \to \mathcal{O}_X \to 0$$

can be expressed as $1 - e^{-D}$ and similarly for D'. Hence we have

$$\text{ch}(\mathcal{E}_{\bullet})\text{ch}(\mathcal{F}_{\bullet}) = (1 - e^{-D})(1 - e^{-D'}).$$

The tensor product $E_{\bullet} \otimes F_{\bullet}$ is the complex

$$0 \to \mathcal{O}_X(-D - D') \to \mathcal{O}_X(-D) \oplus \mathcal{O}_X(-D') \to \mathcal{O}_X \to 0$$

which is locally the Koszul complex in the two elements that locally generate D and D'. We use the computation of the Koszul complex on two elements from the previous section. Let Y be the blow-up of the ideal generated by the ideals that define the divisors D and D', and let $f : Y \to X$ be the projection to X. We recall (see Section 8.9) that in Y the sum of ideals defining D and D' defines a divisor E called the exceptional divisor and that there are divisors $\tilde{D}$ and $\tilde{D}'$ such that $D = \tilde{D} + E$, $D' = \tilde{D}' + E$, and $\tilde{D} \cap \tilde{D}'$ is empty.

Let x and y be local generators of $\mathcal{O}_X(-D)$ and $\mathcal{O}_X(-D')$, respectively. The image of $\mathcal{O}_Y(-D) \oplus \mathcal{O}_Y(-D')$ in $\mathcal{O}_Y$ is the ideal generated by x and y, which is by definition $\mathcal{O}_Y(-E)$. Hence the quotient in the filtration (as computed in the previous section) is the complex

$$0 \to \mathcal{O}_Y(-E) \to \mathcal{O}_Y \to 0.$$

The kernel of the map from $\mathcal{O}_Y(-D) \oplus \mathcal{O}_Y(-D')$ to $\mathcal{O}_Y$ consists of those multiples of $(-y, x)$ by elements of the quotient field of $\mathcal{O}_Y$ whose coordinates lie in the ring. The element $(-y, x)$ itself generates a submodule of $\mathcal{O}_Y(-D) \oplus \mathcal{O}_Y(-D')$ isomorphic to $\mathcal{O}_Y(-D - D')$. In Y, the elements x and y have a common divisor locally that is given by a generator e of E, and the quotients x/e and y/e generate the unit ideal. Hence the kernel of the above map is generated locally by $(-y/e, x/e)$, and thus the kernel is isomorphic to $\mathcal{O}_Y(-D - D' + E)$.

Hence we have shown that the two quotients in the filtration of

$$0 \to \mathcal{O}_Y(-D - D') \to \mathcal{O}_Y(-D) \oplus \mathcal{O}_Y(-D') \to \mathcal{O}_Y \to 0$$

are $0 \to \mathcal{O}_Y(-D - D') \to \mathcal{O}_Y(-D - D' + E) \to 0$ shifted by one and $0 \to \mathcal{O}_Y(-E) \to \mathcal{O}_Y \to 0$. Using the formula for the local Chern character of an elementary complex we must thus show that

$$(1 - e^{-D})(1 - e^{-D'}) = (1 - e^{-E}) - (e^{E-D-D'} - e^{-D-D'}).$$

Multiplying the left-hand side, canceling equal terms, and changing signs, this equation becomes

$$e^{-D} + e^{-D'} = e^{-E} + e^{E-D-D'}.$$

We now substitute $D = \tilde{D} + E$ and $D' = \tilde{D}' + E$, which also gives $E - D - D' = -E - \tilde{D} - \tilde{D}'$, and we must show that

$$e^{-\tilde{D}-E} + e^{-\tilde{D}'-E} = e^{-E} + e^{-E-\tilde{D}-\tilde{D}'}.$$

This equality is an equality of power series, and the equation for the nth term, after canceling $(-1)^n$, is

$$(\tilde{D} + E)^n + (\tilde{D}' + E)^n = E^n + (\tilde{D} + \tilde{D}' + E)^n.$$

This formula holds because $\tilde{D} \cap \tilde{D}'$ is empty, so that $\tilde{D}\tilde{D}' = 0$. Hence if we expand both sides using the binomial theorem and consider the ith term for $i < n$, on the left we have

$$\binom{n}{i} \tilde{D}^i E^{n-i} + \binom{n}{i} \tilde{D}'^i E^{n-i}$$

while on the right, using that $\tilde{D}\tilde{D}' = 0$, we have

$$\binom{n}{i} (\tilde{D} + \tilde{D}')^i E^{n-i} = \binom{n}{i} (\tilde{D}^i + \tilde{D}'^i) E^{n-i}.$$

These expressions are equal, and for $i = n$ the ith term in both sides is $2E^n$. Hence the two sides are equal as was to be shown. ☐

As a corollary, we can compute the local Chern character defined by a Koszul complex.

Corollary 12.3.2 *Let A be a local ring, and let $x_1, \ldots, x_n$ be a sequence of elements in A. Let $K_\bullet(x_1, \ldots, x_n)$ denote the Koszul complex on $x_1, \ldots, x_n$, and let η be a cycle in $A_*(A)$. For each i, let (x_i) denote the divisor defined by x_i. Then*

$$\mathrm{ch}(K_\bullet(x_1, \ldots, x_n))(\eta) = (x_1) \cap \cdots \cap (x_n) \cap (\eta).$$

Proof Since the Koszul complex is a tensor product of Koszul complexes $K_\bullet(x_i)$ on x_i for each i, this result follows from the multiplicativity formula and the computation for a single element in Section 11.1. ☐

12.4 The Todd Class

One of the main theorems that we prove in this chapter is the local Riemann-Roch theorem, which relates the Euler characteristic of a complex to its local Chern character. Given any bounded complex $\mathcal{M}_\bullet$ of coherent sheaves over a projective scheme X, with no condition of being quasi-isomorphic to a bounded complex of locally free sheaves, there is a class in the rational Chow group of the support of the module, called the Todd class of $\mathcal{M}_\bullet$. If $\mathcal{F}_\bullet$ is a perfect complex, then the local Riemann-Roch formula states that the Todd class of $\mathcal{F}_\bullet \otimes \mathcal{M}_\bullet$ is the same as the local Chern character of $\mathcal{F}_\bullet$ applied to the Todd class of $\mathcal{M}_\bullet$. In particular, if $M_\bullet$ is a complex of finitely generated modules over the local ring A in degree 0, we define a Todd class of $M_\bullet$ in $A_*(\mathrm{Spec}(A))_\mathbb{Q}$. If $M_\bullet$ is the ring itself in degree 0, and if $F_\bullet$ is a bounded free complex over A with homology of finite length, this theorem implies that the Todd class of $F_\bullet$ is the local Chern character of $F_\bullet$ applied to the Todd class of A. Now if $F_\bullet$ has homology of finite length, we will show that its Todd class, which is an element of $A_0(\mathrm{Spec}(A/\mathfrak{m})) \cong \mathbb{Q}$, is equal to the Euler characteristic of $F_\bullet$ times the class of $A/\mathfrak{m}$. Thus, we have a formula for the Euler characteristic of $F_\bullet$ in terms of the local Chern character of $F_\bullet$, which is of importance for applications to questions on multiplicities.

If $\mathcal{M}_\bullet$ is a bounded complex of coherent modules over $X = \mathrm{Proj}(A)$, the Todd class will be defined by using the Chern class of a free resolution of $\mathcal{M}_\bullet$ over a polynomial ring over a regular local ring that maps onto A. We recall that we are assuming that all local rings are homomorphic images of regular local rings.

It then follows that if A is a graded ring over a local ring, A is a homomorphic image of a polynomial ring over a regular local ring. For the sake of brevity, we refer to a graded polynomial ring over a regular local ring simply as a *graded polynomial ring*. For the purposes of proving the properties of the Todd class, we make a further simplifying assumption. Suppose that R and R' are graded polynomial rings together with surjective maps $\phi : R \to A$ and $\phi' : R' \to A$. We assume that there is a third graded polynomial ring R'' that maps onto R and R' and such that R'' can be obtained from R and from R' by adjoining indeterminates and elements of degree 0 that form part of a regular sequence of parameters. We refer to the latter process as adjoining regular parameters. For polynomial rings it is clear that this can be done, since, if we have graded polynomial rings $R[S_1, \ldots, S_r]$ and $R[T_1, \ldots, T_s]$ mapping onto A, we can simply take the polynomial ring $R[S_1, \ldots, S_r, T_1, \ldots, T_s]$, which maps onto both rings and can be obtained from either of them by adding indeterminates. If A is a local ring, the same is true if we assume that A is a homomorphic image of a power series ring over a fixed field or discrete valuation ring, or if A is a localization of a polynomial ring over a field. Since this includes most standard rings that arise in practice and includes the case of complete local rings, it is sufficient for the applications in the next chapter.

We first define the Todd class in the local case. Let A be a local ring, let $M_\bullet$ be a bounded complex of finitely generated modules over the ring A, and let A be a homomorphic image of the regular local ring R. Let $F_\bullet$ be a free resolution of $M_\bullet$ over R. We define the Todd class of M to be

$$\tau(M_\bullet) = \mathrm{ch}(F_\bullet)([R]).$$

The Todd class is an element of the Chow group of the support of $M_\bullet$, and it thus also defines an element of the Chow group of A. We must show, of course, that this definition does not depend on the choice of the regular local ring R; we prove this fact below.

In the graded case, the situation is a little more complicated. To see why this is so, we discuss the question raised above of whether the Todd class is well-defined. Consider the local case in which A is itself regular and can also be written in the form R/xR, where R is regular and x is an element of $\mathfrak{m} - \mathfrak{m}^2$, where $\mathfrak{m}$ is the maximal ideal of R. We may define the Todd class of A using A itself, so that we have simply $\tau(A) = [A]$. We may also define the Todd class of A using R, in which case it is the local Chern character of the complex $R \xrightarrow{x} R$ applied to the class $[R]$, since

$$0 \to R \xrightarrow{x} R \to A \to 0$$

is a free resolution of A over R. We computed this example at the end of the first section of Chapter 11, and we showed that this computation also gives $[R/xR] = [A]$. Thus, both procedures give the same result in this case.

However, the situation is different if we apply the same process in the graded case. Let A and R be polynomial rings over a field with the usual grading and assume that $A = R/XR$ where X is a nonzero element of R of degree 1. Let h denote the hyperplane section on $\text{Proj}(R)$ and on $\text{Proj}(A)$. The Chern character of the complex $R \xrightarrow{X} R$ is not simply h in this case, but from Theorem 11.4.6 we see that it is equal to $1 - e^{-h} = h - h^2/2 + \cdots$. (Of course, this formula holds also in the local case, but in that case $h^n = 0$ for $n > 1$.) For this reason, it is necessary to introduce a factor to cancel out the extra terms.

Let $Q(x)$ be the power series defined by

$$Q(x) = \frac{x}{1 - e^{-x}}.$$

We note that

$$1 - e^{-x} = x - \frac{x^2}{2!} + \frac{x^3}{3!} - \cdots = x\left(1 - \frac{x}{2!} + \frac{x^2}{3!} - \cdots\right)$$

and that $Q(x)$ is a power series with constant term 1. Its higher coefficients are defined in terms of Bernoulli numbers (see for example Fulton [17, Example 3.2.4]), but we will not need this fact.

Let A be a $\mathbb{Z}^n$-graded ring, and let $\mathcal{M}_\bullet$ be a bounded complex of coherent sheaves on $\text{Proj}(A)$. Let A be a homomorphic image of a graded polynomial ring B over the regular local ring R. Suppose that B is generated over R by elements X_i of degrees a_i. Let h_{a_i} denote the hyperplane section on X of degree a_i for each i. Let $\mathcal{F}_\bullet$ be a resolution of $\mathcal{M}_\bullet$ by locally free sheaves over B. We define the Todd class of $\mathcal{M}_\bullet$ to be

$$\tau(\mathcal{M}_\bullet) = \text{ch}(\mathcal{F}_\bullet)\left(\prod_i Q\left(h_{a_i}\right)([R])\right).$$

Again it is necessary to prove that this definition is independent of choice of R.

Theorem 12.4.1 *The Todd class does not depend on the choice of the graded polynomial ring R.*

Proof Suppose we have two maps of graded polynomial rings onto A. We can then, as discussed above, find a third ring mapping to both rings by adding variables and adjoining regular parameters. It thus suffices to prove the result when we adjoin one such element. Hence we may assume that we have a graded polynomial ring R and a bounded complex of coherent sheaves $\mathcal{M}_\bullet$

over $\text{Proj}(R)$ and that we extend to the graded ring $R[T]$, where T is an indeterminate of degree i or a regular parameter of degree 0.

Let $\mathcal{F}_\bullet$ be a bounded resolution of $\mathcal{M}_\bullet$ over R by locally free sheaves. Let $X = \text{Proj}(R)$, and let $Y = \text{Proj}(R[T])$, where T is the new variable of degree i. The complex $\mathcal{F}_\bullet \otimes_R R[T]$ will then be a locally free resolution of $\mathcal{M}_\bullet \otimes_R R[T]$ over Y. Hence the complex $\mathcal{M}_\bullet$ can be recovered by tensoring with $R[T]/(T)$, and a free resolution of $\mathcal{M}_\bullet$ over Y can be obtained by tensoring the complex $\mathcal{F}_\bullet \otimes_R R[T]$ with the complex

$$0 \to R[T][-i] \xrightarrow{T} R[T] \to 0$$

which, in the terminology of sheaves, is

$$0 \to \mathcal{O}_Y(-i) \xrightarrow{T} \mathcal{O}_Y \to 0.$$

Let $\mathcal{G}_\bullet$ denote this complex. Let h_i denote the hyperplane section corresponding to i, and let h_{a_j} denote the hyperplane sections corresponding to the generators of R. Then the local Todd class of $\mathcal{M}_\bullet$, defined using its free resolution over R, is

$$\text{ch}(\mathcal{F}_\bullet) \left(\prod_j Q(h_{a_j})[R] \right).$$

The Todd class computed using its resolution over $R[T]$ is

$$\text{ch}(\mathcal{G}_\bullet \otimes (\mathcal{F}_\bullet \otimes_R R[T])) Q(h_i) \prod_j Q(h_{a_j})([R[T]]).$$

Using the commutativity of the operations involved and the mutiplicativity formula, the latter expression can also be written

$$\text{ch}(\mathcal{G}_\bullet) Q(h_i) \text{ch}(\mathcal{F}_\bullet \otimes_R R[T]) \prod_j Q(h_{a_j})([R[T]]).$$

Now Proposition 12.2.2 states that $\text{ch}(\mathcal{G}_\bullet)$ is multiplication by $1 - e^{-h_i}$, where the factor of h_i is intersection with the divisor defined by T. Thus, the product of the two factors on the left is $Q(h_i)(1 - e^{-h_i}) = h_i$, and h_i denotes intersection with (T). On the other hand, the cycle

$$\text{ch}(\mathcal{F}_\bullet \otimes_R R[T]) \prod_j Q(h_{a_j})([R[T]])$$

is the pull-back of

$$\text{ch}(\mathcal{F}_\bullet) \prod_j Q(h_{a_j})([R])$$

to $\text{Proj}(R[T])$. Hence the whole product is the intersection with T of the pull-back of this cycle to $R[T]$, which is the same as the original cycle $\mathrm{ch}(\mathcal{F}_\bullet) \prod_j Q(h_{a_j})([R])$ on $\mathrm{Proj}(R)$. Thus, the Todd class is the same whether defined over R or over $R[T]$.

If T is a regular parameter we follow the same argument, but we do not have to consider the factor $Q(h_i)$. $\quad\square$

In the proof of the local Riemann-Roch formula we will need one further result on the independence of the Todd class. Let $\mathcal{E}$ be a locally free sheaf of rank 1 on X, and let $P(\mathcal{E})$ be the projective scheme defined by the symmetric algebra of $\mathcal{E}$ over X. Then, since $\mathcal{E}$ has rank 1, the map from $P(\mathcal{E})$ to X is an isomorphism. We need to show that the Todd class computed by $P(\mathcal{E})$ is the same as that computed by X under this isomorphism. This fact will follow from the Riemann-Roch theorem, which we prove in the next section.

Let A be a local ring. The Todd class of A is then computed by taking the local Chern character of a resolution of A over a regular local ring. We first prove a part of the main result, the local Riemann-Roch formula, which relates the Todd class to the Euler characteristic.

Theorem 12.4.2 *Let* $M_\bullet$ *be a bounded complex with homology of finite length over a local ring* A. *Then* $\tau(M_\bullet) = \chi(M_\bullet)[A/\mathfrak{m}]$.

Proof We note first that this statement is a statement about regular local rings, since the Todd class is defined as the local Chern character over a regular local ring applied to the class of the ring. We may thus assume that A is regular and that we are computing the Todd class of $M_\bullet$ by taking a resolution over A. Since $M_\bullet$ has homology of finite length, it can be obtained from a sequence of extensions in which the quotients are the module k in various degrees, where k is the residue field of A.

Thus, it suffices to prove the result for a free resolution of k. However, a free resolution of k is the Koszul complex $K_\bullet(x_1, \ldots, x_d)$ on a regular sequence of parameters $x_1, \ldots, x_d$. By Corollary 12.3.2, we have

$$\mathrm{ch}(K_\bullet(x_1, \ldots, x_d))[A] = (x_1) \cap \cdots \cap (x_n) \cap [A].$$

Since this intersection of divisors is $[A/\mathfrak{m}]$ and the Euler characteristic of the Koszul complex is 1, this completes the proof. $\quad\square$

A consequence of this theorem is a general formula for the part of the Todd class of maximal dimension.

Corollary 12.4.3 *Let* $\mathcal{M}_\bullet$ *be a complex with homology of dimension at most* d. *Then the component of* $\tau(\mathcal{M}_\bullet)$ *of dimension* d *is the class*

$$[\mathcal{M}_\bullet]_d = \sum (-1)^i [H_i(\mathcal{M}_\bullet)]_d.$$

Proof The operations involved in defining the Todd class commute with localization. Hence we can reduce to the case in which $\mathcal{M}_\bullet$ has homology of finite length, and in this case the result follows from Theorem 12.4.2. □

Proposition 12.4.4 *Let* A *be a local ring of dimension* d. *Then*

(i) *The Todd class* $\tau(A)$ *is equal to* $[A]_d$ *plus terms of lower dimension.*
(ii) *If* A *is a complete intersection, then* $\tau(A) = [A]_d$.
(iii) *If* A *is Gorenstein, then* $\tau(A)_i = 0$ *if* $d - i$ *is odd.*

Proof The first statement is a special case of Corollary 12.4.3.

If A is a complete intersection, its resolution is a Koszul complex and the result follows from Corollary 12.3.2. If A is Gorenstein, then it can be shown that its free resolution is self dual using the results of Chapter 4. Thus, Proposition 12.2.2 implies that we have $\tau_i(A) = -\tau_i(A)$ for all i such that $d - i$ is odd, and since we are working over $\mathbb{Q}$ these components must be zero. □

12.5 The Riemann-Roch Theorem

In this section we prove a version of the Riemann-Roch theorem. This theorem states that the function that takes a complex to its Todd class commutes with push-forward by projective maps. In Chapter 9 we defined the push-forward of a bounded complex $\mathcal{M}_\bullet$ of coherent sheaves by a projective map and showed that the push-forward is quasi-isomorphic to a bounded complex of coherent sheaves. We also computed the result for a map $\phi : X \to Y$ where X is defined by a graded polynomial algebra over $\mathcal{O}_Y$, and $\mathcal{M}_\bullet$ is the sheaf $\mathcal{O}_X(i)$ for some $i \in \mathbb{Z}^n$. The form of the Riemann-Roch theorem we prove states that the Todd class of $\phi_*(\mathcal{M}_\bullet)$ is the push-forward of the Todd class of $\mathcal{M}_\bullet$.

Theorem 12.5.1 (The Riemann-Roch Theorem) *Let* $\phi : X \to Y$ *be a projective map, and let* $\mathcal{M}_\bullet$ *be a bounded complex of coherent sheaves on* X. *Let* Z *be the support of* $\mathcal{M}_\bullet$ *and let* $W = \phi(Z)$. *Then*

$$\phi_*(\tau(\mathcal{M}_\bullet)) = \tau(\phi_*(\mathcal{M}_\bullet))$$

in $A_*(W)$.

Proof In the statement of the theorem the ϕ_* on the left is push-forward on the Chow group, while the ϕ_* on the right is push-forward of complexes.

Let $Y = \mathrm{Proj}(A)$ and $X = \mathrm{Proj}(B)$, where A is a $\mathbb{Z}^n$-graded ring and B is a $\mathbb{Z}^{n+1}$-graded ring generated over A by elements of degree $(a_i, 1)$ for various $a_i \in \mathbb{Z}^n$.

We first note that both sides of the formula are well-defined up to quasi-isomorphism and are additive on short exact sequences. We can replace $\mathcal{M}_\bullet$ by a quasi-isomorphic complex such that the sheaf $\mathcal{M}_i$ has support contained in Z for each i. To see this, let k be the largest integer such that $\mathcal{M}_k \neq 0$, and consider the map $d_k : \mathcal{M}_k \to \mathcal{M}_{k-1}$. The kernel of d_k is supported in Z, so there exists a subsheaf $\mathcal{N}$ of $\mathcal{M}_k$ such that $\mathcal{N}$ meets the kernel of d_k trivially and the quotient $\mathcal{M}_k/\mathcal{N}$ is supported in Z. We may thus divide by the complex $0 \to \mathcal{N} \to d_k(\mathcal{N}) \to 0$; since this complex is exact, the quotient is quasi-isomorphic to $\mathcal{M}_\bullet$. Thus, we may produce a complex $\overline{\mathcal{M}}_\bullet$ that is a quotient of $\mathcal{M}_\bullet$ and quasi-isomorphic to $\mathcal{M}_\bullet$ and such that the support of $\overline{\mathcal{M}}_k$ is contained in Z. We then continue this process with $\overline{\mathcal{M}}_{k-1}$; since $\mathcal{M}_\bullet$ is bounded, we can thus obtain a complex with component sheaves supported in Z after a finite number of steps.

Taking a filtration and using additivity, we may reduce to the case where $\mathcal{M}$ is a single sheaf defined by a module of the form $B/\mathfrak{q}[m]$ for some graded prime ideal $\mathfrak{q}$ and some index m. Since shifting the degrees of a complex will change both sides of the above equality by multiplying by ± 1, we may assume that $\mathcal{M}_\bullet$ is the complex consisting of $B/\mathfrak{q}[m]$ in degree 0. Let $\mathfrak{p} = \phi(\mathfrak{q})$ be the inverse image of $\mathfrak{q}$ in A. Then $\mathrm{Proj}(A/\mathfrak{p})$ is contained in $W = \phi(Z)$, and it suffices to show that the image of $\tau(B/\mathfrak{q}[m])$ under the restriction of ϕ_* is equal to the result of applying τ to the push-forward of $B/\mathfrak{q}$ in $\mathrm{Proj}(A/\mathfrak{p})$. Dividing by $\mathfrak{p}$ and $\mathfrak{q}$, we may assume that $\mathfrak{p}$ and $\mathfrak{q}$ are zero.

We now assume that ϕ is a map from $X = \mathrm{Proj}(B)$ to $Y = \mathrm{Proj}(A)$ and we want to prove that $\phi_*(\tau(B[m])) = \tau(\phi_*(B[m]))$. Let T_i be indeterminates of the appropriate degrees and let B be a homomorphic image of $A[T_1, \ldots, T_k]$. Localizing at $\mathfrak{p}$, we can take a locally free resolution of $(B/\mathfrak{q})_{(\mathfrak{p})}$ over $k(\mathfrak{p})[T_1, \ldots, T_k]$. We can then clear denominators to get a complex of graded free $A/\mathfrak{p}[T_1, \ldots, T_k]$-modules and a map to $B/\mathfrak{q}$ that is an isomorphism up to a complex $\mathcal{N}_\bullet$ such that the image of the support of $\mathcal{N}_\bullet$ in Y is strictly smaller than $V(\mathfrak{p})$. By induction on the dimension of the image of the support of $\mathcal{M}_\bullet$ in Y, it suffices to prove the result when B is a polynomial ring $A[T_1, \ldots, T_k]$ over A.

Thus, we have $B = A[T_1, \ldots, T_k]$, where T_i has degree $(a_i, 1)$ for some $a_i \in \mathbb{Z}^n$. Let R be a $\mathbb{Z}^n$-graded polynomial ring over a regular local ring, and let f be a surjective map from R to A. The map f extends to a surjective map from $R[T_1, \ldots, T_k]$ to B. Let $\mathcal{F}_\bullet$ be a free resolution of A over R. Then $\phi^*(\mathcal{F}_\bullet)$

is a locally free resolution of B over $R[T_1, \ldots, T_k]$ and $\mathcal{O}_X(m) \otimes \phi^*(\mathcal{F}_\bullet)$ is a locally free resolution of $B[m]$. Let h_{b_j} denote the hyperplane sections corresponding to generators of R; we denote the hyperplane section of T_i by $h_{(a_i,1)}$. We must then show that

$$\phi_* \left(\mathrm{ch}(\mathcal{O}_X(m)) \mathrm{ch}(\phi^*(\mathcal{F}_\bullet)) \prod_j Q\left(h_{b_j}\right) \prod_i Q\left(h_{(a_i,1)}\right)[X] \right)$$

is equal to

$$\mathrm{ch}(\mathcal{F}_\bullet) \prod_j Q\left(h_{b_j}\right) \mathrm{ch}(\phi_*(\mathcal{O}_X(m)))[Y].$$

The factor $\mathrm{ch}(\phi^*(\mathcal{F}_\bullet)) \prod_j Q(h_{b_j})[X]$ in the first product is the pull-back of the class $\mathrm{ch}(\mathcal{F}_\bullet) \prod_j Q(h_{b_j})[Y]$ on Y. Hence it suffices to show that for any class η on Y, we have

$$\phi_*(\mathrm{ch}(\mathcal{O}_X(m))) \prod_i Q\left(h_{(a_i,1)}\right)(\phi^*(\eta)) = \mathrm{ch}(\phi_*(\mathcal{O}_X(m)))(\eta).$$

We note that $p_*(\mathcal{O}_X(m))$ is locally free and also that, if $m = (m_1, m_2)$ with $m_1 \in \mathbb{Z}^n$ and $m_2 \in \mathbb{Z}$, we can replace m_1 by any other element, since the difference is pulled back from Y. To prove this result we first prove a Lemma, taken from Lang and Fulton [40].

Lemma 12.5.2 *Let* $T, X_1, \ldots, X_{r+1}$ *be indeterminates, and let* F *be the power series*

$$F(T, X_1, \ldots, X_{r+1}) = Q(T - X_1) \cdots Q(T - X_{r+1})$$

in $A[[X_1, \ldots, X_{r+1}]][T]$, *where* $Q(x)$ *is defined to be the power series* $x/1 - e^{-x}$ *as above. Consider the image of* F *in the quotient ring*

$$A[[X_1, \ldots, X_{r+1}]][T]/((T - X_1) \cdots (T - X_{r+1})).$$

Then F *has a unique representation* $G(T, X_i)$ *in this quotient as a polynomial in* T *of degree at most* r *with coefficients that are power series in the* X_i. *Furthermore, the coefficient of* r *in* G *is* 1.

Proof We first note that if F were a polynomial it would follow immediately from the fact that $(T - X_1) \cdots (T - X_{r+1})$ is a monic polynomial in T of degree $r + 1$ that we could represent F in the quotient uniquely as a polynomial of degree at most r in T. Thus, for each $i \geq 1$ we can find a unique polynomial $\alpha_i(T)$ of degree at most r that is congruent to T^{i+r} modulo $(T - X_1) \cdots (T - X_{r+1})$. Let $\mathfrak{a}$ be the ideal generated by $X_1, \ldots, X_{r+1}$. Then the coefficient of T^j in $(T - X_1) \cdots (T - X_{r+1})$ lies in $\mathfrak{a}^{r+1-j}$, and it thus follows from the division algorithm for polynomials that α_i will have coefficients in $\mathfrak{a}^i$. Hence the coefficients in $\alpha_i(T)$ lie in higher and higher powers of $\mathfrak{a}$ as i increases, and we may use this construction to represent any power series in $X_1, \ldots, X_{r+1}, T$ as a polynomial in T of degree at most r with coefficients that are power series in the X_i.

It remains to determine the coefficient of T^r in the case of the power series F. Let $a_0, \ldots, a_r$ be power series in $X_1, \ldots, X_{r+1}$ such that $a_0 + a_1 T + \cdots + a_r T^r$ is congruent to F modulo $(T - X_1) \cdots (T - X_{r+1})$. For each i we let $F(X_i)$ denote the result of substituting T with X_i in F, so that $F(X_i) = F(X_i, X_1, \ldots, X_{r+1})$. We can determine the coefficients a_i by using the fact that they satisfy the conditions

$$a_0 + a_1 X_i + \cdots + a_r X_i^r \cong F(X_i) \, \text{modulo} \, \prod (T - X_i) \qquad (*)$$

for $i = 1$ to $r + 1$.

Consider the equations $(*)$ as a system of linear equations for a_i with coefficients X_i^j for $j = 0, \ldots r$. Then a_r, the coefficient we are trying to compute, can be found by Cramer's rule as the quotient of determinants, where the denominator is the Vandermonde determinant and the numerator is the determinant of the matrix obtained by replacing X_i^r by $F(X_i)$ for each i. We now compute this determinant.

The matrix whose determinant we must compute is the matrix with X_i^{j-1} in the ij position except for $F(X_i)$ in the last column. Thus, we must compute

$$\begin{vmatrix} 1 & X_1 & X_1^2 & \cdots & X_1^{r-1} & F(X_1) \\ 1 & X_2 & X_2^2 & \cdots & X_2^{r-1} & F(X_2) \\ \vdots & & \vdots & & \vdots & \vdots \\ 1 & X_{r+1} & X_{r+1}^2 & \cdots & X_{r+1}^{r-1} & F(X_{r+1}) \end{vmatrix}.$$

To evaluate this determinant, we first compute $F(X_i)$. Substituting X_i into the formula for F, and noting that the factor $Q(X_i - X_i)$ is $Q(0) = 1$, we have

$$F(X_i) = \prod_{j \neq i} \frac{(X_i - X_j)}{\left(1 - e^{-(X_i - X_j)}\right)}.$$

We now evaluate this determinant by using the cofactor expansion along the last column. For each i the corresponding cofactor is $(-1)^{r+1+i}$ times a Vandermonde determinant, which is equal to the product of $X_k - X_m$ for $k > m$ and $k, m \neq i$. If we multiply this cofactor by the numerator of the above expression for $F(X_i)$, we obtain, up to sign, the expression for the Vandermonde determinant corresponding to $X_1, \ldots, X_{r+1}$, which occurs in the denominator of the formula for a_r. The sign that occurs is $(-1)^k$, where k is the number of

j between 1 and $r + 1$ greater than i, which is $(-1)^{r+1-i}$. Hence this sign is the same as the sign coming from the cofactor, so they cancel in the product. Thus, we have

$$a_r = \sum_{i=1}^{r+1} \prod_{j \neq i} \frac{1}{1 - e^{-(X_i - X_j)}} = \sum_{i=1}^{r+1} \prod_{j \neq i} \frac{1}{1 - e^{X_j - X_i}}.$$

We now show that this expression is equal to 1. We first multiply the numerator and denominator of each factor in each term by e^{-X_j} to obtain

$$a_r = \sum_{i=1}^{r+1} \prod_{j \neq i} \frac{e^{-X_j}}{e^{-X_j} - e^{-X_i}}.$$

For each j, let $Y_j = e^{-X_j}$. Then we have

$$a_r = \sum_{i=1}^{r+1} \prod_{j \neq i} \frac{Y_j}{Y_j - Y_i}.$$

We claim that this expression is equal to 1. To see this, multiply the entire expression by $G(Y_i) = \prod_{k>m}(Y_k - Y_m)$. The product is a polynomial of degree equal to the degree of $G(Y_i)$, and we claim that it is equal to $G(Y_i)$. It is not difficult to check that the product is divisible by $Y_i - Y_j$ for all $i \neq j$, so that it is a constant multiple of $\prod_{k>m}(Y_k - Y_m)$, and then to check that the constant is 1. We leave the details to the reader.   □

We now complete the proof of the Riemann-Roch theorem. We must show that

$$\phi_* \left(\mathrm{ch}(\mathcal{O}_X(m)) \prod_i Q\big(h_{(a_i,1)}\big)(\phi^*(\eta)) \right) = \mathrm{ch}(\phi_*(\mathcal{O}_X(m))(\eta)). \qquad (*)$$

We recall that we have $B = A[T_1, \ldots, T_k]$ and ϕ is the projection from $X = \mathrm{Proj}(B)$ to $Y = \mathrm{Proj}(A)$. Let $k = r+1$ in the above Lemma. We map the power series ring $\mathbb{Q}[[T, X_1, \ldots, X_k]]$ to the ring of operators on $A_*(X)$ by sending T to $h_{(0,1)}$ and X_i to $h_{(-a_i,0)}$. Since $\prod_i h_{(a_i,1)}$ represents the intersection of the ideals generated by the T_i, and together they generate the irrelevant ideal, this product is zero. Thus, $\prod_i (T - X_i)$ is mapped to zero. Furthermore, the factor $\prod_i Q(h_{(a_i,1)})$ in the above formula is the image of $\prod_i Q(T - X_i)$. By the Lemma, this product is equal to $h_{(0,1)}^r$ plus terms of lower degree in $h_{(0,1)}$ modulo a polynomial that maps to zero. Thus, Proposition 9.2.1 implies that it projects to η.

If m is of the form $(m_1, 0)$, this completes the proof. In fact, in this case we may assume that $m = (0, 0)$, and the formula $(*)$ becomes

$$\phi_* \left(\prod_i Q(h_{(a_i,1)})(\phi^*(\eta)) \right) = \eta$$

which is what we just proved.

For arbitrary $m = (m_1, m_2)$, we use induction on the absolute value of m_2 and on k. We note that the operator $ch(\mathcal{O}_X(m))$ is multiplication by e^{h_m}, where h_m is the hyperplane section corresponding to m. Write $(m_1, m_2) = (m_1 + a_k, m_2 + 1) + (-a_k, -1)$. We then have

$$e^{h_{(m_1,m_2)}} = e^{h_{(m_1+a_k,m_2+1)}} e^{h_{(-a_k,-1)}} = e^{h_{(m_1+a_k,m_2+1)}} e^{-h_{(a_k,1)}}$$

$$= e^{h_{(m_1+a_k,m_2+1)}}((e^{-h_{(a_k,1)}} - 1) + 1).$$

Mutiplying this expression by $\prod(Q(h_{(a_i,1)}))$, the denominator of the factor $Q(h_{(a_k,1)})$ cancels $e^{-h_{(a_k,1)}} - 1$ and gives $-h_{(a_k,1)}$. Thus, we can write the expression

$$\phi_* \left(e^{h_{(m_1,m_2)}} \prod_i Q(h_{(a_i,1)})(\phi^*(\eta)) \right)$$

as a sum

$$-\phi_* \left(e^{h_{(m_1+a_k,m_2+1)}} \prod_{i<k} Q(h_{(a_i,1)}) h_{(a_k,1)}(\phi^*(\eta)) \right)$$

$$+ \phi_* \left(e^{h_{(m_1+a_k,m_2+1)}} \prod_i Q(h_{(a_i,1)})(\phi^*(\eta)) \right).$$

We can interpret $h_{(a_k,1)}$ as intersection with the divisor defined by T_k. Then $h_{(a_k,1)}\phi^*$ is the pull-back from Y to the subscheme $X' = \text{Proj}(A[T_1, \ldots, T_{k-1}])$ of X. If m_2 is negative, we can use the above decomposition and induction on the absolute value of m_2. If m_2 is positive, we use combined induction on the absolute value of m_2 and on k and use the above expression to write

$$\phi_* \left(e^{h_{(m_1+a_k,m_2+1)}} \prod_i Q(h_{(a_i,1)})(\phi^*(\eta)) \right)$$

as two terms, one of which has a lower value of m_2 and the other a lower value of k.

Thus, we can write the left-hand side of equation $(*)$ as a sum of terms in which the absolute value of m_2 or k is smaller. To complete the proof, we must show that there is a corresponding decomposition of the right-hand side.

Assume first that $m_2 > 0$. Let ϕ' be the map from X' to Y. Then the push-down of $\mathcal{O}_X(m_1, m_2)$ has nonzero components only in degree 0, and, moving the negative term to the other side in the above equation, we must show that

$$\phi_*(\mathcal{O}_X(m_1, m_2)) \oplus \phi'_*(\mathcal{O}_X(m_1 + a_k, m_2 + 1)) = \phi_*(\mathcal{O}_X(m_1 + a_k, m_2 + 1)).$$

This formula follows from the explicit computation of the push-forward of sheaves of the form $\mathcal{O}_X(m_1, m_2)$ in Theorem 9.8.3. We showed there that

$$H^0(\phi_*(\mathcal{O}_X(m))) = \oplus \mathcal{O}_Y\left(m_1 - \sum i_j a_j\right)$$

where the sum is taken over all k-tuples $i_1, \ldots, i_k$ with $i_1 + \cdots + i_k = m_2$ and $i_j \geq 0$. If we divide the expression for $\phi_*(\mathcal{O}_X(m_1 + a_k, m_2 + 1))$ into two summands, one of which contains those with $i_k > 0$ and the other those with $i_k = 0$, we obtain the desired decomposition into a direct sum.

If m_2 is negative, then the push-forward is in degree $k - 1$ and we have a change in sign between ϕ and ϕ'. Hence we have to show that

$$\phi_*(\mathcal{O}_X(m_1, m_2)) = \phi'_*(\mathcal{O}_X(m_1 + a_k, m_2 + 1)) \oplus \phi_*(\mathcal{O}_X(m_1 + a_k, m_2 + 1)).$$

This follows from Theorem 9.8.3 in the same way, decomposing the sum into those terms for which $i_k = -1$ and those for which $i_k < -1$. $\square$

12.6 The Local Riemann-Roch Formula

The local Riemann-Roch formula is the main formula that connects local Chern characters to Euler characteristics. It states that if $\mathcal{F}_\bullet$ is a complex of locally free sheaves and $\mathcal{M}_\bullet$ is a bounded complex of coherent sheaves, then

$$\tau(\mathcal{F}_\bullet \otimes \mathcal{M}_\bullet) = \mathrm{ch}(\mathcal{F}_\bullet)(\tau(\mathcal{M}_\bullet)).$$

This formula can be considered a generalization of the multiplicativity formula proven in Section 12.3. The proof makes use of the multiplicativity formula, together with the splitting principle and the Riemann-Roch theorem proved in the last section.

Theorem 12.6.1 (Local Riemann-Roch Formula) *Let $X = \mathrm{Proj}(A)$ be a projective scheme, let $\mathcal{F}_\bullet$ be a bounded complex of locally free sheaves on X, and let $\mathcal{M}_\bullet$ be a bounded complex of coherent sheaves on X. Then*

$$\tau(\mathcal{F}_\bullet \otimes \mathcal{M}_\bullet) = \mathrm{ch}(\mathcal{F}_\bullet)(\tau(\mathcal{M}_\bullet))$$

in $A_*(Z)_\mathbb{Q}$, where Z is the support of $\mathcal{F}_\bullet \otimes \mathcal{M}_\bullet$.

Proof Let $\phi : Y \to X$ be a splitting map constructed as in Theorem 12.2.1. Let $Y = \mathrm{Proj}(B)$. To reduce to the case in which $\mathcal{F}_\bullet$ is an elementary complex, we need to show that we can replace $\mathcal{M}_\bullet$ with the push-forward of a complex of sheaves on Y. Using the additivity of both sides of the formula, we may reduce to the case in which $\mathcal{M}_\bullet$ is a sheaf of the form $A/\mathfrak{p}[m]$ for some m. Since the map induced by ϕ on rational Chow groups is surjective, some positive multiple of $[A/\mathfrak{p}]$ is in the image of an element in the Chow group of Y, and the proof of Theorem 12.2.1 shows that a multiple of the cycle $[A/\mathfrak{p}]$ is the image of a cycle of the form $[B/\mathfrak{q}]$. Let $\mathcal{N}_\bullet$ be the push-forward of the sheaf $B/\mathfrak{q}[m]$. Then $\mathcal{N}_\bullet$ is isomorphic to a direct sum of copies of $A/\mathfrak{p}[m]$ up to a complex with support of dimension strictly less than the dimension of Z. Hence, by induction on the dimension of Z, we can assume that $\mathcal{M}_\bullet$ is the push-forward of a complex of sheaves $\tilde{\mathcal{M}}_\bullet$ on Y.

Since both pull-back and push-forward of sheaves are defined by tensor product, the associativity of the tensor product implies that we have an isomorphism

$$\phi_*(\phi^*(\mathcal{F}_\bullet) \otimes \tilde{\mathcal{M}}_\bullet) \cong \mathcal{F}_\bullet \otimes \phi_*(\tilde{\mathcal{M}}_\bullet).$$

Assuming that the theorem holds on Y, we now apply Theorem 12.5.1 (the Riemann-Roch theorem) and use the projection formula for local Chern characters to obtain

$$\tau(\mathcal{F}_\bullet \otimes \mathcal{M}_\bullet) = \tau(\phi_*(\phi^*(\mathcal{F}_\bullet) \otimes \tilde{\mathcal{M}}_\bullet)) = \phi_*(\tau(\phi^*(\mathcal{F}_\bullet) \otimes \tilde{\mathcal{M}}_\bullet))$$

$$= \phi_*(\mathrm{ch}(\phi^*(\mathcal{F}_\bullet))\tau(\tilde{\mathcal{M}}_\bullet)) = \mathrm{ch}(\mathcal{F}_\bullet)(\phi_*(\tau(\tilde{\mathcal{M}}_\bullet)))$$

$$= \mathrm{ch}(\mathcal{F}_\bullet)(\tau(\phi_*(\tilde{\mathcal{M}}_\bullet))) = \mathrm{ch}(\mathcal{F}_\bullet)(\tau(\mathcal{M}_\bullet)).$$

Thus, we may assume that $\mathcal{F}_\bullet$ has a filtration with quotients that are elementary complexes. Using additivity, we may then assume that $\mathcal{F}_\bullet$ is an elementary complex.

Let R be a graded polynomial ring that maps onto A, let $M = \mathrm{Proj}(R)$, and let i denote the embedding from X into M. Suppose that the complex $\mathcal{F}_\bullet$ is the restriction to X of a complex of locally free sheaves $\mathcal{G}_\bullet$ on M. Let $\mathcal{H}_\bullet$ be a locally free resolution of $\mathcal{M}_\bullet$ on M. Then $\mathcal{G}_\bullet \otimes \mathcal{H}_\bullet$ is a locally free resolution of $\mathcal{F}_\bullet \otimes \mathcal{M}_\bullet$. Let η be the product $\prod_i Q(h_i)[M]$ used in the definition of the Todd class. Then by the multiplicativity formula we have

$$\tau(\mathcal{G}_\bullet \otimes \mathcal{M}_\bullet) = \mathrm{ch}(\mathcal{G}_\bullet \otimes \mathcal{H}_\bullet)(\eta) = \mathrm{ch}(\mathcal{G}_\bullet)\mathrm{ch}(\mathcal{H}_\bullet)(\eta)$$

$$= \mathrm{ch}(\mathcal{G}_\bullet)(\tau(\mathcal{M}_\bullet)) = \mathrm{ch}(\mathcal{F}_\bullet)(\tau(\mathcal{M}_\bullet))$$

since $\tau(\mathcal{M}_\bullet)$ is a cycle in $A_*(X)$ and the restriction of $\mathcal{G}_\bullet$ to X is $\mathcal{F}_\bullet$. The remainder of the proof shows that we can reduce to this case.

We first assume that $\mathcal{F}_\bullet$ is a locally free sheaf of rank 1 $\mathcal{L}$ in one degree, and we may take the degree to be 0. Let notation be as above, and consider the sheaf $\mathcal{L}$ as a module over R, so that it defines a coherent sheaf on M, which we denote $i_*(\mathcal{L})$. Let $\mathcal{G}$ be a locally free sheaf defined by a sum of copies of $R[i]$ for various i that maps onto $i_*(\mathcal{L})$. Then, tensoring with A, a direct sum of copies of $A[i]$ maps onto $\mathcal{L}$. Let $\tilde{M}$ be $\text{Proj}(R[T_i])$, where $R[T_i]$ is a polynomial ring with generators of degree i, so that it is the symmetric algebra on $\mathcal{G}$ over M. Let $R(\mathcal{L})$ denote the Rees algebra and let $Y = \text{Proj}(R(\mathcal{L}))$. Then, since $\mathcal{L}$ is locally free of rank 1, the map from Y to X is an isomorphism. Furthermore, there is a surjective map from $R[T_i]$ to $R(\mathcal{L})$ defined by the surjection from $\mathcal{G}$ to $i_*(\mathcal{L})$. We thus have an embedding from Y into $\tilde{M}$, and under this embedding the pull-back of $\mathcal{E}$ is equal to the pull-back of $\mathcal{L}$ from X. Using the Riemann-Roch theorem again, we may compute the Todd classes on Y instead of X. But now the sheaf $\mathcal{L}$ is a pull-back of a sheaf on $\tilde{M}$, so the result follows from the multiplicativity formula.

We next consider the case of an elementary complex with nonzero sheaves in two positions. A complex of this form is of the form

$$\mathcal{L} \otimes (0 \to \mathcal{O}_X(-D) \to \mathcal{O}_X \to 0)$$

for some locally free sheaf of rank 1 $\mathcal{L}$, and using the multiplicativity of local Chern characters and the previous case we may assume that $\mathcal{L} = \mathcal{O}_X$. We first do the case where $X = \text{Proj}(\text{Sym}_B(B \oplus \mathcal{L}))$ and D is defined by $\mathcal{L}$ in degree 1, where $\mathcal{L}$ defines a locally free sheaf of rank 1 on $\text{Proj}(B) = W$. Applying the argument of the previous paragraph, we can replace W by an isomorphic projective scheme in which $\mathcal{L}$ is the pull-back from $M = \text{Proj}(R)$, where R is a graded polynomial ring mapping onto B, and where $\mathcal{L}$ is the pull-back of $\mathcal{O}_M(1)$ on M. We then take a further extension, adjoining generators S_1 and S_2 of degrees 0 and -1, respectively, and mapping $R[S_1, S_2]$ to the pull-back of $\mathcal{O}_M \oplus \mathcal{O}_M(1)$ on W, which is isomorphic to the sheaf defined by $B \oplus \mathcal{L}$. Let $N = \text{Proj}(R[S_1, S_2])$. Now the complex

$$0 \to \mathcal{O}_X(-D) \to \mathcal{O}_X \to 0$$

is the pull-back of the complex

$$0 \to \mathcal{O}_N(1) \to \mathcal{O}_N \to 0$$

defined by the ideal generated by S_2 in $R[S_1, S_2]$, and again the result follows from the multiplicativity formula.

We now deform to the general case to the one we just considered using Fulton's "reduction to the normal cone," which is an argument very similar to the graph construction. Let $\mathcal{O}_X(-D)$ be defined by the locally principal ideal $\mathfrak{a}$ in $X = \text{Proj}(A)$, and let $\mathcal{O}_D$ denote the sheaf corresponding to $A/\mathfrak{a}$. As usual, we may assume that A is an integral domain and that $\mathcal{M}_\bullet$ is of the form $\mathcal{O}_X(m)$

for some m. Then the formula we are proving can be written

$$\tau(\mathcal{O}_D(m)) = \text{ch}(\mathcal{O}_X(-D) \to \mathcal{O}_X)(\tau(\mathcal{O}_X(m))).$$

We adjoin an indeterminate s and take the ideal generated by $\mathfrak{a}$ and s in $A[s]$. We then take $Y = \text{Proj}(R)$, where R is the Rees ring of the ideal generated by $\mathfrak{a}$ and s over $A[s]$. Let U denote the generator of the Rees ring corresponding to s in degree 1, and let $\tilde{\mathfrak{a}}$ be the ideal of R generated by $\mathfrak{a}$ in degree 1. We denote the complexes of the form

$$0 \to \mathcal{O}_X(-E) \to \mathcal{O}_X \to 0$$

for E equal to the divisors D and $\tilde{D}$ and those defined by s and $s-1$, respectively, by $\mathcal{F}_\bullet(\mathfrak{a})$, $\mathcal{F}_\bullet(\tilde{\mathfrak{a}})$, $\mathcal{F}_\bullet(s)$, and $\mathcal{F}_\bullet(s-1)$. Since the divisors (s) and $(s-1)$ are principal generated by elements of degree 0, they can be pulled back to any graded polynomial ring and the local Riemann-Roch theorem holds for them; in addition, $\text{ch}(\mathcal{F}_\bullet(s))$ is intersection with the divisor (s) and similarly for $s-1$.

By the argument of reducing to projective space of Theorem 12.1.2, the classes

$$\text{ch}(\mathcal{F}_\bullet(s))\text{ch}(\mathcal{F}_\bullet(\tilde{\mathfrak{a}}))(\tau(\mathcal{M}_\bullet))$$

and

$$\text{ch}(\mathcal{F}_\bullet(s-1))\text{ch}(\mathcal{F}_\bullet(\tilde{\mathfrak{a}}))(\tau(\mathcal{M}_\bullet))$$

are rationally equivalent. We now compute each of these in reverse order. If we first intersect with (s), then by the local computation of Section 11.1 we have two components, one generated by U and the other generated by $\mathfrak{a}$ in degree 0. The divisor $\tilde{\mathfrak{a}}$ is generated by the image of $\mathfrak{a}$ in degree 1. If we divide by U, then this image generates the irrelevant ideal and we get zero. On the other hand, the image of $\tilde{\mathfrak{a}}$ in the other component is the ideal generated by $\mathfrak{a}/\mathfrak{a}^2$ in degree 1 in the symmetric algebra on $A/\mathfrak{a} \oplus \mathfrak{a}/\mathfrak{a}^2$ over $A/\mathfrak{a}$, and that is the case we did above. We can thus deduce that

$$\tau(\mathcal{O}_D(m)) = \text{ch}(\mathcal{F}_\bullet(\tilde{\mathfrak{a}}))\text{ch}(\mathcal{F}_\bullet(s))(\tau(\mathcal{O}_Y(m))).$$

If we first intersect with $(s-1)$, the ideal $(s, \mathfrak{a})$ becomes principal generated by 1, and the Rees ring is just $A[T]$ for T an indeterminate of degree $(0, 1)$. Thus, the associated projective scheme is isomorphic to X. Furthermore, the ideal generated by $\tilde{\mathfrak{a}}$ is equivalent to the pull-back of $\mathfrak{a}$ under this isomorphism. We can conclude that

$$\text{ch}(\mathcal{F}_\bullet(\tilde{\mathfrak{a}}))\text{ch}(\mathcal{F}_\bullet(s-1))(\tau(\mathcal{O}_Y(m))) = \text{ch}(\mathcal{O}_X(-D) \to \mathcal{O}_X)(\tau(\mathcal{O}_X(m))).$$

Thus, combining these two equations and using that the results are rationally equivalent, we have that

$$\tau(\mathcal{O}_D(m)) = \text{ch}(\mathcal{O}_X(-D) \to \mathcal{O}_X))(\tau(\mathcal{O}_X(m))).$$

This completes the proof.   □

12.7 Local Chern Characters and Dutta Multiplicity

Let A be a ring of positive characteristic p and dimension d, and let F denote the Frobenius map on A. We assume that the residue field of A is perfect and that the Frobenius is a finite map. Let $F_\bullet$ be a bounded free complex over A with homology of finite length, and for each n let $F_\bullet^{[n]}$ be the result of tensoring $F_\bullet$ n times with the Frobenius map. In this situation we defined the Dutta multiplicity $\chi_\infty(F_\bullet)$ in Section 7.3 to be the limit of the Euler characteristic of $F_\bullet^{[n]}$ divided by p^{nd}. We show here that the Dutta multiplicity of $F_\bullet$ can be interpreted in terms of local Chern characters.

Theorem 12.7.1 *Let A and $F_\bullet$ be as above. Then we have*

$$\chi_\infty(F_\bullet) = \text{ch}(F_\bullet)([A]_d).$$

Proof By the local Riemann-Roch formula, we have

$$\chi\left(F_\bullet^{[n]}\right) = \text{ch}\left(F_\bullet^{[n]}\right)(\tau(A))$$

for each integer n. Let ϕ_n be the map from $\text{Spec}(A)$ to itself induced by F^n; we are assuming that ϕ_n is a finite map. We apply the projection formula to ϕ_n. The pull-back of $F_\bullet$ under ϕ_n is $F_\bullet^{[n]}$. On the other hand, we proved in Section 7.3 that the image of a cycle η of dimension i under $(\phi_n)_*$ is equal to $p^{ni}\eta$. Thus, the projection formula states that for every cycle η of dimension i we have

$$\text{ch}\left(F_\bullet^{[n]}\right)(\eta) = \text{ch}\left(\phi_n^*(F_\bullet)\right)(\eta) = \text{ch}(F_\bullet)((\phi_n)_*(\eta))$$
$$= \text{ch}(F_\bullet)(p^i\eta) = p^i\text{ch}(F_\bullet)(\eta).$$

On the other hand, we have

$$\chi_\infty(F_\bullet) = \lim_{n\to\infty} \frac{\chi\left(F_\bullet^{[n]}\right)}{p^{nd}}$$

$$= \lim_{n\to\infty} \frac{\text{ch}\left(F_\bullet^{[n]}\right)(\tau(A))}{p^{nd}}.$$

Writing $\tau(A)$ in terms of its components $\tau_i(A)$ in each dimension i, we conclude that this limit is equal to

$$\lim_{n\to\infty} \frac{\mathrm{ch}\big(F_\bullet^{[n]}\big)\big(\sum_i \tau_i(A)\big)}{p^{nd}} = \sum \lim_{n\to\infty} \frac{p^{ni}\mathrm{ch}(F_\bullet)(\tau_i(A))}{p^{nd}}$$

$$= \mathrm{ch}(F_\bullet)(\tau_d(A)) = \mathrm{ch}(F_\bullet)([A]_d).$$

$\square$

Exercises

12.1 Prove commutativity of local Chern characters using the splitting principle.

12.2 Carry out the details of the proof that

$$\sum_{i=1}^{r+1} \prod_{j\neq i} \frac{Y_j}{Y_j - Y_i} = 1.$$

12.3 Let $X = \mathbb{P}^3 = \mathrm{Proj}(k[X, Y, Z, W])$, and let Y be the subscheme $\mathrm{Proj}(k[X, Y, Z, W]/(XY - ZW))$ of X.

(a) Show that the ideal generated by X and Z defines a divisor on Y that cannot be extended to a divisor on X.

(b) Let $\mathcal{O}_D$ be the sheaf defined on Y by the graded module $k[X, Y, Z, W]/(X, Z)$. Compute $\tau(\mathcal{O}_Y)$ and $\tau(\mathcal{O}_D)$ by taking resolutions over X and verify that $\tau(\mathcal{O}_D) = \mathrm{ch}(\mathcal{F}_\bullet)\tau(\mathcal{O}_Y)$, where $\mathcal{F}_\bullet$ is the complex

$$0 \to \mathcal{O}_Y(-D) \to \mathcal{O}_Y \to 0.$$

13
Applications and Examples

This chapter is devoted to several applications of the theory of local Chern characters to local algebra. In the first section we show how to extend the definition of intersection multiplicities in the nonregular case. The second section consists of two examples of negative intersection multiplicities, and in the third we discuss their implications to the theory of local Chern characters. In the fourth section we prove the Peskine-Szpiro intersection theorem in mixed characteristic.

13.1 Intersection Multiplicities

As mentioned in the introduction, one of the motivations behind much of the theory discussed in this book is the problem of defining intersection multiplicities of two subschemes that meet at a point. While there is still not a complete solution to this question, the use of local Chern characters makes it possible to extend the definitions to more general situations. Essentially, it allows a definition of intersection multiplicities for two cycles in the Chow group when both cycles are associated to the homology of a complex of finite length of locally free sheaves. We limit ourselves here to the case of bounded complexes of free modules over a local ring.

Let $F_\bullet$ be a bounded free complex over a local integral domain A of dimension d, and suppose that the support of $F_\bullet$ has dimension at most k. We then define a cycle $[F_\bullet]_k$ of dimension k by letting

$$[F_\bullet]_k = \sum_{\mathfrak{p};\dim(\mathfrak{p})=k} \chi((F_\bullet)_\mathfrak{p})[A/\mathfrak{p}].$$

We note that this is the same as the cycle $[M]_k$ if $F_\bullet$ is a free resolution of a module M of dimension k.

We now define an intersection product on pairs of cycles defined by bounded free complexes of complementary dimension meeting in the closed point of

286

Spec(A). Let $F_\bullet$ and $G_\bullet$ be bounded free complexes defining cycles η and ω of dimensions k and m, respectively, with $k + m = d$ and such that the intersection of the supports of $F_\bullet$ and $G_\bullet$ is the closed point. We then define

$$[F_\bullet]_k \cap [G_\bullet]_m = \eta \cap \omega = \mathrm{ch}_m(F_\bullet)(\omega).$$

We first show that this definition is in fact symmetric in η and ω. In fact, we note that, by Corollary 12.4.3, we have that ω is equal to $\mathrm{ch}_k([A])$, so that the above intersection can also be written $\mathrm{ch}_m(F_\bullet)\mathrm{ch}_k(G_\bullet)([A])$. It thus follows from the commutativity of local Chern characters that the definition is symmetric.

We now list several important properties of this product.

(i) The product is defined modulo rational equivalence. This is true since the local Chern characters are defined modulo rational equivalence.

(ii) The definition is independent of choice of complexes with the given cycles. This follows since the original definition makes no mention of the complex $G_\bullet$ and the product is commutative.

(iii) The definition agrees with Serre's intersection multiplicity in the case of regular local rings. In this case, every module has a finite free resolution, and we may assume that our modules are of the form $A/\mathfrak{p}$ and $A/\mathfrak{q}$ of dimensions k and m, respectively. By Serre's results [66], if $\mathfrak{p} + \mathfrak{q}$ is primary to the maximal ideal, then $\dim(A/\mathfrak{p}) + \dim(A/\mathfrak{q}) \leq \dim(A)$. Let $F_\bullet$ and $G_\bullet$ be resolutions of $A/\mathfrak{p}$ and $A/\mathfrak{q}$. Then Serre's intersection multiplicity is given by

$$\chi(A/\mathfrak{p}, A/\mathfrak{q}) = \chi(F_\bullet \otimes G_\bullet).$$

But the local Riemannn-Roch and multiplicativity formulas state that

$$\chi(F_\bullet \otimes G_\bullet) = \mathrm{ch}_d(F_\bullet \otimes G_\bullet)([A]) = \sum_{i=0}^{d} \mathrm{ch}_i(F_\bullet)\mathrm{ch}_{d-i}(G_\bullet)([A]).$$

Now since the support of $G_\bullet$ has dimension $m = d - k$, all terms are zero for $d - i < k$. Using commutativity again, we similarly obtain that $\mathrm{ch}_i(F_\bullet)\mathrm{ch}_{d-i}(G_\bullet)$ $([A]) = \mathrm{ch}_{d-i}(G_\bullet)\mathrm{ch}_i(F_\bullet)([A])$ for $i < m$. Thus, all terms are zero except $\mathrm{ch}_m(F_\bullet)\mathrm{ch}_k(G_\bullet)([A])$, and this definition agrees with the one given above.

One corollary of this is Serre's vanishing conjecture.

Corollary 13.1.1 *If A is a regular local ring of dimension d, and if M and N are finitely generated modules such that $M \otimes_A N$ has finite length and $\dim(M) + \dim(N) < d$, then $\chi(M, N) = 0$.*

Proof Let dim$(M) = k$ and $m = d - k$. Let $F_\bullet$ and $G_\bullet$ be finite free resolutions of M and N, respectively. Then

$$\chi(M, N) = [F_\bullet]_k \cap [G_\bullet]_m = [F_\bullet] \cap 0 = 0$$

since the dimension of N is less than m.   □

On the other hand, Serre's positivity conjecture is still not known for rings of mixed characteristic. The fact that intersection multiplicities are non-negative was recently proven by Gabber (see Berthelot [6]). A special case (where the subvarieties are not tangent) was proven by Tennison [71].

13.2 Negativity of Intersection Multiplicities

In this section we construct two examples of nonpositive intersection multiplicities. The first example shows that the definition of multiplicities using Euler characteristics cannot be extended to the most general case where it could conceivably be defined. The second shows that even where intersection multiplicities can be defined, the intersection multiplicity of two positive cycles can be negative in the singular case.

The first example is due to Dutta, Hochster and MacLaughlin [12], and it had numerous consequences for the theory. We recall the general situation. Let M and N be modules over a local ring A such that $M \otimes_A N$ has finite length and N has finite projective dimension. It then makes sense to define the Euler characteristic $\chi(M, N) = \sum (-1)^i \operatorname{length}(\operatorname{Tor}_i^A(M, N))$. Suppose that we wish to define intersection multiplicities of the associated cycles using this definition; it is then necessary that we have $\chi(M, N) = 0$ if $\dim(M) + \dim(N) < \dim(A)$, since otherwise we would have a situation where we could modify a module of dimension k by something of lower dimension that would define the same cycle in dimension k but give a different answer for the multiplicity. The example we present shows that we do not have $\chi(M, N) = 0$ in this situation in general.

Let $A = k[X, Y, Z, W]/(XW - YZ)$. We show that there exists a module M of finite length and finite projective dimension such that $\chi(M, A/\mathfrak{p}) = -1$, where $\mathfrak{p}$ is the prime ideal generated by X and Y. Thus, $\chi(M, A/\mathfrak{p}) \neq 0$ while $\dim(M) + \dim(A/\mathfrak{p}) = 0 + 2 < 3 = \dim(A)$.

We first prove some general facts about modules of finite projective dimension over this particular ring A. First, a module of finite length is a finite dimensional vector space over k with four commuting nilpotent endomorphisms, which we denote X, Y, Z, W, such that $XW = YZ$. We assume that we have such a vector space M with four commuting endomorphisms. In this situation the length of M as an A-module is the same as its dimension as a vector space over k.

A minimal free resolution of $A/\mathfrak{p}$ is given by the following complex:

$$\cdots \longrightarrow A \oplus A \xrightarrow{\begin{pmatrix} W & -Y \\ -Z & X \end{pmatrix}} A \oplus A \xrightarrow{\begin{pmatrix} X & Y \\ Z & W \end{pmatrix}}$$

$$\rightarrow A \oplus A \xrightarrow{\begin{pmatrix} W & -Y \\ -Z & X \end{pmatrix}} A \oplus A \xrightarrow{(X \quad Y)} A \rightarrow A/\mathfrak{p} \rightarrow 0.$$

Lemma 13.2.1 *The module M has finite projective dimension if and only if the sequence*

$$M \oplus M \xrightarrow{\begin{pmatrix} X & Y \\ Z & W \end{pmatrix}} M \oplus M \xrightarrow{\begin{pmatrix} W & -Y \\ -Z & X \end{pmatrix}} M \oplus M$$

is exact.

Proof Since $A/\mathfrak{p}$ is regular, the A-module k has a finite resolution by sums of copies of $A/\mathfrak{p}$ (these modules are, of course, not free A-modules). The module M has finite projective dimension if and only if $\operatorname{Tor}^i_A(M, k) = 0$ for all sufficiently large i, and the existence of a finite resolution of k by sums of copies of $A/\mathfrak{p}$ implies that $\operatorname{Tor}^i_A(M, k) = 0$ for all sufficiently large i if and only if $\operatorname{Tor}^i_A(M, A/\mathfrak{p}) = 0$ for all sufficiently large i. Now we have just seen that $A/\mathfrak{p}$ has a free resolution with matrices alternating between the two in the statement of the Lemma. If we tensor with M, we then have the above complex, and its homology is $\operatorname{Tor}^i_A(M, A/\mathfrak{p}) = 0$. Thus, if M has finite projective dimension, the above sequence is exact. On the other hand, if the sequence is exact, then the sum of ranks of the two maps in the sequence is twice the dimension of M as a vector space over k. This in turn implies that the homology of the sequence

$$M \oplus M \xrightarrow{\begin{pmatrix} W & -Y \\ -Z & X \end{pmatrix}} M \oplus M \xrightarrow{\begin{pmatrix} X & Y \\ Z & W \end{pmatrix}} M \oplus M$$

is also zero. Thus, $\operatorname{Tor}^i_A(M, A/\mathfrak{p}) = 0$ for all large i, and M has finite projective dimension. $\qquad \square$

Lemma 13.2.2 *If M has finite projective dimension, then the multiplicity $\chi(M, A/\mathfrak{p})$ is equal to $r - s$, where r is the rank of the map from $M \oplus M$ to itself given by the matrix $\begin{pmatrix} W & -Y \\ -Z & X \end{pmatrix}$, and s is the length of M.*

Proof We have shown that $\mathrm{Tor}^i_A(M, A/\mathfrak{p}) = 0$ for $i > 1$, so the intersection multiplicity is the difference between the dimensions of the first and second homology modules in the sequence

$$M \oplus M \xrightarrow{\begin{pmatrix} W & -Y \\ -Z & X \end{pmatrix}} M \oplus M \xrightarrow{(X\ Y)} M \to 0.$$

Let N be the image of $\begin{pmatrix} W & -Y \\ -Z & X \end{pmatrix}$ in $M \oplus M$. We then have a short exact sequence

$$0 \to \mathrm{Tor}_1(M, A/\mathfrak{p}) \to (M \oplus M)/N \to M \to \mathrm{Tor}_0(M, A/\mathfrak{p}) \to 0.$$

Hence

$$\chi(M, A/\mathfrak{p}) = \mathrm{length}(\mathrm{Tor}_0(M, A/\mathfrak{p})) - \mathrm{length}(\mathrm{Tor}_1(M, A/\mathfrak{p}))$$

$$= \mathrm{length}(M) - \mathrm{length}((M \oplus M)/N) = s - (2s - r) = r - s.$$

$\square$

Thus, the problem is to find a vector space with four maps satisfying the above properties such that the rank of the matrix above is not equal to the dimension of M. It is shown in [12] that the minimum dimension of an example with these properties is 15. We outline their construction in dimension 15.

We represent the endomorphisms that define X, Y, Z, W by 15-by-15 matrices in block form with sides 5, 4, 6, so they are 3-by-3 matrices with entries given by various size blocks. We denote these matrices A_1, A_2, A_3, A_4. They are assumed to be upper triangular and of the form

$$A_i = \begin{pmatrix} 0 & a_i & c_i \\ 0 & 0 & b_i \\ 0 & 0 & 0 \end{pmatrix}.$$

Let $a_2 = b_2 = a_4 = b_4 = 0$. Further, let

$$a_1 = \begin{matrix} & 2 & 2 \\ 2 \\ 3 \end{matrix}\begin{pmatrix} 1 & 0 \\ 0 & 0 \end{pmatrix} \qquad a_3 = \begin{matrix} & 2 & 2 \\ 2 \\ 3 \end{matrix}\begin{pmatrix} 0 & 1 \\ 0 & 0 \end{pmatrix}$$

$$b_1 = \begin{matrix} & 2 & 2 & 2 \\ 2 \\ 2 \end{matrix}\begin{pmatrix} 1 & 0 & 0 \\ 0 & 1 & 0 \end{pmatrix} \qquad b_3 = \begin{matrix} & 2 & 2 & 2 \\ 2 \\ 2 \end{matrix}\begin{pmatrix} 0 & 1 & 0 \\ 0 & 0 & 1 \end{pmatrix}.$$

Here the 1s denote identity matrices and the 0s denote matrices of zeros of sizes given by the numbers above and to the left of the matrices.

Multiplying $A_i A_j$ gives a matrix that is zero except for $a_i b_j$ in the upper corner. To check commutativity, we need to show that $a_i b_j = a_j b_i$ for all i

and j, which is trivial except when $i = 1$ and $j = 3$, where it is easy to check. Similarly, the relation $A_1 A_4 = A_2 A_3$ holds since both sides are zero.

For the c_i we let:

$$c_1 = \begin{pmatrix} 0 & 0 & 0 & 0 & 0 & 0 \\ 0 & 0 & 0 & 0 & 0 & 0 \\ 0 & 0 & 0 & 0 & 0 & 0 \\ 0 & 0 & 0 & 0 & 1 & 0 \\ 0 & 0 & 0 & 0 & 0 & 1 \end{pmatrix} \qquad c_2 = \begin{pmatrix} 0 & 0 & 0 & 0 & 0 & 0 \\ 0 & 0 & 0 & 0 & 0 & 0 \\ 1 & 0 & 0 & 0 & 0 & 0 \\ 0 & 1 & 0 & 0 & 0 & 0 \\ 0 & 0 & 1 & 0 & 0 & 0 \end{pmatrix}$$

$$c_3 = \begin{pmatrix} 0 & 0 & 0 & 0 & 0 & 0 \\ 0 & 0 & 0 & 0 & 0 & 0 \\ 0 & 0 & 0 & 0 & 0 & 1 \\ 0 & 0 & 0 & 0 & 0 & 0 \\ 0 & 0 & 0 & 0 & 0 & 0 \end{pmatrix} \qquad c_4 = \begin{pmatrix} 1 & 0 & 0 & 0 & 0 & 0 \\ 0 & 1 & 0 & 0 & 0 & 0 \\ 0 & 0 & 0 & 1 & 0 & 0 \\ 0 & 0 & 0 & 0 & 1 & 0 \\ 0 & 0 & 0 & 0 & 0 & 1 \end{pmatrix}$$

It is straightforward to verify that the 30-by-30 matrices $\begin{pmatrix} A_1 & A_2 \\ A_3 & A_4 \end{pmatrix}$ and $\begin{pmatrix} A_4 & -A_2 \\ -A_3 & A_1 \end{pmatrix}$ have ranks 16 and 14, respectively, so that they define a module with negative intersection multiplicity. For more details, and a somewhat more conceptual proof of this last statement, we refer to Dutta, Hochster, MacLaughlin [12].

Thus, we have constructed a module M of finite length and finite projective dimension such that $\chi(M, N) \neq 0$ for a module N of dimension less than the dimension of the ring, showing that intersection multiplicities cannot be defined using Euler characteristics in this generality. In the next section we discuss further consequences of this example.

The next example shows that, even in cases where the intersection multiplicities can be defined properly, the intersection multiplicity of two positive cycles can be negative. We construct a local ring of dimension 4 together with two bounded free complexes $F_\bullet^1$ and $F_\bullet^2$ with support of dimension 2 meeting only in the closed point of $\mathrm{Spec}(A)$, which define positive cycles, and such that the intersection multiplicity defined as in Section 1 is negative. This example appeared in Roberts [63].

Let U, V, W, X, Y, Z be indeterminates, and let A be the polynomial ring $k[U, V, W, X, Y, Z]$ modulo the ideal generated by maximal minors of the matrix

$$\begin{pmatrix} U & V & W \\ X & Y & Z \end{pmatrix}$$

localized at the maximal ideal (U, V, W, X, Y, Z). Let $\mathfrak{p}_1 = (U, V, X, Y)$ and $\mathfrak{p}_2 = (V, W, Y, Z)$. We wish to find bounded complexes of free modules $F_\bullet^i$ with homology of dimension 2 such that $[F_\bullet^i]_2 = [A/\mathfrak{p}_i]$. We carry out the construction for $\mathfrak{p}_1$; the construction of $\mathfrak{p}_2$ is the same. (These ideals do not themselves have finite projective dimension.)

For positive integers k and n we consider the ideal $I_{k,n}$ generated by X^k and $Y - U^n$. We will show that the given generators of $I_{k,n}$ form a regular sequence, so that $I_{k,n}$ has finite projective dimension and so that the support of $A/I_{k,n}$ consists of $\mathfrak{p}_1$ together with another ideal $\mathfrak{q}_1$ also of height 2. Furthermore, the lengths of $(A/I_{k,n})_{\mathfrak{p}_1}$ and $(A/I_{k,n})_{\mathfrak{q}_1}$ are k and $k(n+1)$, respectively. We will then put two resolutions of the quotients $A/I_{k,n}$ together for appropriate choices of k and n to produce a complex with Euler characteristic $[A/\mathfrak{p}_1]$ as desired.

To study the ideal $I_{k,n}$, we take the quotient of the polynomial ring $k[U, V, W, X, Y, Z]$ by the ideal generated by the maximal minors of the above matrix plus the two generators of $I_{k,n}$. We then have the quotient

$$k[U, V, W, X, Y, Z]/(UY - XV, UZ - XW, VZ - YW, X^k, Y - U^n).$$

Suppose now that $\mathfrak{p}$ is a minimal prime ideal in this quotient ring. Then $\mathfrak{p}$ contains X^k, so $\mathfrak{p}$ contains X. Thus, the minimal prime ideals of this ring are the same as the minimal prime ideals of the ring

$$k[U, V, W, X, Y, Z]/(UY - XV, UZ - XW, VZ - YW, X, Y - U^n)$$

$$\cong k[U, V, W, Z]/(U^{n+1}, UZ, VZ - U^n W).$$

A minimal prime ideal of this latter ring must contain U, so the minimal prime ideals are the same as the minimal prime ideals of the ring $k[V, W, U, Z]/(U, VZ - U^n W) \cong k[V, W, Z]/(VZ)$. The minimal prime ideals of this ring are the ideals generated by V and Z. Pulling these ideals back to A, we have the ideals generated by (X, Y, U, V), which is $\mathfrak{p}_1$, and by (X, Y, Z, U), which we denote $\mathfrak{q}_1$.

Both of these prime ideals have height 2, and the fact that the generators of $I_{k,n}$ form a regular sequence follows from the fact that A is Cohen-Macaulay.

It remains to compute the lengths of $(A/I_{k,n})_{\mathfrak{p}_1}$ and $(A/I_{k,n})_{\mathfrak{q}_1}$; these lengths are the coefficients of $[A/\mathfrak{p}_1]$ and $[A/\mathfrak{q}_1]$ in the cycle $[A/I_{k,n}]_2$. The rings $A_{\mathfrak{p}_1}$ and $A_{\mathfrak{q}_1}$ are regular, and their maximal ideals are generated by (X, Y) and (U, Y), respectively. In $A_{\mathfrak{p}_1}$, X is a unit times U, so there is a unit α in $A_{\mathfrak{p}_1}$ with $I_{k,n}A_{\mathfrak{p}_1} = (X^k, Y - \alpha X^n)$, and it is clear that the length of the quotient is k. In the ring $A_{\mathfrak{q}_1}$, the element V is a unit, so, since $XV = YU$, we have $I_{k,n}A_{\mathfrak{q}_1} = (Y^k U^k, Y - U^n)$. Substituting U^n for Y in $Y^k U^k$ we conclude that the length of the quotient is $k(n+1)$.

We now define a complex by taking a resolution $F_\bullet^1$ of $I_{2,1}$ and a resolution $G_\bullet^1$ of $I_{1,3}$. Then the complex $H_\bullet^1 = F_\bullet^1 \oplus G_\bullet^1[-1]$ is such that $[H_\bullet^1]_2 = [A/\mathfrak{p}_1]$.

Up to this point we have constructed a complex representing the cycle $[A/\mathfrak{p}_1]$; replacing U, V, X, Y by V, Y, W, Z, respectively, and replacing the ideal $I_{k,n}$

by $(W^k, Z - V^n)$, we obtain a complex $F_\bullet^2$ representing the cycle $[A/\mathfrak{p}_2]$. Note that the supports of $F_\bullet^1$ and $F_\bullet^2$ are larger than the sets of prime ideals containing $\mathfrak{p}_1$ and $\mathfrak{p}_2$, respectively, but their supports still intersect only at the maximal ideal of A. Thus, to find the multiplicity we must calculate the local Chern character of $F_\bullet^1$ applied to the class of $A/\mathfrak{p}_2$. However, since $F_\bullet^1$ is a direct sum of Koszul complexes, we can compute the multiplicity by computing the Euler characteristic $\chi(F_\bullet^1 \otimes F_\bullet^2)$. This calculation is straightforward; since each complex is the sum of one Koszul complex and another Koszul complex shifted by one, and since the ring A is Cohen-Macaulay, the Euler characteristic reduces to

$$\text{length}(A/(X^2, Y - U, W^2, Z - V)) - \text{length}(A/(X^2, Y - U, W, Z - V^3))$$

$$- \text{length}(A/(X, Y - U^3, W^2, Z - V))$$

$$+ \text{length}(A/(X, Y - U^3, W, Z - V^3)).$$

This computation gives

$$\chi(H_\bullet^1 \otimes H_\bullet^2) = 12 - 14 - 14 + 15 = -1.$$

Thus, we have shown that there exist two prime ideals $\mathfrak{p}_1$ and $\mathfrak{p}_2$ of dimension 2 in a ring of dimension 4 such that $[A/\mathfrak{p}_1] \cap [A/\mathfrak{p}_2]$ is defined and is equal to -1. The question of the existence of two modules (rather than complexes) of finite projective dimension whose cycles meet in the correct codimension but have negative intersection multiplicity is still open. However, if there is a module of finite projective dimension whose support is $\mathfrak{p}_1$ (together with possibly a set of dimension at most one), then by the independence of the definition of multiplicity given above on the complex defining the cycle, such a module, together with a similar module with support $\mathfrak{p}_2$, would provide an example.

These examples show that some of the properties of intersection multiplicities for regular local rings do not carry over to the singular case. We remark also that Mumford [47] has defined intersection multiplicities on normal surfaces; in this case the results are always positive for positive cycles, but they take values in the rational numbers and are not necessarily integers.

13.3 Further Consequences of the Examples

The counterexamples to general vanishing conjectures, when interpreted in terms of Chern characters via the local Riemann-Roch formula, give rise to other examples concerning the Euler characteristics defined by perfect complexes.

Let $A = k[X, Y, Z, W]/(XW - YZ)$ be the ring defined in the first part of the last section, and let $F_\bullet$ be a free resolution of the module of negative intersection multiplicity M constructed there. Let $\mathfrak{p} = (X, Y)$ and let $N = A/\mathfrak{p}$. First, the local Riemann-Roch theorem gives

$$\mathrm{ch}(F_{\bullet})(\tau(N)) = \chi(F_{\bullet} \otimes N) = -1.$$

Now $A/\mathfrak{p}$ is a regular local ring, so it follows from the invariance of the local Todd class that $\tau(N) = [A/\mathfrak{p}] = [N]$. Thus, it follows in particular not only that the class $[A/\mathfrak{p}]$ is not zero in the Chow group (a fact we proved in a much more elementary manner in Chapter 1) but that we have a complex supported at the maximal ideal whose Chern character does not vanish on this element.

We next wish to give further examples constructed from this one. We construct a family of finite extensions of A as follows.

Assume that the field k is infinite and not of characteristic 2, and let $a_1, \ldots a_{2n}$ be distinct elements of k. Let $\eta = (X - a_1 Y)(X - a_2 Y) \cdots (X - a_{2n} Y)$, and let B_n be the ring $A[\sqrt{\eta}]$ obtained by adjoining the square root of η to A. Let C_n be the integral closure of B_n in its quotient field.

Since we are assuming that the characteristic of k is not 2, the ring C_n is easy to describe. First, the trace map gives a splitting $C_n = A \oplus I \sqrt{\eta}$, where I is the fractional ideal of elements $a \in A$ such that $a\sqrt{\eta}$ is integral over A, and the condition that $a\sqrt{\eta}$ is integral over A means that $a^2\eta$ is in A. The quotient $A/(Z - a_i W)$ is isomorphic to $k[Y, W, Z]/(a_i Y W - Y Z)$, and from this fact we can conclude that the ideal generated by $X - a_i Y$ is the intersection of the prime ideals (X, Y) and $(X - a_i Y, Z - a_i W)$. Each of these is a height 1 prime in the integrally closed domain A, so the condition that $u \in I$ is equivalent to the condition that u can be written as t/X^n with $t \in (X, Z)^n$. Thus, we obtain that $C_n = A \oplus I \sqrt{\eta}$, where I is a fractional ideal isomorphic as a module to $(X, Z)^n$. Thus, for each positive integer n we have constructed an integral domain C_n containing A such that C_n is isomorphic to $A \oplus (X, Z)^n$ as an A-module.

Let $F_{\bullet}$ be the free resolution of the module M constructed in the previous section as above. Recall that the length of M as an A-module is 15. It follows from the fact that $\chi(F_{\bullet} \otimes A/(X, Z)) = -1$ and that the length of the localization $(A/(X, Z)^n)_{(X,Z)}$ is n that $\chi(F_{\bullet} \otimes A/(X, Z)^n) = -n$. Finally, the short exact sequence

$$0 \to (X, Z)^n \to A \to A/(X, Z)^n \to 0$$

shows that we must have $\chi(F_{\bullet} \otimes (X, Z)^n) = \chi(M, A) - n = 15 - n$. Thus, if we take n greater than 30 in this construction, and let $G_{\bullet} = F_{\bullet} \otimes C_n$, we obtain a complex of free C_n-modules of length equal to the dimension of C_n with Euler characteristic

$$\chi(G_{\bullet}) = \chi(M, A \oplus (X, Z)^n) = 15 + (15 - n) = 30 - n < 0.$$

We recall that in Chapter 7 we showed that the Dutta multiplicity of a bounded free complex of length equal to the dimension of the ring with nonzero homology of finite length is always positive. This example shows that the corresponding result does not hold for Euler characteristics.

In the previous chapter we defined the Todd class $\tau(A)$ of a ring A and showed that if the dimension of A is d, then the part of $\tau(A)$ of dimension d is equal to $[A]_d$. We also showed that if A is a complete intersection, then $\tau(A) = [A]_d$, and if A is Gorenstein, then $\tau(A)_i = 0$ for all i such that $d - i$ is odd. We consider now the Todd class of the example C_n of dimension 3 above. We show not only that $\tau_2(C_n)$ is nontrivial but that there exists a bounded free complex $F_\bullet$ with homology of finite length such that $\mathrm{ch}_2(F_\bullet)(\tau(C_n)) \neq 0$. In fact, the class of C_n in dimension 3 pushes down to twice the class of A in the Chow group of A, so we have $\mathrm{ch}_3(G_\bullet)([C_n]) = 2\mathrm{ch}_3(F_\bullet)([A]) = 30$. On the other hand, the local Riemann-Roch formula gives

$$\chi(G_\bullet) = \mathrm{ch}(G_\bullet)(\tau(C_n)) = \mathrm{ch}_2(G_\bullet)(\tau_2(C_n)) + \mathrm{ch}_3(G_\bullet)(\tau_3(C_n))$$

$$= \mathrm{ch}_2(G_\bullet)(\tau_2(C_n)) + 30.$$

Putting this together with the fact that $\chi(G_\bullet) = 30 - n$, we see that $\tau_2(C_n) \neq 0$ and that $\mathrm{ch}_2(G_\bullet)(\tau_2(C_n)) = -n$. As shown at the end of the last chapter, the Dutta multiplicity is equal to $\mathrm{ch}_3(G_\bullet)([C_n]) = 30$.

The final example is when $n = 1$ in the above example. Then the ring C_1 is Cohen-Macaulay, and $\chi(G_\bullet) = 29$. This example shows first of all that, for a Cohen-Macaulay ring C of dimension 3, it is possible for $\tau_2(C)$ not to vanish. We showed in Chapter 12 that if $F_\bullet$ is a complex of length $d = \dim(A)$ with homology of finite length, then

$$\chi(\mathrm{Hom}(F_\bullet, A)) = \sum_i (-1)^i \mathrm{ch}_i(F_\bullet)(\tau_i(A)).$$

In particular, if A is Gorenstein, then $\tau_i(A) = 0$ when $d - i$ is odd, so we conclude that $\chi(\mathrm{Hom}(F_\bullet, A)) = (-1)^d \chi(F_\bullet)$ (we note that this last equality can also be deduced from the fact that A is dualizing since A is Gorenstein). It had been asked whether such a result might hold for Cohen-Macaulay rings (see Szpiro [68]). However, in this example, we have that $\chi(G_\bullet) = 29$ while $\chi(\mathrm{Hom}(G_\bullet, A)) = -31$.

Kurano [38] has constructed an example of a Gorenstein ring A of dimension 5 such that $\tau_3(A) \neq 0$. It is not known, however, whether there is a bounded free complex with local Chern character that does not vanish on this class.

13.4 The Intersection Theorem in Mixed Characteristic

In Chapter 6 we gave a proof of the Peskine-Szpiro intersection theorem for rings of positive characteristic and outlined how the theorem can be extended to the case of equal characteristic. Here we prove the theorem in mixed characteristic. The proof is by computing the Chern character of the complex reduced modulo p and comparing the result to the answer given by Dutta multiplicity. We recall the statement of the "new" version of the theorem.

Theorem 13.4.1 (New Intersection Theorem) *Let*

$$F_\bullet = 0 \to F_k \to \cdots \to F_0 \to 0$$

be a nonexact bounded free complex over a local ring A with homology of finite length. Then

$$k \geq \dim(A).$$

Proof From previous results we may assume that A is a ring of mixed characteristic p, and dividing by a minimal prime ideal we may assume that A is a domain. By reducing modulo p and applying the result in the positive characteristic case, we conclude that $F_\bullet$ is a complex of length at least $d - 1$. If the theorem is false, we can find an example of length exactly $d - 1$; we will show that this assumption leads to a contradiction.

We apply the local Chern character associated to the complex $F_\bullet$ to the class $[A]$ of dimension d. Denote the complex $F_\bullet \otimes_A A/pA$ by $\overline{F}_\bullet$. Since local Chern characters commute with intersection with divisors, we have

$$\mathrm{ch}_{d-1}(\overline{F}_\bullet)([A/pA]) = \mathrm{ch}_{d-1}(\overline{F}_\bullet)((p) \cap [A]) = (p) \cap \mathrm{ch}_{d-1}(F_\bullet)([A])).$$

On the other hand, the cycle $\mathrm{ch}_{d-1}(F_\bullet)([A/pA])$ is supported at the maximal ideal of A, and p is in the maximal ideal, so intersecting with the divisor (p) gives zero. However, Theorem 12.7.1 comparing Chern characters to Dutta multiplicity implies that $\mathrm{ch}_{d-1}(\overline{F}_\bullet)([A/pA]) = \chi_\infty(\overline{F}_\bullet)$ over the ring A/pA. In Theorem 7.3.5 we proved that the Dutta multiplicity of a complex of length $d - 1$ over a ring of dimension $d - 1$ is positive. Thus, the quantity $\mathrm{ch}_{d-1}(\overline{F}_\bullet)([A/pA]) = \chi_\infty(\overline{F}_\bullet)$ would have to be both zero and positive, so this contradiction proves the theorem. □

We remark that this method does not suffice to prove the improved new intersection conjecture in mixed characteristic.

Bibliography

[1] M. F. Atiyah and I. G. Macdonald, *Introduction to Commutative Algebra*, Addison-Wesley, Reading, MA, 1969.

[2] M. Auslander, Modules over unramified regular local rings, *Proc. Intern. Congress Math.* (1962), 230–233.

[3] L. Avramov and H.-B. Foxby, *Cohen-Macaulay properties of ring homomorphisms*, Københavns Universitet, Matematisk Institut, Preprint series no. 20 (1993).

[4] H. Bass, On the ubiquity of Gorenstein rings, *Math. Z.* **82** (1963), 8–28.

[5] P. Baum, W. Fulton, and R. MacPherson, Riemann-Roch for singular varieties, *Publ. Math. IHES* **45** (1975), 101–145.

[6] P. Berthelot, *Altérations de variétés algébriques [d'après A. J. de Jong]* Séminaire Bourbaki n. 815 (1996).

[7] P. L. Bhattacharya, The Hilbert function of two ideals, *Proc. Cambridge Phil. Soc.* **53** (1956), 568–575.

[8] W. Bruns and U. Vetter, Determinantal Rings, *Lecture Notes in Mathematics* **1327**, Springer-Verlag, New York, 1988.

[9] D. A. Buchsbaum and D. S. Rim, A generalized Koszul complex II. Depth and multiplicity, *Trans. Amer. Math. Soc.* **111** (1964), 197–224.

[10] A. de Jong, Smoothness, stability, and alterations, *Publ. Math. IHES* **83** (1996), 51–93.

[11] S. P. Dutta, On the canonical element conjecture, *Trans. Amer. Math. Soc.* **299** (1987), 803–811.

[12] S. P. Dutta, M. Hochster, and J. E. McLaughlin, Modules of finite projective dimension with negative intersection multiplicities, *Invent. Math.* **79** (1985), 253–291.

[13] S. P. Dutta, Frobenius and multiplicities, *J. Algebra* **85** (1983), 424–448.

[14] S. P. Dutta, Ext and Frobenius, *J. Algebra* **127** (1989), 163–177.

[15] G. Evans and P. Griffith, The Syzygy problem, *Ann. Math.* **114** (1981), 323–333.

[16] H.-B. Foxby, The MacRae Invariant, *Commutative Algebra (Durham 1981)*, London Math Soc. Lecture Note Series **72** (1982), 121–128.

[17] W. Fulton, *Intersection Theory*, Springer-Verlag, Berlin, 1984.

[18] H. Gillet and C. Soulé, K-théorie et nullité des multiplicités d'intersection, *C. R. Acad. Sci. Paris Série I no. 3, t.* **300** (1985), 71–74.

[19] H. Gillet and C. Soulé, Intersection theory using Adams operations, *Invent. Math.* **90** (1987), 243–277.

297

298 *Bibliography*

[20] A. Grothendieck, Le théorie des classes de Chern, *Bull. Soc. Math. France* **86** (1958), 137–154.
[21] R. Hartshorne, *Algebraic Geometry*, Graduate Texts in Mathematics, Springer-Verlag, New York, 1977.
[22] R. Heitmann, A counterexample to the rigidity conjecture for rings, *Bull. Amer. Math. Soc. New Series* **29** (1993), 94–97.
[23] M. Hochster, The equicharacteristic case of some homological conjectures on local rings, *Bull. Amer. Math. Soc.* **80** (1974), 683–686.
[24] M. Hochster, Topics in the Homological Theory of Modules over Commutative Rings, *Regional Conference Series in Mathematics* **24**, Amer. Math. Soc. Providence, RI (1975).
[25] M. Hochster, Canonical elements in local cohomology modules and the direct summand conjecture, *J. Algebra* **84** (1983), 503–553.
[26] M. Hochster and C. Huneke, Tight closure, invariant theory, and the Briançon-Skoda Theorem, *J. Amer. Math. Soc.* **1** (1990), 31–116.
[27] M. Hochster and C. Huneke, Infinite Integral Extensions and Big Cohen-Macaulay Algebras, *Ann. Math.* **135** (1992), 53–89.
[28] C. Huneke, Tight Closure and its Applications, *Regional Conference Series in Mathematics* **88**, Amer. Math. Soc., Providence, RI (1996).
[29] C. Huneke, A remark concerning multiplicities, *Proc. Amer. Math. Soc.* **85** (1982), 331–332.
[30] B. Iversen, *Local Chern Classes*, Ann. Sc. Éc. Norm. Sup. **9** (1976), 155–169.
[31] D. Katz, Complexes acyclic up to integral closure, *Math. Proc. Cambridge Phil. Soc.* **116** (1994), 310–314.
[32] D. Katz and J. K. Verma, Extended Rees algebras and mixed multiplicities, *Math. Z.* **202** (1989), 111–128.
[33] S. L. Kleiman, A generalized Teissier-Plücker formula, *Contemp. Math.* **162** (1994).
[34] S. L. Kleiman, The transversality of a general translate, *Compositio Math.* **38** (1974), 287–297.
[35] D. Kirby, On the Buchsbaum-Rim multiplicity associated with a matrix, *J. London Math. Soc.* **32**(2) (1985), 57–61.
[36] D. Kirby and D. Rees, Multiplicities in Graded Rings 1: The General Theory, *Contemp. Math.* **159** (1994), 209–267.
[37] J. Koh, Degree p extensions of an unramified regular local ring of mixed characteristic p, *J. Algebra* **99** (1986), 310–323.
[38] K. Kurano, A remark on the Riemann-Roch formula on affine schemes associated with Noetherian local rings, *Tohoku Math. J.* **48** (1996), 121–138.
[39] S. Lang, *Algebra*, Third edition, Addison-Wesley, Reading, MA, 1993.
[40] S. Lang and W. Fulton, *Riemann-Roch Algebra*, Springer-Verlag, New York, 1985.
[41] C. Lech, On the associativity formula for multiplicities, *Ark. Math.* **3** (1956), 301–314.
[42] S. Lichtenbaum, On the vanishing of Tor in regular local rings, *Ill. J. Math.* **10** (1966), 220–226.
[43] F. S. Macaulay, *The Algebraic Theory of Modular Systems*, Cambridge University Press, Cambridge, England, 1916.
[44] H. Matsumura, *Commutative Ring Theory*, Cambridge University Press, Cambridge, England, 1986.
[45] J. Milnor and J. Stasheff, Characteristic Classes, *Annals of Math. Studies* **76**, Princeton Univ. Press, Princeton, NJ, 1974.

[46] P. Monsky, The Hilbert-Kunz function, *Math. Ann.* **263** (1983), 43–49.

[47] D. Mumford, Topology of normal singularities and a criterion for simplicity, *Publ. Math. IHES* **9** (1961), 5–22.

[48] M. Nagata, *Local Rings*, Interscience, New York, 1962.

[49] C. Peskine and L. Szpiro, Sur la topologie des sous-schémas fermés d'un schéma localement noethérien, définis comme support d'un faisceau cohérent localement de dimension projective finie, *C. R. Acad. Sci. Paris, Sér. A* **269** (1969), 49–51.

[50] C. Peskine and L. Szpiro, Dimension projective finie et cohomologie locale, *Publ. Math. IHES* **42** (1973), 47–119.

[51] C. Peskine and L. Szpiro, Syzygies et multiplicités, *C. R. Acad. Sci. Paris, Sér. A* **278** (1974), 1421–1424.

[52] D. Rees, Transforms of local rings and a theorem on multiplicities of ideals, *Math. Proc. Camb. Philos. Soc.* **57** (1961), 8–17.

[53] D. Rees, Reduction of modules, *Math. Proc. Camb. Philos. Soc.* **101** (1987), 431–449.

[54] D. Rees and R. Sharp, On a theorem of B. Teissier on multiplicities of ideals in local rings, *J. London Math. Soc.* **13** (1978), 449–463.

[55] P. Roberts, *Homological Invariants of Modules over Commutative Rings*, Les Presses de l'Université de Montréal, Montréal, Québec, Canada (1980).

[56] P. Roberts, An infinitely generated symbolic blow-up in a power series ring and a new counterexample to Hilbert's fourteenth problem, *J. Algebra* **132** (1990), 461–473.

[57] P. Roberts, Chern classes of matrices over Noetherian rings, *Math. Z.* **216** (1994), 157–168.

[58] P. Roberts, Cohen-Macaulay complexes and an analytic proof of the new intersection conjecture, *J. Algebra* **66** (1980), 220–225.

[59] P. Roberts, The vanishing of intersection multiplicities of perfect complexes, *Bull. Amer. Math. Soc.* **13** (1985), 127–130.

[60] P. Roberts, Local Chern characters and intersection multiplicities, *Proc. Sympos. Pure Math.* **46**(2) (1987), 389–400.

[61] P. Roberts, Le théorème d'intersection, *C. R. Acad. Sc. Paris, Sér. I no. 7, t.* **304** (1987), 177–180.

[62] P. Roberts, *Intersection Theorems*, Commutative Algebra, Proceedings of an MSRI Microprogram, Springer-Verlag, New York, 1989, pp. 417–436.

[63] P. Roberts, Negative intersection multiplicities on singular varieties, *Proceedings of the Zeuthen Conference*, Copenhagen, 1989, Contemporary Mathematics **123** (1991), 213–222.

[64] J. Rotman, *An Introduction to Homological Algebra*, Academic Press, New York, 1979.

[65] G. Seibert, Complexes with homology of finite length and Frobenius functors, *J. Algebra* **125** (1989), 278–287.

[66] J.-P. Serre, Algèbre Locale – multiplicités. *Lecture Notes in Mathematics* **11**, Springer-Verlag, New York, 1961.

[67] I. Swanson, Mixed multiplicites, joint reductions, and quasi-unmixed local rings, *J. London Math. Soc.* **48** (1993), 1–14.

[68] L. Szpiro, Sur la théorie des complexes parfaits, *Commutative Algebra (Durham 1981)*, *London Math Soc. Lecture Note Series* **72** (1982), 83–90.

[69] B. Teissier, Cycles évanescents, sections planes, et condition de Whitney, *Singularités à Cargèse 1972, Astérisque* **7–8** (1973), 285–362.

[70] B. Teissier, Sur une inégalité à la Minkowski pour les multiplicités, *Ann. Math.* **106** (1977), 38–44.

[71] B. R. Tennison, Intersection multiplicities and tangent cones, *Math. Proc. Camb. Philos. Soc.* **85** (1979), 33–42.

[72] J. K. Verma, Rees algebras and mixed multiplicities, *Proc. Amer. Math. Soc.* **104** (1988), 1036–1044.

[73] J. K. Verma, Multigraded Rees algebras and mixed multiplicities, *J. Pure Appl. Algebra* **77** (1992), 219–228.

Index

acyclicity lemma, 75
affine scheme, 142
annihilator ideal, 4
Artin approximation theorem, 137
Artin-Rees lemma, 27
associated graded ring, 26
associated prime, 4–6
Auslander-Buchsbaum theorem, 78
Auslander-Buchsbaum-Serre theorem, 80

basic open set, 148
Bass numbers, 85
Betti numbers, 76
Bézout's theorem, 174

canonical element conjecture, 113, 123
canonical sheaves of the Grassmannian, 210,
    215, 219
Chern classes of, 180, 182, 220
    homological computation of, 220, 223
    global definition, 212
catenary ring, 72
Chern character, 196
    additive property, 198
    of an m-primary matrix, 249
    multiplicative property, 198
Chern class, 180, 182
    commutativity of, 192
    composition of, 190
    definition of, 186
    functorial properties of, 187
    independence modulo rational equivalence,
        190
    of an m-primary matrix, 232, 234, 235
    of a rank one sheaf, 181
Chow group, 9
    of Proj of a symmetric algebra, 183
    rational, 196
Cohen-Macaulay
    module, 66

ring, 66, 70, 87, 250, 295, 229
coherent sheaf, 200
commutativity of intersection with divisors,
    171, 172, 251
complete intersection, 70
    Todd class of, 274
complex
    bounded, 43
    double, 52
    dualizing, 86, 99
    elementary, 262
    exact sequence of, 43
    homology of, 43
    Koszul, 57, 78, 171, 201, 265, 269
    minimal, 76
    of modules, 42
    total, of a double complex, 52
condition on generators of a graded ring, 146
cycle
    defined by a module, 9
    defined by a complex, 237, 239, 240, 243
    group of cycles, 8

degree
    of an element, 19
    of a graded module, 115, 175
depth, 64
determinantal ideals, 221
dimension, 8, 28, 29, 31, 72
    of a graded module, 144
    of a $\mathbb{Z}^n$-graded module, 154
divisor, 189
    effective, 167
    intersection with, 94, 167, 169
    sum of divisors, 170
divisor class group, 12
divisorial ideal, 10
dualizing complex, 86, 99
Dutta multiplicity, 121, 132, 258, 284, 294–96

301